The Law of Electronic Commerce

Written specifically for legal practitioners and students, this book examines the concerns, laws and regulations involved in electronic commerce.

In just a few years, commerce via the World Wide Web and other online platforms has boomed, and a new field of legal theory and practice has emerged. Legislation has been enacted to keep pace with commercial realities, cyber-criminals and unforeseen social consequences, but the ever-evolving nature of new technologies has challenged the capacity of the courts to respond effectively.

This book addresses the legal issues relating to the introduction and adoption of various forms of electronic commerce. From intellectual property, to issues of security and privacy, Alan Davidson looks at the practical challenges for lawyers and commercial parties while providing a rationale for the underlying legal theory.

Alan Davidson is Senior Lecturer in the School of Law, University of Queensland. He is also a solicitor and barrister of the Supreme Court of New South Wales and the High Court of Australia.

The Law of Electronic Commerce

Alan Davidson

CAMBRIDGE
UNIVERSITY PRESS

CAMBRIDGE UNIVERSITY PRESS
Cambridge, New York, Melbourne, Madrid, Cape Town, Singapore, São Paulo, Delhi

Cambridge University Press
477 Williamstown Road, Port Melbourne, VIC 3207, Australia

www.cambridge.org
Information on this title: www.cambridge.org/9780521678650

First published 2009

Cover design by MAPG
Edited by Sarah Shrubb
Typeset by Aptara
Printed in Australia by Ligare

National Library of Australia Cataloguing in Publication data
Davidson, Alan.
The law of electronic commerce / Alan Davidson.
9780521678650 (pbk.)
Law series (Port Melbourne, Vic.)
Includes index
Electronic commerce – Law and legislation – Australia
Electronic data interchange – Law and legislation – Australia
Business records – Law and legislation – Australia
Internet (Computer network) – Law and legislation – Australia.
Computer crimes – Law and Legislation – Australia.
343.940999

ISBN 978-0-521-67865-0 paperback

Contents

Acknowledgements

On one analysis electronic commerce emerged in mid 19th century with the invention of the telegraph and telephone. But it was not until the creation and growth of computers and the internet, and the development of this realm called cyberspace, that the legal system faced real challenges in private, public and criminal law. Speaking about cyberspace, 'the new home of the mind', John Perry Barlow declared:

> Ours is a world that is both everywhere and nowhere, but it is not where bodies live . . . In our world, whatever the human mind may create can be reproduced and distributed infinitely at no cost. The global conveyance of thought no longer requires your factories to accomplish. We will create a civilization of the Mind in Cyberspace. May it be more humane and fair than the world your governments have made before.[1]

In the past two decades many of the legal challenges have been answered. The milestones in the past 20 years have been many. Some of the major ones are the global content of the internet (in *Reno v American Civil Liberties Union* 521 US 844 (1997)), jurisdiction in cyberspace (in *Dow Jones v Gutnick* [2002] HCA 56), and the functional equivalence of writing and signatures (by the UNCITRAL Models Law of Electronic Commerce and the Electronic Transaction Acts). Other areas that have been dealt with include the regulation of spam, internet gambling, identity theft, digital privacy, email and cybersquatting.

My interest in this area arose some three decades ago when I undertook a degree in computing science while practising law. At the time the combination was most unusual, but the world of cyberspace and the regulation of that world have now become part of the landscape. The aim of this work is to define the law relating to electronic commerce within Australia as determined by the legislature, judicial interpretations and the common law. It is intended for legal practitioners and students of what has broadly become known as cyberlaw.

I would like to thank my colleagues Russell Hinchy, Paul O'Shea and my research assistant William Hickey for their feedback, suggestions and assistance. I would especially like to thank my assistant and colleague Garth Wooler for his

1 John Perry Barlow, "Declaration of Independence of Cyberspace".

patience, proofreading and contributions. Finally I would thank my late father and my wonderful family Dianne, Taylor and Chelsea.

Dr Alan Davidson
Senior Lecturer, TC Beirne School of Law University of Queensland
Solicitor and Barrister
April 2009

Table of Cases

Table of statutes

1

The law of electronic commerce

Electronic commerce refers to all commercial transactions based on the electronic processing and transmission of data, including text, sound and images. This involves transactions over the internet, plus electronic funds transfers and Electronic Data Interchange (EDI).

On one level, electronic commerce began in the mid-1800s, when the first contract was entered into using the telegraph or telephone. However, the expression 'electronic commerce' is typically used in connection with the expansion of commerce using computers and modern communications, most notably the internet and cyberspace. The development of security protocols has aided the rapid expansion of electronic commerce by substantially reducing commercial risk factors.

The advantages of electronic commerce to commercial parties include ease of access, anonymous browsing of products, larger choice, the convenience of shopping from the computer and enormous efficiencies. The disadvantages include the potential for invasion of privacy and security risks. There are also questions regarding jurisdiction, standards, protection of intellectual property, taxation, trade law and many other issues. Nevertheless, acceptance of electronic products and services has grown substantially.

Security is of paramount importance in electronic commerce. Public key cryptology was invented in response to security concerns and has revolutionised electronic commerce. Communications are now relatively secure: digital signatures or certificates permit the authentication of the sender of a message or of an electronic commerce product.

This book addresses legal issues relating to the introduction and adoption of various forms of electronic commerce. Whether it is undertaking a commercial transaction on the World Wide Web, sending electronic communications to enter into commercial arrangements, downloading material subject to copyright or privacy concerns about our digital personas, there are legal considerations. Parties must consider the risks of electronic commerce, whether electronic writing and signatures are equivalent to paper writing or wet ink signatures; which jurisdiction and which law governs a dispute between parties, if the parties are in different countries from the servers. This book addresses intellectual property, cybercrime, surveillance and domain name usage and disputes.

Electronic commerce law

An examination of the law of electronic commerce must begin with a fundamental understanding of the law and its role in society as it has evolved over the centuries. It necessitates understanding terrestrial norms, social behaviour and the application of the rule of law. These principles must be applied to new circumstances, infrastructure and contexts, even if this challenges such foundations of society as sovereignty and human rights. It is an exciting time to be charting the course and watching as legislators, courts, merchants and the populace wrestle with this new epoch. In the 18th and 19th centuries there may have been a similar opportunity to observe principles evolving, as there were new developments in relation to consideration (in contract law) and the postal acceptance rule (also in contract law), and as principles of equity matured. But such development was at a snail's pace compared with the eruption of the law of electronic commerce over the last two decades.

The majority of legal problems arising through the use of electronic commerce can be answered satisfactorily by the application of standard legal principles. Contract law, commercial law and consumer law, for example, all apply to the internet, email communications, electronic banking and cyberspace generally. However, cyberspace gives rise to unique and unusual circumstances, rights, privileges and relationships that are not adequately dealt with by traditional law. This has necessitated legislation, international agreements and a plethora of cases before the courts to resolve myriad questions. The expression 'electronic commerce law' is used to describe all changes and additions to the law that are a result of the electronic age.

Justice Fryberg,[1] in an address to the Australian Conference on the Law of Electronic Commerce, asked, 'Is there such a thing as Electronic Commerce Law? I suggest there is not',[2] although he acknowledged that he had completed a keynote address on precisely that topic. Joseph Sommer[3] argued that 'cyberlaw' was non-existent as a separate body of law, and that cyberspace 'is a delightful new playground for old games':

1 Justice of the Supreme Court of Queensland.
2 Supreme Court of Queensland publications: archive.sclqld.org.au/judgepub/2003/fry070403.pdf.
3 'Against cyberlaw', (2000) 15 *Berkeley Tech L J* 1145.

> [N]ot only is 'cyberlaw' nonexistent, it is dangerous to pretend that it exists. A lust to define the future can be very dangerous, especially when we cannot even agree on the present. A lust to define the law of the future is even worse, since law tends to evolve through an inductive accretion of experience. It is much safer to extract first principles from a mature body of law than to extract a dynamic body of law from timeless first principles. An overly technological focus can create bad taxonomy and bad legal analysis, at least. At worst, it can lock us into bad law, crystallizing someone's idea of a future that will never be.[4]

Judge Frank Easterbrook[5] initiated the debate with his article 'Cyberspace and the law of the horse'. Easterbrook argued that cyberlaw is unimportant because it invokes no first principles.[6] He made reference to the comment of a former Dean of the University of Chicago Law School that a course in the law of the horse was not offered: 'Lots of cases deal with sales of horses; others deal with people kicked by horses; still more deal with the licensing and racing of horses, or with the care veterinarians give to horses, or with prizes at horse shows.' Nevertheless there is no discrete body of horse law.[7] Judge Easterbrook argued that there was no reason to teach the 'law of cyberspace', any more than there was reason to teach the 'law of the horse', because neither, he suggested, would 'illuminate the entire law'. By analogy he proclaimed that cyberlaw did not exist.[8]

In his article 'The law of the horse: What cyberlaw might teach', Lawrence Lessig[9] responded to Easterbrook's assertions. Through a series of examples he demonstrated that cyberlaw or electronic commerce law, however described, forms a unique area of legal discourse. Lessig referred to privacy and spam in cyberspace. He argued that any lesson about cyberspace requires an understanding of the role of law, and that in creating a presence in cyberspace, we must all make choices about whether the values we embed there will be the same values we espouse in our real space experience. Understanding how the law applies in cyberspace in conjunction with demands, social norms and mores, and the rule of cyberspace, will be valuable in understanding and assessing the role of law everywhere.

Easterbrook and Lessig's disquisitions are now dated by a decade in a field which has advanced more quickly than any other field of law. The law of electronic commerce has increasingly become a distinct class of study, with legal specialists, dedicated monographs and courses in every law school. Legislation has been deemed necessary for several cyber issues. Those who scorned words like 'cyberlaw' and 'cybercrime' perhaps winced at the introduction of the Australian *Cybercrime Act 2001* (Cth). Traditional laws proved inadequate,

4 Ibid.
5 Now Chief Judge of the US Court of Appeals for the Seventh Circuit.
6 'Cyberspace and the law of the horse', (1996) *U Chi Legal F* 207.
7 Interestingly, the US law firm of Miller, Griffin and Marks advertises that it specialises in 'commercial, corporate and equine matters'.
8 See also James Boyle, 'Foucault in cyberspace: Surveillance, sovereignty, and hard-wired censors', (1997) *University of Cincinnati Law Review* 66, 177.
9 'The law of the horse: What cyberlaw might teach', (1999) 113 *Harv L Rev* 501.

necessitating legislation on computer-related crime, credit card fraud, bank card fraud, computer forgery, computer sabotage, unauthorised access to computer systems, unauthorised copying or distribution of computer programs, cyber stalking, theft of intellectual property and identity theft.

Spam has become a real economic waste for virtually all business, resulting in legislation and international agreements.[10] The digitalisation of data results in real privacy concerns. In response to this the Australian *Privacy Act 1988* (Cth) was overhauled in 2000 to make the private sector accountable. The Australian Law Reform Commission (ALRC) is currently undertaking a further review, with recommendations to expand privacy laws so that they deal with technological developments.

Domain names are valuable business identifiers, traded in the millions of dollars and subject to numerous disputes. Most national domain name administrators have introduced dispute resolution procedures. The courts have dramatically expanded the tort of passing off, in a manner not contemplated until recently, in an attempt to provide remedies.

Conflict of laws principles in cyberspace have been inadequately served by traditional principles established over centuries. The courts have formed new approaches. Online defamation, for instance, is unlike the static occurrences. Now, a defamation statement can be published continuously worldwide 24 hours a day. The courts have had to reconsider the single publication rule, and the applicability of local laws to a website intended for another jurisdiction, but with global reach.

Child pornography, terrorism, suicide materials, spyware and censorship are issues on which laws vary dramatically internationally, and yet each website is typically available globally. Nations have different ages at which a person is no longer regarded as a child; freedom of speech issues arise with terrorism issues (plans to make a bomb) and suicide information, but the law must address the easy reach of such material in the digital age, in ways that in other contexts may be considered draconian. Censorship laws for print and television are ineffective for online materials. What is electronic writing, and an electronic signature? The range of issues related to electronic contracting has resulted in the Electronic Transactions Acts internationally. Evidentiary issues arise in relation to digitising paper documents and printing out electronic documents. The internet raises a range of intellectual property issues, such as peer-to-peer file sharing (music and videos in particular), digital rights management, time shifting and format shifting. Electronic commerce by its nature does not recognise borders and it raises questions regarding security of transactions, standards and protection, legally and otherwise, in an international context.

10 For example the *Spam Act 2003* (Cth) and the Memoranda of Understanding between Australia, South Korea and the US. See Chapter 16.

Many international organisations have spent considerable time and resources on resolving legal issues and difficulties in electronic commerce: the UN Commission on International Trade Law (UNCITRAL), the International Chamber of Commerce (ICC), the Asia-Pacific Economic Cooperation (APEC), the Organisation for Economic Cooperation and Development (OECD) and the World Trade Organization (WTO) are some examples.[11]

The law of electronic commerce (or cyberlaw) has emerged as a new, disparate and coherent body of law.

Internet use in Australia

According to the Australian Bureau of Statistics (ABS), as at the end of the March quarter 2007, there were 6.43 million active internet subscribers in Australia, comprised of 761 000 business and government subscribers and 5.67 million household subscribers. The number of non dial-up subscribers was 4.34 million; the number of dial-up subscribers was 2.09 million. Non dial-up subscribers increased by 16 per cent between September 2006 and March 2007, while dial-up dropped by 16 per cent. The growth in non dial-up was driven mainly by household subscribers. Non dial-up subscribers represented 67 per cent of total internet subscribers in Australia at the end of March 2007, compared with 60 per cent at the end of September 2006. Digital Subscriber Line (DSL) continued to be the dominant access technology used by non dial-up subscribers (3.36 million or almost 78 per cent of total non dial-up subscribers were connected this way). Connections with download speeds of 1.5 Mbps or greater had increased by 43 per cent by March 2007 to 1.56 million (there were 1.09 million at the end of September 2006).

Home internet access across Australia reached 67 per cent in 2007, up from 35 per cent in 2001. In 2006, 66 per cent of homes in major cities had internet access, compared with 42 per cent for very remote Australia. Broadband was used by 46 per cent of homes in major cities and 24 per cent in very remote Australia.[12]

The ABS report also found that income and education were key factors in people's internet access. Households with an income of $2000 or more per week were three times more likely to have broadband than households on less than $600 per week. Families with children under 15 (or dependant students) were three to four times more likely to have internet access than other families. People in low-skill occupations were about a quarter less likely to have broadband than those with higher skills. People not in the labour force were 18 per cent less likely to have broadband than those in the labour force. Unemployed people

11 Other bodies include the Free Trade Agreement of the Americas, the International Telecommunications Union and the International Organisation for Standardisation. See Chapter 19.

12 In 2006 the ACT had the highest connection rate, with 75% of all homes connected and 53% of these on broadband connections. Similar rates were seen in New South Wales (63% total and 42% broadband), Victoria (63% and 42%), Queensland (64% and 41%) and Western Australia (65% and 41%). The lowest connection rate was in Tasmania (55% and 28%).

were 12 per cent less likely to have broadband than employed people. Indigenous households are about half as likely to have broadband as non-indigenous households.

Judicial consideration in Australia

The High Court of Australia has had few opportunities to consider the impact of electronic commerce, cyberspace and the operation of the internet. *Dow Jones v Gutnick*[13] in 2002 was one such opportunity. The court considered defamation on the World Wide Web and whether it was appropriate for the Supreme Court of Victoria to exercise jurisdiction over a US-based website. This was the first real opportunity for the court to consider its role in law making and the common law in the context of cyberspace and electronic commerce.

The nature and essence of the common law makes it amenable to development, subject to the Constitution and statute.[14] The judiciary, scholars and commentators debate the length and breadth of acceptable developments using various forms of legal reasoning to justify their individual approaches. Whether the approach is principle-based, a coherence-based incremental method or policy-based, development is an integral part of the common law and of our socio-legal structure.[15]

In a joint majority judgment Gleeson CJ, McHugh, Gummow and Hayne JJ accepted the evidence before the judge at first instance, Hedigan J, regarding 'the unusual features of publication on the internet and the World Wide Web'. The majority accepted that the internet is 'a telecommunications network that links other telecommunication networks . . . [that] enables inter-communication using multiple data-formats . . . among an unprecedented number of people using an unprecedented number of devices [and] among people and devices without geographic limitation'.[16] The majority expressed concern regarding the lack of evidence adduced to reveal what electronic impulses pass or what electronic events happen in the course of passing or storing information on the internet. Nevertheless the majority took the opportunity to define a broad range of internet terms:

> 14. The World Wide Web is but one particular service available over the Internet. It enables a document to be stored in such a way on one computer connected to the Internet that a person using another computer connected to the Internet can request and receive a copy of the document . . . the terms conventionally used to refer to the materials that are transmitted in this way are a 'document' or a 'web

13 *Dow Jones v Gutnick* [2002] HCA 56.

14 See *Theophanous v Herald and Weekly Times Limited* [1994] HCA 46, especially Brennan J at para 4. See also *D'Orta-Ekenaike v Victorian Legal Aid* [2005] HCA 12; (2005) 223 CLR 1 and *Arthur J S Hall & Co v Simmons* [2000] UKHL 38; [2002] 1 AC 615.

15 For a considered and valuable discourse see Russell Hinchy, *The Australian legal system: History, institutions and method*, Pearson Education Australia, Sydney, 2008, in particular, Part Three.

16 *Dow Jones v Gutnick* [2002] HCA 56, para 13.

page' and a collection of web pages is usually referred to as a 'web site'. A computer that makes documents available runs software that is referred to as a 'web server'; a computer that requests and receives documents runs software that is referred to as a 'web browser'.

15. The originator of a document wishing to make it available on the World Wide Web arranges for it to be placed in a storage area managed by a web server. This process is conventionally referred to as 'uploading'. A person wishing to have access to that document must issue a request to the relevant server nominating the location of the web page identified by its 'uniform resource locator (URL)'. When the server delivers the document in response to the request the process is conventionally referred to as 'downloading'.

In the same case Kirby J was more philosophical in discussing the ramifications of technological developments, quoting Lord Bingham of Cornhill, who said that the impact of the internet on the law of defamation will require 'almost every concept and rule in the field . . . to be reconsidered in the light of this unique medium of instant worldwide communication'.[17]

In any reformulation of the common law, Kirby J continued, many factors would need to be balanced: the economic implications of any change, valid applicable legislation, the pros and cons of imposing retrospective liability on persons, and social data and public consultation.[18] Most significantly, reform is the purvey of government; it is not the primary role of the courts. Nevertheless, as Kirby J pointed out, courts have reversed long-held notions of common law principle when 'stimulated by contemporary perceptions of the requirements of fundamental human rights'.[19]

As they have recognised the enormity of the impact of the internet as a revolutionary communications giant, courts all over the world have been forced to reconsider basic principles. Kirby J appropriately quoted noted US jurist Billings Learned Hand:

> The respect all men feel in some measure for customary law lies deep in their nature; we accept the verdict of the past until the need for change cries out loudly enough to force upon us a choice between the comforts of further inertia and the irksomeness of action.[20]

Of this passage Brennan J, in *Theophanous v Herald and Weekly Times Limited*,[21] remarked, 'Other judges find the call to reform more urgent.'[22] To varying degrees the judiciary acknowledges its role in the law-making process, balancing consistency, cohesion and precedent in an analytical ballet of deductive, inductive and abductive reasoning. On the one hand, the common law is the

17 Matthew Collins, *The law of defamation and the internet*, Oxford University Press, Oxford, 2001.
18 *Dow Jones v Gutnick* [2002] HCA 56, paras 75 and 76.
19 *Dow Jones v Gutnick* [2002] HCA 56, para 77.
20 Hand, 'The contribution of an independent judiciary to civilisation', in Irving Dillard (ed.), *The spirit of liberty: Papers and addresses of Learned Hand*, 3rd edn, Alfred A Knopf, New York, 1960.
21 [1994] HCA 46.
22 [1994] HCA 46, para 5.

rock and the foundation of modern English and colonial law (including Australia's law). On the other hand, it has remained sufficiently flexible to adapt to modern developments. And so it should. With the advent of new and novel factual circumstances the judiciary must find solutions for the benefit, stability and protection of society and commerce. This has been described as the 'genius of the common law': that the courts may adapt principles of past decisions, by analogical reasoning, 'to the resolution of entirely new and unforeseen problems'.[23] When the new problem is as novel, complex and global as Gutnick's case, an opportunity – indeed a duty – arises to fashion and build the common law.[24]

Two examples illustrate the best and worst of judicial malleability. The common law crime of larceny dictates that the prosecution must prove certain elements of the crime to elicit a conviction. One key element is to prove that the accused intended to deprive another of property. This is problematic for intangibles. Where a person accesses a computer and 'steals' a computer file a curious thing happens: the information is both stolen and left behind. The development of common law larceny failed to predict or consider this. Attempts at prosecution failed. The courts could not or would not modify the well-established common law offence of larceny to accommodate this development. It was left to the legislature to enact cybercrime legislation to encompass offences that could not be imagined only a matter of years earlier.[25] A reading of the US Supreme Court decision of *Reno v American Civil Liberties Union*[26] provides a similar missed opportunity in the US context. On the other hand, the English Court of Appeal, in the domain name case *Marks & Spencer v One in a Million*,[27] moulded, twisted and interpreted the five established elements of the tort of passing off to provide a remedy where one previously did not exist. The result is such a departure from traditional passing off that it is referred to as 'domain name passing off'.[28]

This extension of principle when faced with domain name piracy is to be applauded. In the words of Kirby J, 'When a radically new situation is presented to the law it is sometimes necessary to think outside the square.'[29] Kirby J reflected on the specific issue of defamation in the Dow Jones appeal and theorised a re-evaluation and formation of a new 'paradigm', and a 'common sense' approach to change. His Honour questioned the wisdom of refusing to find a new remedy in new circumstances, describing it as 'self-evidently unacceptable'.[30]

23 *Dow Jones v Gutnick* [2002] HCA 56, para 91, per Kirby J.
24 For a view of the role and limitations of the judiciary see the judgment of Brennan J in *Mabo v Queensland (No. 2)* [1992] HCA 23; (1992) 175 CLR 1.
25 See Chapter 16.
26 521 US 844 (1997). See also *American Civil Liberties Union v Reno* 929 F. Supp 824 (1996), particularly the joint judgment of Sloviter CJ, Buckwalter and Dalzell JJ.
27 [1999] 1 WLR 903; [1998] EWCA Civ 1272.
28 See Chapter 9.
29 *Dow Jones v Gutnick* [2002] HCA 56, para 112.
30 *Dow Jones v Gutnick* [2002] HCA 56, para 115.

Change is not new. The *lex mercatoria*, for example, emerged from the customs and practices of merchants from the Middle Ages. Rules governing bills of exchange and letters of credit were crafted by commercial parties long before the law makers had the opportunity to legislate. Their aim was commerce, but the result was that commercial customs and practice became recognised by the courts and then the legislature, both institutions responding to commercial realities.

In the context of reconceptualisation, Kirby J (in Gutnick's case) dabbles with new rules for a 'unique technology' and the 'urgency' of considering such changes:

> To wait for legislatures or multilateral international agreement to provide solutions to the legal problems presented by the Internet would abandon those problems to 'agonizingly slow' processes of lawmaking. Accordingly, courts throughout the world are urged to address the immediate need to piece together gradually a coherent transnational law appropriate to the 'digital millennium'. The alternative, in practice, could be an institutional failure to provide effective laws in harmony, as the Internet itself is, with contemporary civil society – national and international. The new laws would need to respect the entitlement of each legal regime not to enforce foreign legal rules contrary to binding local law or important elements of local public policy. But within such constraints, the common law would adapt itself to the central features of the Internet, namely its global, ubiquitous and reactive characteristics. In the face of such characteristics, simply to apply old rules, created on the assumptions of geographical boundaries, would encourage an inappropriate and usually ineffective grab for extra-territorial jurisdiction . . .
>
> Generally speaking, it is undesirable to express a rule of the common law in terms of a particular technology. Doing so presents problems where that technology is itself overtaken by fresh developments. It can scarcely be supposed that the full potential of the Internet has yet been realised. The next phase in the global distribution of information cannot be predicted. A legal rule expressed in terms of the Internet might very soon be out of date.[31]

In some instances, such as electronic banking and finance, commercial parties have embraced electronic commerce, forging new paths and boldly leading the way, even in the absence of any adequate laws. In other instances the law makers have taken the initiative: the UN Commission on International Trade Law (UNCITRAL) in 1996, with the Model Law of Electronic Commerce, is an important example.[32]

There are many pitfalls, warnings and surprises in the regulation of electronic commerce, both practical and legal. These voyages are examined in this book. The law relating to electronic commerce, the internet and cyberspace requires careful management by legislators and structured application and development of existing legal principles by the courts.

31 [2002] HCA 56, paras 119 and 125.
32 For details see Chapter 3.

Further reading

Justice George Fryberg, Keynote Address to the Australian Conference on the Law of Electronic Commerce, Brisbane 2003, Supreme Court of Queensland publications: archive.sclqld.org.au/judgepub/2003/fry070403.pdf.

Billings Learned Hand, 'The contribution of an independent judiciary to civilisation', in Irving Dillard (ed.), *The spirit of liberty: Papers and addresses of Learned Hand*, 3rd edition, Alfred A Knopf, New York, 1960.

Russell Hinchy, *The Australian legal system: History, institutions and method*, Pearson Education Australia, Sydney, 2008, Part Three.

Neil MacCormick, *Legal reasoning and legal theory*, Clarendon Law Series, Oxford University Press, Oxford, 1979.

2

The rule of cyberspace

This chapter examines and contemplates law and culture in cyberspace. The role of law and indeed the rule of law have different dynamics in cyberspace as a consequence of the architecture of cyberspace and its anonymous, pseudonymous and borderless features. The resultant structural balance among technology, law and culture may be expressed as the 'rule of cyberspace'. This spatial dimension, economic influence on human culture and the role of law and regulation together form a subculture which both impacts on and moulds electronic commerce.

First, the nature of cyberspace is examined. This is followed by consideration of theoretical bases for law and order in cyberspace. The rule of cyberspace emerges, by processes known as 'spontaneous order', from the environmental factors fashioning cyberspace. It is spontaneous order which best describes and to a limited extent predicts regulation for electronic commerce.

This chapter examines the juxtaposition of culture and cyberspace, a modern application of spontaneous order, and then uses a discussion of libertarian and classical approaches to predict the future of cyberspace.

Cultural and environmental juxtaposition with cyberspace

Human interaction tends towards order and has an aversion to chaos. Culture brings about communities, law, order and stability. And so is it for cyberspace and the rule of cyberspace.[1]

1 See Sir Edward Burnett Tylor, *Primitive culture: Researches into the development of mythology, philosophy, religion, language, art and custom*, Gordon Press, London, (1871) 1974.

Cyberspace is infused with a kind of spontaneous order, and has thus evolved protocols through public participation. No one controls cyberspace. There are many stakeholders and users, all with their own agendas, impacts and influences. Customs, usages and structure have emerged from human action and interaction, but not human command. Organising bodies do not know the exigencies of the diverse predilections and demands of the participants. The size, direction, extent and use of cyberspace have challenged forecasters. By incalculable actions and inputs – spontaneous order – cyberspace has gained structure and presence.

Through a process of exploration and learning humankind has begun to understand the limits of its physical world. With curiosity and wonder we develop an appreciation of real space as a home and vessel for life. Similarly, cyberspace is a mystery; at least initially. Some dip their toe in and play in this new universe. To many cyberspace is a nebulous, unsafe and ungoverned place.[2] Cyberspace poses problems and challenges to visitors, and to owners, regulators, consumers, merchants and tourists.

The legend of King Canute[3] is of a king believed to be so powerful that he could command the tides to stop. Concerned that his subjects believed him to be almighty to the point of immortality, Canute undertook a practical demonstration at Thorney Island and ordered the tide to stop. This was of course futile, and the tide proceeded to disobey him. The demonstration merely confirmed the limits of his sovereignty. The sea was beyond mere human governance. Many have drawn a parallel with cyberspace: any attempt to regulate it would be equally futile.[4]

Cyberspace

Cyberspace is an illusion. It has no physical presence. Yet its users visit it, send messages and transact business through it. Electrical, magnetic and optical forces with storage facilities permit users to carry out steps that produce results in real space. Whether users find information or goods and services via Google, listen online to radio around the world, or send and receive email, every step is carefully planned and choreographed. This is not new. Technology has fooled human senses for years. The dot matrices on newspaper photographs give the illusion of people and places. Tiny pixels on a television or computer screen give the illusion of words, drawings and movement. Flashing still pictures 24 times a second gives the illusion of movement. Users become as immersed in cyberspace as they are engrossed in the narrative of television and movies. The illusion is real enough; it takes on an ethereal quality and allows escape

2 Tom Bell, *The internet: Heavily regulated by no one in particular*, (1997), www.tomwbell.com/writings/InterReg.html.

3 Canute (994–1035AD) was king of England, Denmark and Norway and overlord of Schleswig and Pomerania.

4 For example, see Michael Kirby, 'Privacy in cyberspace', (1998) *UNSWLJ* 47 and Graham Greenleaf, 'An endnote on regulating cyberspace: Architecture vs law?', (1998) *UNSWLJ* 52.

from terrestrial shackles. In the end, though, it is all real people dealing with real people, and sometimes their relationships and actions become convoluted and conflicted enough for them to look to the law for resolution, to regulation or a form of rule in cyberspace.

In *Reno v American Civil Liberties Union* the US Supreme Court has defined 'the internet' and 'cyberspace' in the following terms:

> The Internet is an international network of interconnected computers . . . [which] now enable[s] tens of millions of people to communicate with one another and to access vast amounts of information from around the world . . . 'cyberspace' – located in no particular geographical location but available to anyone, anywhere in the world, with access to the Internet . . . Cyberspace undeniably reflects some form of geography; chat rooms and Web sites, for example, exist at fixed 'locations' on the Internet. Since users can transmit and receive messages on the Internet without revealing anything about their identities or ages . . . cyberspace is malleable. Thus, it is possible to construct barriers in cyberspace and use them to screen for identity, making cyberspace more like the physical world and, consequently, more amenable to zoning laws. This transformation of cyberspace is already underway.[5]

The High Court of Australia described the described the internet and cyberspace in the following terms:

> [A] decentralised, self-maintained telecommunications network. It is made up of interlinking small networks from all parts of the world. It is ubiquitous, borderless, global and ambient in its nature. Hence the term 'cyberspace'. This is a word that recognises that the interrelationships created by the Internet exist outside conventional geographic boundaries and comprise a single interconnected body of data, potentially amounting to a single body of knowledge. The Internet is accessible in virtually all places on Earth where access can be obtained either by wire connection or by wireless (including satellite) links. Effectively, the only constraint on access to the Internet is possession of the means of securing connection to a telecommunications system and possession of the basic hardware.[6]

Science fiction author William Gibson first coined the term 'cyberspace' in his 1982 novelette 'Burning Chrome' in *Omni* magazine.[7] He depicts cyberspace as a 'consensual hallucination experienced daily by billions of legitimate operators'.[8] Michael Benedikt[9] describes cyberspace as: 'A new universe, a parallel universe created and sustained by the world's computers and communication lines . . . The tablet become a page become a screen become a world, a virtual world . . . Its corridors form wherever electricity runs with intelligence . . . The realm of pure information.' He believes that cyberspace, 'like cityspace, can be inhabited,

5 521 US 844 (1997).
6 *Dow Jones v Gutnick* [2002] HCA 56, para 80 per Kirby J.
7 The concept was expanded and popularised in his 1984 novel *Neuromancer* (Ace Books, New York). Many sources incorrectly cite Gibson's 1984 book as the first use of the word 'cyberspace'. Gibson has stated, 'When I came up with the term [cyberspace] in *Burning Chrome*, I used it to define a kind of navigable, iconic, three-dimensional representation of data.'
8 William Gibson, *Neuromancer*, Ace Books, New York, 1984.
9 Michael Benedikt, *Cyberspace: First steps*, MIT Press, Cambridge MA, 1991.

explored, and designed'.[10] Benedikt puts forward the notion that we drift into cyberspace every time we speak on the telephone or become absorbed in a book.

It is not a precondition for living, working and engaging in commerce in real space that its participants understand real space's underlying physics. And so it is for cyberspace. Users, vendors and consumers embrace cyberspace however it is defined. The law of cyberspace emerges as a consequence of their actions. Yet real space norms remain a starting point. Appreciating the difference gives meaning to the application of law and regulation. The conservative approach is to apply all real space laws, rules and behavioural norms in cyberspace. The libertarian view is to remove the shackles of real space and create a utopia, free for any form of regulation or encumbrance.

Author and artist John Perry Barlow falls into the latter group. He describes the action of visiting cyberspace as 'like having had your everything amputated'.[11] Physical existence in cyberspace remains impossible. Nevertheless consciousness becomes meshed and even lost in computer depictions. Barlow was an early advocate of freedom of expression and freedom from regulation in cyberspace. His response to the US *Telecom Reform Act* of 1996 was to write and issue a 'Declaration of the Independence of Cyberspace',[12] in which he colourfully proclaims that cyberspace cannot be ruled by terrestrial norms, legal or otherwise. This declaration galvanised the growing anxiety regarding rights and liberties within the cyberspace community:

> **A Declaration of the Independence of Cyberspace**
>
> Governments of the Industrial World, you weary giants of flesh and steel, I come from Cyberspace, the new home of Mind. On behalf of the future, I ask you of the past to leave us alone. You are not welcome among us. You have no sovereignty where we gather.
>
> We have no elected government, nor are we likely to have one, so I address you with no greater authority than that with which liberty itself always speaks. I declare the global social space we are building to be naturally independent of the tyrannies you seek to impose on us. You have no moral right to rule us nor do you possess any methods of enforcement we have true reason to fear.
>
> Governments derive their just powers from the consent of the governed. You have neither solicited nor received ours. We did not invite you. You do not know us, nor do you know our world. Cyberspace does not lie within your borders. Do not think that you can build it, as though it were a public construction project. You cannot. It is an act of nature and it grows itself through our collective actions.
>
> You have not engaged in our great and gathering conversation, nor did you create the wealth of our marketplaces. You do not know our culture, our ethics, or the

10 *Cityspace, cyberspace, and the spatiology of information*, University of Texas at Austin, 1993: www.utexas.edu/architecture/center/benedikt_articles/cityspace.html.
11 Michael E Doherty Jr, 'Marshall McLuhan meets William Gibson in "Cyberspace"', (1995) *Computer-Mediated Communication Magazine*: www.ibiblio.org/cmc/mag/1995/sep/doherty.html.
12 Available at www.eff.org/Misc/Publications/John_Perry_Barlow/barlow_0296.declaration.txt.

unwritten codes that already provide our society more order than could be obtained by any of your impositions.

You claim there are problems among us that you need to solve. You use this claim as an excuse to invade our precincts. Many of these problems don't exist. Where there are real conflicts, where there are wrongs, we will identify them and address them by our means. We are forming our own Social Contract. This governance will arise according to the conditions of our world, not yours. Our world is different.

Cyberspace consists of transactions, relationships, and thought itself, arrayed like a standing wave in the web of our communications. Ours is a world that is both everywhere and nowhere, but it is not where bodies live.

We are creating a world that all may enter without privilege or prejudice accorded by race, economic power, military force, or station of birth.

We are creating a world where anyone, anywhere may express his or her beliefs, no matter how singular, without fear of being coerced into silence or conformity.

Your legal concepts of property, expression, identity, movement, and context do not apply to us. They are based on matter. There is no matter here.

Our identities have no bodies, so, unlike you, we cannot obtain order by physical coercion. We believe that from ethics, enlightened self-interest, and the commonweal, our governance will emerge. Our identities may be distributed across many of your jurisdictions. The only law that all our constituent cultures would generally recognize is the Golden Rule. We hope we will be able to build our particular solutions on that basis. But we cannot accept the solutions you are attempting to impose.

In the United States, you have today created a law, the Telecommunications Reform Act, which repudiates your own Constitution and insults the dreams of Jefferson, Washington, Mill, Madison, De Toqueville, and Brandeis. These dreams must now be born anew in us.

You are terrified of your own children, since they are natives in a world where you will always be immigrants. Because you fear them, you entrust your bureaucracies with the parental responsibilities you are too cowardly to confront yourselves. In our world, all the sentiments and expressions of humanity, from the debasing to the angelic, are parts of a seamless whole, the global conversation of bits. We cannot separate the air that chokes from the air upon which wings beat.

In China, Germany, France, Russia, Singapore, Italy and the United States, you are trying to ward off the virus of liberty by erecting guard posts at the frontiers of Cyberspace. These may keep out the contagion for a small time, but they will not work in a world that will soon be blanketed in bit-bearing media.

Your increasingly obsolete information industries would perpetuate themselves by proposing laws, in America and elsewhere, that claim to own speech itself throughout the world. These laws would declare ideas to be another industrial product, no more noble than pig iron. In our world, whatever the human mind may create can be reproduced and distributed infinitely at no cost. The global conveyance of thought no longer requires your factories to accomplish.

These increasingly hostile and colonial measures place us in the same position as those previous lovers of freedom and self-determination who had to reject the authorities of distant, uninformed powers. We must declare our virtual selves immune to your sovereignty, even as we continue to consent to your rule over our bodies. We will spread ourselves across the Planet so that no one can arrest our thoughts.

> We will create a civilization of the Mind in Cyberspace. May it be more humane and fair than the world your governments have made before.
>
> John Perry Barlow Cognitive Dissident, Co-Founder, Electronic Frontier Foundation

Barlow co-founded the Electronic Frontier Foundation, an apt metaphor as cyberspace genuinely is a frontier. Some parallel the electronic frontier with the Western frontier of North America in its openness and initial lawlessness in the 1800s. David Post[13] is one such who saw the potential of the new frontier, as did his role model, Thomas Jefferson, in relation to an earlier frontier. Post advocates unfettered movement in cyberspace, arguing that the internet 'should be a free-flowing cauldron of creative energy unencumbered by copyright, trademark, jurisdictional, and other concerns'.[14] Andrew Morriss continues the analogy in his article 'The Wild West meets cyberspace'.[15] He argues that the internet is the product of spontaneous order in the same manner as the American Wild West, which was 'far away from government because there [we]re no centralised control mechanisms'. Nevertheless the miners and cattlemen in the West established protocols and developed norms of behaviour. In both the Wild West and the internet, order spontaneously emerges. Morriss predicted that the internet would not need outside assistance to reach its potential. Governmental control, regulation or even taxes, he said, would only hamper the free flow of information and design by the millions of stakeholders and users. However, the social norm of netiquette, a form of etiquette on the internet, has been judicially recognised.[16]

The rule of law and the rule of cyberspace

The 'rule of law' is a basic precept of legal systems. It is enshrined in the common law tradition, and in continental Europe is known as Rechtsstaat. The concept of impartial rule of law is found in the Chinese political philosophy of legalism. The rule of cyberspace is not and cannot be based on the rule of law. Nevertheless there are parallels between the development of the rule of law and the emergence and adoption of the rule cyberspace. Order and regulation emerge from varied parties, unplanned and unexpected. The rule of cyberspace morphs, melds and moulds in response to its own inherent dynamic forces. An understanding of the rule of law allows the cyber user to appreciate the rule of cyberspace.

13 See David Post, 'Napster, Jefferson's moose, and the law of cyberspace', (2000) www.temple.edu/lawschool/dpost/Napster.html; 'What Larry doesn't get: Code, law, and liberty in cyberspace', www.temple.edu/lawschool/dpost/Code.pdf; and '"The free use of our faculties": Jefferson, cyberspace, and the languages of social life', papers.ssrn.com/sol3/papers.cfm?abstract_id=264201.

14 Elizabeth Greene, 'David Post: Freeing cyberspace from the rule of law', (2000) *The Chronicle*, 20 October.

15 Andrew Morriss, 'The Wild West meets cyberspace', (1998) 48 *The Freeman: Ideas on liberty* 427.

16 *Ontario Inc. v Nexx Online Inc.* [1999] OJ No. 2246 (Sup Ct), dealing with bulk emailing at a time when such practices were not legislatively prohibited.

The rule of law

The rule of law ordains that all people come before the law equally. No one is above the law. Many writers have espoused subsets of legal principles in an attempt to define this concept. Aristotle posited the rule of law as a system of rules inherent in the natural order of things. More than 2000 years ago Aristotle said, 'The rule of law is preferable . . . to that of any individual.'[17] It ensures fairness to all, the right to be heard by an unbiased medium and similar outcomes in similar circumstances.

In the British common law tradition, Albert Venn Dicey, in his treatise Law of the Constitution in 1895, explained the rule of law thus:

> [E]very official, from the Prime Minister down to a constable or a collector of taxes, is under the same responsibility for every act done without legal justification as any other citizen. The Reports abound with cases in which officials have been brought before the courts, and made, in their personal capacity, liable to punishment, or to the payment of damages, for acts done in their official character but in excess of their lawful authority. [Appointed government officials and politicians alike] . . . and all subordinates, though carrying out the commands of their official superiors, are as responsible for any act which the law does not authorise as is any private and unofficial person.[18]
>
> [The 'rule of law' ensures] . . . the absolute supremacy or predominance of regular law as opposed to the influence of arbitrary power . . . Englishmen are ruled by the law, and by the law alone; a man may with us be punished for a breach of law, but he can be punished for nothing else. It means, again, equality before the law or the equal subjection of all classes to the ordinary law of the land administered by the ordinary law courts; the 'rule of law' in this sense excludes the idea of any exemption of officials or others from the duty of obedience to the law which governs other citizens or from the jurisdiction of the ordinary tribunals.[19]

In a similar vein, in American law, John Adams, in a draft of the constitution of the Commonwealth of Massachusetts, in justification and enunciation of the principle of separation of powers, wrote: 'to the end it may be a government of laws and not of men'.[20] The concept of rule of law does not relate to morality, justice or efficacy of laws, but to how the legal system upholds the law. It is often perceived as a prerequisite for democracies, though non-democratic nations and dictatorships can also be said to adhere to the rule of law.

Dicey outlined the three 'kindred conceptions' underpinning the English constitution,[21] namely:

- No man could be punished or lawfully interfered with by the authorities except for breaches of law.

17 Aristotle, *The Politics*, Book Three, Part XVI, trans. Benjamin Jowett, http://classics.mit.edu/Aristotle/politics.3.three.html.
18 AV Dicey, *Law of the Constitution*, 9th edn, Macmillan, London, 1950, p. 194.
19 AV Dicey, *Introduction to the study of the Law of the Constitution*, 10th edn, Macmillan, London, 1959, pp. 202–03.
20 Massachusetts Constitution, First Part, art. XXX (1780). The quote has been cited with approval by the US Supreme Court and every State Supreme Court in the United States.
21 Dicey, *Introduction to the study of the Law of the Constitution*.

- No man is above the law and everyone, regardless of rank, is subject to the ordinary laws of the land.
- There is no need for a bill of rights because the general principles of the constitution are the result of judicial decisions determining the rights of the private person.[22]

Bradford Smith describes the internet as a third industrial revolution, based on the joint and corresponding technological advances in software, hardware and telecommunications.[23] He examines the characteristics of the rule of law in real space and compares them with the rule of law in cyberspace.

The rule of cyberspace

Cyberspace is crammed with multitudes of data, text, images, sounds and videos; from priceless gems of information, knowledge and reasoning to abhorrent rubbish and trivia. This data is not stored or presented with any order or structure. There are no global editors, data checkers or barriers. While certain nation states do control and restrict access and there are laws which address certain materials and behaviour, neither directly impacts the ability of a user to place any data online, nor attempts any structural control. Cyberspace is chaos.

However, powerful search engines give order and structure to the data. Forms of etiquette emerge in social and commercial situations. Order spontaneously results from such chaos. The rule of cyberspace is the natural, emergent order arising from data chaos. Demand yields an effective order.

Spontaneous (or endogenous) order

> Good order results spontaneously when things are let alone.
>
> Zhuangzi, 4th century BCE

The concept of spontaneous order has been used to explain the development of language, science and technology, markets and economies, agriculture and eco-systems. It is a natural mechanism, operating without any specific intent or design: given an environment with rules, multiple behaviours and structures, a coherent order emerges. In all human culture, language emerged without design. Given human relations, emerging intelligence, multiple behavioural inputs and a community, language became ordered. A rainforest has a particular hierarchical order because of laws of science and physical attributes (gravity, light, energy transference, water flow, soil nutrients, air composition and so forth). The rainforest has an order and stability which may have evolved over eons. And yet, should any factor alter, so do countless other

22 Ibid., pp. 175–84.
23 Bradford Smith, 'The third industrial revolution: Policymaking for the internet', (2001) 3 *Colum Sci & Tech L Rev* 1.

things, and a new order emerges with such speed that it is said to be effectively spontaneous. Cyberspace, like the rainforest, is complex, with a multitude of independent factors. And yet, from what begins as disorder, an order emerges.

However, the expression 'spontaneous' is misleading. On one level, order in cyberspace is developing continuously. Rather than arising 'spontaneously', order arises naturally, even automatically. An alternative expression might be 'endogenous order': that is, order which originates naturally from within.

Friedrich Hayek was a Nobel Laureate economist whose theories and principles relating to spontaneous order have been applied to the rule of law. For Hayek, the rule of law means that government is bound by rules fixed in advance; this make how those in command of the rules will use (or abuse) their coercive powers in given factual situations foreseeable.

Hayek defined order as a reaction to given stimuli and circumstances. He contrasts 'cosmos' and 'taxis' types of order, the former being spontaneous and self-generating, the latter forced or engineered. Hayek argued that human beings lived within a social order that was more 'taxis' than 'cosmos'. He believed that the rule of law could not arise artificially. Instead it is a spontaneous development for societies structured around a belief in the virtues of the free market.

For example, Hayek recognised and accepted that language emerged not as a conscious human phenomenon; it emerged as an ordered form, but spontaneously. Equally, he regarded the economic concept of a 'price mechanism' as a spontaneous form of order rather than a conscious human invention. He discussed and theorised about evolved behaviour. In later years he attributed the birth of civilisation to private property:[24]

> the spontaneous interaction of a number of people, each possessing only bits of knowledge, brings about a state of affairs in which prices correspond to costs, etc., and which could be brought about by deliberate direction only by somebody who possessed the combined knowledge of all those individuals.[25]

Andrew Morriss[26] argues that the regulation of the internet component of cyberspace arose spontaneously in the same manner as did regulation in the Wild West of the United States.[27] In both the internet and the Wild West, acceptable behaviour is maintained by rules which have appeared and then developed spontaneously. There is not necessarily any law-making authority to determine and enforce appropriate rules. Morriss provides a range of insights into natural order in the Wild West and parallels this to examples in cyberspace. Through the mechanism of spontaneous order, he argues that cyberspace is well equipped to govern and rule itself.

An alternative approach to the concepts of spontaneous order and natural order is Adam Smith's 'invisible hand' theory:

24 Friedrich Hayek, *The Fatal Conceit*, University of Chicago Press, Chicago IL, 1988.
25 Friedrich Hayek, 'Economics and knowledge', *Economica* (New Series), 1937, vol. IV, p. 33.
26 See note 15.
27 Morriss, 'The Wild West meets cyberspace'.

> Every individual . . . generally, indeed, neither intends to promote the public interest, nor knows how much he is promoting it . . . he intends only his own gain, and he is in this, as in many other cases, led by an invisible hand to promote an end which was no part of his intention.[28]

Smith's *Theory of Moral Sentiments*,[29] an early work, contains the philosophical underpinnings and statements of methodology that underlie his later broader works on political economics, such as *The Wealth of Nations*.[30] It commences with the following declaration regarding the dual facilities of human existence and natural influences:

> *How selfish soever man may be supposed, there are evidently some principles in his nature,* which interest him in the fortunes of others, and render their happiness necessary to him, though he derives nothing from it, except the pleasure of seeing it. Of this kind is pity or compassion, the emotion we feel for the misery of others, when we either see it, or are made to conceive it in a very lively manner. That we often derive sorrow from the sorrows of others, is a matter of fact too obvious to require any instances to prove it; for this sentiment, like all the other original passions of human nature, is by no means confined to the virtuous or the humane, though they perhaps may feel it with the most exquisite sensibility. The greatest ruffian, *the most hardened violator of the laws of society, is not altogether without it*.[31]

Notwithstanding the philosophical differences between Smith's early and later writings, each can be extended to human existence in cyberspace. *The Theory of Moral Sentiments* accentuates the union of human intention and behaviour in a benevolent sense. *The Wealth of Nations* argues that an organised economic society results from the individual acting selfishly in a capitalistic manner, as if guided by an 'invisible hand', which results in an overall good for the community.

Cyberspace structure, organisation and continuation can be successfully explained by such spontaneous order notions, and by moral and capitalistic principles. Cyberspace, like real space, operates on multiple levels: social, commercial, domestic, educational and for entertainment. Each section of the community, by investing both time and effort into cyberspace, builds on the tools and structure laid down by predecessors. Cyberspace is dynamically evolving, pulsating with checks and balances, and subject to the vicissitudes of human actions.

A code of cyberspace

Lawrence Lessig espouses a new and separate code applicable in cyberspace; a version of the rule of cyberspace. Lessig is an academic who has contemplated and theorised about the nature of cyberspace from its infancy. In response to a

28 Adam Smith, *The Wealth of Nations*, Book IV Chapter II, Bantam Classic, (1776) 2003.
29 Adam Smith published *The Theory of Moral Sentiments* in 1759.
30 Adam Smith published *The Wealth of Nations* in 1776, at the beginning of the Industrial Revolution.
31 Smith, *The Theory of Moral Sentiments*, emphasis added.

remark by Judge Frank Easterbrook that there was no more a 'law of cyberspace' than there was a 'law of the horse',[32] Lessig published a rebuttal: 'The law of the horse: What cyberlaw might teach'.[33] He began by considering particular problems of regulation that cyberspace might present. He expanded these examples to propose models for applying regulation in cyberspace and real space, then examined how law in real space can regulate cyberspace, and finally proposed that cyberspace could in turn regulate law. At heart, he claimed, there is a systematic competition between cyberspace and real space which 'illuminate[s] the entire law'.

Lessig believed that cyberspace has a parallel 'architecture' to real space.[34] He describes the rule of cyberspace as a code. The code of cyberspace comprises components real and ethereal, 'the software and hardware that makes this part of cyberspace as it is':[35]

> Once it is plain that code can replace law, the pedigree of the codewriters becomes central. Code in essence becomes an alternative sovereign – since it is in essence an alternative structure of regulation. But who authors this sovereign authority? And with what legitimacy?[36]

One of Lessig's examples relates to copying. Cyberspace has impacted on the technology of copying and on the power of law to protect against illegal copying.[37] Lessig explores the notion that computer code may regulate conduct in much the same way that legal codes regulate real space. He argues that the code in cyberspace supplants copyright law in real space, and that such intellectual property concepts are inappropriate in the new medium.

Lessig proposes four factors which regulate real space and cyberspace: law, norms, markets and code.[38] In real space, law demands and requires that citizens behave in a certain community-minded manner, with sanctions for non-compliance. Social norms involve the threat of punishment (for breaches) by fellow citizens. Markets restrain behaviour fiscally and code establishes regulation on the individual and collective behaviour by the sheer 'architecture', meaning simply the 'physical world as we find it'. These four factors apply validly and equally to the unique yet dynamic architecture of cyberspace. But we may be surprised by the manner in which they are manifested. Lessig gives this example: 'Talk about democratic politics in the *alt.knitting* newsgroup, and you open yourself to flaming; "spoof" someone's identity in a MUD, and you might find

32 'Cyberspace and the law of the horse', 1996 *U Chi L Forum* 207. Judge Easterbrook's reference is to an argument by Gerhard Casper, a former Law Dean at University of Chicago, who boasted that the law school did not offer a course in 'the law of the horse'.

33 Lawrence Lessig, 'The law of the horse: What cyberlaw might teach', (1999) 113 *Harvard Law Review* 501.

34 Lawrence Lessig, *Code: The future of ideas*, Basic Books, New York, 2002. See Lawrence Lessig's website code-is-law.org. For example, in *Code: The future of ideas* Lessig argues that too much long-term copyright protection hampers the creation of new ideas, and he advocates the importance of existing works entering the public domain quickly.

35 Lessig, 'The law of the horse: What cyberlaw might teach', p. 502.

36 Lawrence Lessig, *Code, and other laws of cyberspace*, Basic Books, New York, 1999.

37 Lessig, *Code: The future of ideas*, pp. 125–7.

38 See also Lawrence Lessig, 'The new Chicago School', (1998) 27 *J Legal Stud* 661.

yourself toaded.'[39] Understanding the forces inherent in the rule of cyberspace assists in an appreciation of the rule of law in real space by providing a new and unexpected perspective.

Lawrence Lessig does not support Smith's 'invisible hand' approach:

> Constitutions in this sense are built; they are not found. Foundations are laid; they don't magically appear. Just as the founders of our nation learned from the anarchy that followed the revolution (remember: our first constitution, the Articles of Confederation, was a miserable failure of do-nothingness), so too are we beginning to see in cyberspace that this building is *not the work of an invisible hand*. There is *no reason to believe that the grounding for liberty in cyberspace will simply emerge*. In fact, quite the opposite is the case. As our framers learned, and as the Russians saw, we have reason to believe that cyberspace, left to itself, will not fulfill the promise of freedom. Left to itself, cyberspace will become a perfect tool of control.
>
> Control. Not necessarily control by government, and not necessarily control for some evil, fascist end. The invisible hand of cyberspace is building an architecture that's quite the opposite of what it was at cyberspace's birth. The invisible hand, through commerce, is constructing an architecture that perfects control – an architecture that makes possible highly efficient regulation.[40]

Lessig's libertarian writings may be explained as stemming from a deep concern and passion for the protection of the growth of cyberspace as a new and significant medium. He does not want the future of the internet to be left to an invisible hand or the whims of some natural or spontaneous ethereal force.

David Post submits an impressive rebuttal in his article 'What Larry doesn't get'.[41] Post argues that Lessig's calls for collective action are unlikely to entice libertarians into action. Collective action, in Post's view, is another way of using coercive force to achieve directed change for the common good. Post questions any structured direction, expressing concern at who makes such determinations and at what expense. Post states that the 'conscientious libertarian recognizes that there are times when collective action is required to promote the common welfare', but the 'architectures of liberty are of fundamental importance'. He concludes that while cyberspace needs architecture, there remains disagreement about the extent to which coercive power of the state needs to be invoked to get those communities built and to get people to live there.[42]

Information wants to be free

The dynamics of cyberspace and the resultant order give the impression that an overall design and structure was planned, much like Smith's 'invisible hand' concept. But spontaneous order is a process akin to evolution, with no planned

39 Ibid.
40 Ibid.
41 'What Larry doesn't get: Code and other laws of cyberspace', (2000) 52 *Stanford Law Review* 1439.
42 Ibid., pp. 1439, 1458–9.

direction or apparent purpose. Such a process is reflected in the phrase 'information wants to be free', taken from a hackers conference in 1984:

> On the one hand information wants to be expensive, because it's so valuable. The right information in the right place just changes your life. On the other hand, *information wants to be free*, because the cost of getting it out is getting lower and lower all the time. So you have these two fighting against each other.[43]

'Information', of itself, has no mind or direction. Subjectively, to human users certain information may be considered valuable, and other information may be regarded as rubbish. But it is of different value to each individual. And yet, in cyberspace, where multiple access and the copying of digital data (information) has virtually no cost, information appears to have a life of its own, as it is copied, sent, read, posted and downloaded. The infrastructure of cyberspace provides both order and freedom given the demands of its human users.

Conclusion

Spontaneous order facilitates diversity and flexibility and arises in many and varied ways: online protocols, hypertext mark-up language, the commercial presence of the private sector, intellectual property tensions, business identifiers, and speedy communications are just a few of the factors that shape and contribute to – and equally are shaped and developed by – the rich fabric of cyberspace.

The architecture of the internet interprets attempts at censorship as damage and routes around it.[44] This is a by-product of the utilisation of mass networks, which include in their initial design a way to route around damage, should there be, for example, massive actual damage from blackouts, war or even a nuclear catastrophe. Cyberspace has become a medium beyond the control of terrestrial jurisdictions. The behaviour of individuals in real space can be targeted, but the control of cyberspace globally is beyond reach.[45]

Spontaneous or endogenous order is not regulation and it is not control. It is a process. The cyberspace community is impacted first by the human condition and human use, and our penchant for information and entertainment; second by legislation and regulation, the human predilection to control for particular purposes (this targets aspects of cyberspace such as child pornography, defamation and crime); third by protocols, either innate in the architecture or in the form of etiquette, netiquette and rules of human interaction. These factors coexist and coalesce. A variation in any one aspect may result in a change in cyberspace, but the process engenders order. Those who attempt control will cause shifts in the structure of cyberspace and fluctuations in the order; but control of cyberspace is

43 Stewart Brand, *Whole Earth Review*, Point Foundation, San Diego CA, 1984, p. 49, emphasis added.
44 'The internet interprets censorship as damage, and routes around it': John Gilmore, *The Electronic Frontier Foundation quotes collection*, 2005, w2.eff.org/Misc/EFF/?f=quotes.eff.txt.
45 See J Goldsmith, 'Regulation of the internet: Three persistent fallacies', (1998) 73 *Chi-Kent L Rev* 1112.

not possible. Local communities may filter or restrict access: parents and schools may utilise filters, for instance, and North Korea and China may severely regulate their population's use and entry into cyberspace. However the rule of cyberspace has evolved to a point where no person, no body and no government can regulate cyberspace. Behavioural traits and social norms inhibit abhorrent behaviour to a substantial degree, but no less so than in real space. Cyberspace and the internet are exquisite examples of spontaneous or endogenous order.

The rule of cyberspace is not simply the rule of law in cyberspace; it is the amalgamation and the fraternity of factors of spontaneous endogenous order. The rule of cyberspace is both the juxtaposition of hardware and structure with verisimilitude, and human behaviour within the cyberspace environment.

The backbone of a democratic society is the concept of the rule of law. The backbone of a robust community in cyberspace is the rule of cyberspace.

Further reading

Michael Benedikt, *Cyberspace: First steps*, MIT Press, Cambridge MA, 1991.

James Boyle, 'Foucault in cyberspace: Surveillance, sovereignty, and hardwired censors', (1997) *University of Cincinnati Law Review* 66.

AV Dicey, *Introduction to the study of the Law of the Constitution*, Macmillan Press, London, 1885.

William Gibson, *Neuromancer*, Ace Books, New York, 1984.

Graham Greenleaf, 'An endnote on regulating cyberspace: Architecture vs Law?', [1998] *UNSWLJ* 52.

Friedrich Hayek, *Rules and Order, Volume 1: Law, Legislation and Liberty*, 1973.

David R Johnson and David Post, 'Law and borders – The rise of law in cyberspace', (1996) 48 *Stanford Law Review* 1367.

Lawrence Lessig, *Code: The future of ideas – the fate of the commons in a connected world*, First Vintage Books, 2002.

Lawrence Lessig, *Code, and other laws of cyberspace*, Basic Books, 1999.

Lawrence Lessig, 'The law of the horse: What cyberlaw might teach', (1999) 113 *Harvard Law Review* 501.

Lawrence Lessig, 'Reading the Constitution in cyberspace', (1996) 45 *Emory LJ* 869.

AP Morriss, 'The Wild West meets cyberspace', (1998) 48 *The Freeman: Ideas on Liberty* 427.

David Post, 'What Larry doesn't get: Code and other laws of cyberspace', (2000) *Stanford Law Review*, Vol. 52, 1439.

Adam Smith, *The Wealth of Nations*, Book IV Chapter II, Mass Market Paperback, 2003.

3

Electronic commerce and the law of contract

Commerce is typically about profit. The history and development of commerce involves the element of risk. Risk assessment in commerce involves consideration of such factors as the law of contracts, the parties, the goods or services and legal forum. Many commercial parties have braved the new electronic commerce world without knowing or understanding the legal implications.

In an attempt to ensure confidence, many international organisations have proposed treaties, model laws and protocols to encourage certainty and stability for these international electronic commercial practices and in relation to laws of contract.[1] The UNCITRAL Model Law of Electronic Commerce (Model Law) has proved the most popular, with significant international acceptance by national legislatures, including Australia and New Zealand.

This chapter addresses the regulation of, and legislative responses to, electronic contracting. It challenges the wisdom of and necessity for legislation based on the Model Law, and the introduction of the concept of consent as a precondition for the application of selected legislative provisions.

UNCITRAL Model Law of Electronic Commerce

In 1996 the UN Commission on International Law Trade (UNCITRAL) released what is now the most popular model for consumer and commercial protection

1 These organisations have included the Organisation for Economic Cooperation and Development (OECD) www.oecd.org/subject/e_commerce/; the United Nations; the Asia-Pacific Economic Cooperation forum (APEC) www.apec.org; the International Chamber of Commerce (ICC) www.iccwbo.org; the World Trade Organisation (WTO) www.wto.org/english/tratop_e/ecom_e/ecom_e.htm; and the UN Commission on International Trade Law (UNCITRAL) www.uncitral.org.

in an electronic environment. The UNCITRAL Model Law on Electronic Commerce[2] (Model Law) was intended to provide national legislatures with a template of internationally acceptable rules that would remove legal obstacles and create a more secure legal environment for electronic commerce. The Model Law was intended to facilitate the use of electronic communication and storage of information such as electronic data interchange and electronic mail. It provided standard ways to assess the legal value of electronic messages and legal rules for electronic commerce in specific areas such as carriage of goods.

The Model Law has gained significant international acceptance.[3] The drafting process was attended by representatives of over 50 nations and 10 international organisations.

The Model Law does not specifically refer to contract law. Instead it deals with the principle of functional equivalence of electronic media in commercial transactions.[4] That is, where the electronic form is functionally equivalent to the traditional form, it should be treated equally by the law. This principle permeates all legislation based on the Model Law. A second principle underlying the Model Law is that of technology neutrality (the term was chosen in response to the recognition that technology is constantly developing). For example, as 'electronic mail' connotes a certain medium, the Model Law uses the general expression 'data message'.

The Model Law addresses:

- legal recognition of data messages;
- writing;
- signatures;
- originals;
- admissibility and evidentiary weight of data messages;
- retention of data messages;
- formation and validity of contracts;
- recognition by parties of data messages;[5]
- attribution of data messages;

2 General Assembly Resolution 51/162 of 16 December 1996, amended in 1998. Available at www.uncitral.org/pdf/english/texts/electcom/05–89450_Ebook.pdf.

3 Legislation implementing provisions of the Model Law has been adopted in Australia (1999), China (2004), Colombia (1999), the Dominican Republic (2002), Ecuador (2002), France (2000), India (2000), Ireland (2000), Jordan (2001), Mauritius (2000), Mexico (2000), New Zealand (2002), Pakistan (2002), Panama (2001), the Philippines (2000), the Republic of Korea (1999), Singapore (1998), Slovenia (2000), South Africa (2002), Sri Lanka (2006), Thailand (2002) and the United Arab Emirates (2006), Venezuela (2001), and Vietnam (2005). The Model Law has also been adopted in the Bailiwick of Guernsey (2000), the Bailiwick of Jersey (2000) and the Isle of Man (2000), all Crown Dependencies of the United Kingdom of Great Britain and Northern Ireland; in Bermuda (1999), the Cayman Islands (2000), and the Turks and Caicos Islands (2000), overseas territories of the United Kingdom of Great Britain and Northern Ireland; and in the Hong Kong Special Administrative Region of China (2000). Uniform legislation influenced by the Model Law prepared in the United States (Uniform Electronic Transactions Act, adopted in 1999 by the National Conference of Commissioners on Uniform State Law) and Canada. See www.uncitral.org.

4 See *Official Guide to Enactment of the UNCITRAL Model Law on Electronic Commerce* para 15 et al.: www.uncitral.org/pdf/english/texts/electcom/05–89450_Ebook.pdf.

5 For criticism of this approach see *Criticisms* supra.

- acknowledgment of receipt; and
- time and place of dispatch and receipt of data messages.[6]

Legislation

The *Electronic Transactions Act 1999* (Cth) is based on the Model Law. All Australian states and territories have enacted parallel legislation.[7] The Federal Attorney General stated that the 'enactment of the uniform Bill will achieve the Commonwealth's goal of national uniform legislation to remove the legal impediments facing electronic transactions'.[8] The stated objects within the Act are to provide a regulatory framework that:

- recognises the importance of the information economy to the future economic and social prosperity of Australia;
- facilitates the use of electronic transactions;
- promotes business and community confidence in the use of electronic transactions; and
- enables business and the community to use electronic communications in their dealings with government.[9]

Australia

In 1997 the Wallis Report on the financial industry in Australia recognised the need 'to adopt appropriate internationally recognised standards for electronic commerce, including for electronic transactions over the Internet and the recognition of electronic signatures'.[10] Specifically, the report recommended:

- amendments to legislation and industry codes to permit the appropriate use of digital signatures, electronic notices and documents to improve the efficiency of financial transactions and reduce costs;
- endorsement by industry and government of the Public Key Authentication Framework; and
- amendments to the Evidence Acts to take account of electronic transactions and record keeping.

In May 1999, at a meeting of the Standing Committee of Attorneys General, the states agreed to enact parallel legislation based on the Commonwealth Bill. All

6 See Model Law, Arts 5 to 15.
7 *Electronic Transactions Act 2000* (NSW); *Electronic Transactions (Queensland) Act 2001* (Qld) ; *Electronic Transactions Act 2000* (SA) ; *Electronic Transactions Act 2000* (Tas) ; *Electronic Transactions (Victoria) Act 2000* (Vic); *Electronic Transactions Act 2003* (WA); *Electronic Transactions Act 2001* (ACT) ; *Electronic Transactions (Northern Territory) Act 2000* (NT). References to Acts in this chapter will be to the corresponding Acts listed here.
8 News release by Federal Attorney General Daryl Williams, April 2000.
9 Section 3.
10 Chapter 2, para 2.67.

states and territories have prepared legislation in accordance with the agreement. The nine pieces of legislation are:

Cth	*Electronic Transactions Act 1999*
NSW	*Electronic Transactions Act 2000*
Qld	*Electronic Transactions (Queensland) Act 2001*
SA	*Electronic Transactions Act 2000*
Tas	*Electronic Transactions Act 2000*
Vic	*Electronic Transactions (Victoria) Act 2000*
WA	*Electronic Transactions Act 2003*
ACT	*Electronic Transactions Act 2001*
NT	*Electronic Transactions (Northern Territory) Act 2000*

The *Electronic Transactions Act 1999* (Cth) came into full operation on 1 July 2001 and applies to all laws of the Commonwealth except those specifically exempted by its regulations.[11] For constitutional reasons the Commonwealth Act applies only 'for the purposes of a law of the Commonwealth'.[12] 'Laws of the Commonwealth' are specified in the regulations.[13] The federal legislature relies primarily upon the corporations power, the trade and commerce power and the territories power to give it the ability to deal with electronic commerce issues.[14] The corresponding state and territory legislation has no need for such a restriction.

Because there are minor changes in the legislation across the states, territories and the Commonwealth, choice of law provisions will play an important part in the application of the rules in this area. All these Acts bind the Crown.[15]

New Zealand

The New Zealand *Electronic Transactions Act 2002* and the Electronic Transactions Regulations came into force on 21 November 2003. The stated purpose is to facilitate the use of electronic technology by reducing uncertainty regarding the legal effect of information that is in electronic form or that is communicated by electronic means, and the time and place of dispatch and receipt of electronic communications and by functional equivalence. The Act provides that in its interpretation, reference may be made to the UNCITRAL Model Law and any document that relates to the Model Law.[16]

11 More than 150 exemptions have been proclaimed by regulations. See Electronic Transactions Regulations 2000 (Cth).
12 Section 8.
13 *Electronic Transactions Act 1999* (Cth) s5.
14 *Australian Constitution*, ss51(xx), 51(i) and 122 respectively.
15 Cth – s6; NSW – s6; Qld – s7; SA – s6; Tas – s4; Vic – s6; WA – s6; NT – s6. The ACT Act omits the provision.
16 Section 6.

Provisions of the Electronic Transactions Acts

Provisions	Cth	ACT	NSW	NT	Qld	SA	Tas	Vic	WA	NZ[17]
Short title	1	1	1	1	1	1	1		1	1
Commencement	2		2	2	2	2	2	2	2	2
Object	3	3	3	3	3	3		1, 4	3	3
Simplified outline	4	4	4	4	4, 5	4		5		4
Definitions	5	5	5	5	6	5	3	3	5	5–6
Crown to be bound	6		6	6	7	6	4	6	6	7
External Territories	7									
Validity of electronic transactions	8	7	7	7	8	7	5	7	7	8
Writing	9	8	8	8	9–13	8	6	8	8	18–21
Signature	10	9	9	9	14–15	9	7	9	9	22–24
Production of document	11	10	10	10	16–18	10	8	10	10	28–29
Retention	12	11	11	11	19–21	11	9	11	11	25–26
Exemptions	13	12	12	12		12	10	12	12	14
Time and place of dispatch and receipt of electronic communications	14	13	13	13	22–25	13	11	13	13	9–13
Attribution of electronic communications	15	14	14	14	26	14	12	14	14	
Regulations	16	15	15	15	27	15	13	15	15	36
Administration of Act							14			

United States

The United States *Uniform Electronic Transactions Act 1999* is the product of the National Conference of Commissioners on Uniform State Laws and uses the UNCITRAL Model Law as its template. The Act applies only to transactions between parties who have agreed to conduct transactions by electronic means.[18] The Act defines an 'electronic record' as 'a record created, generated, sent, communicated, received, or stored by electronic means'.[19]

United Kingdom

In the United Kingdom the two relevant pieces of legislation are the *Land Registration Act* 2002 (UK) and the *Electronic Communications Act* 2000 (UK).

The *Land Registration Act 2002* replaced 1925 legislation relating to registered land and dealings with unregistered land in England and Wales. Part 8 implements electronic conveyancing processes. An electronic document to which section 91 applies is to be regarded for the purposes of any enactment as a deed. Section 91 applies to certain dispositions (including documents used in conveyancing) in an electronic form which make provision for the time and date when they take effect and include certified electronic signatures of each person by whom they purport to be authenticated. In these circumstances the

17 The New Zealand Act also includes provisions dealing with access, originals, copyright and a separate provision dealing with consent.
18 See *Consent*, below.
19 The US Act is available at: www.law.upenn.edu/bll/archives/ulc/fnact99/1990s/ueta99.htm.

dispositions are to be regarded as in writing and signed by each individual, and sealed by each corporation whose electronic signature they have. Where notice of assignment made by means of a document pursuant to section 91 is given in electronic form, it is to be regarded for the purposes of any enactment as having been given in writing.[20]

Subsequent provisions of the Land Registration Act allow the registrar to arrange an electronic communications network for a range of purposes such as electronic registration and electronic settlement. Separate rules deal with the communication of documents in electronic form to the registrar and the electronic storage of documents communicated to the registrar in electronic form.

The Land Registration Act refers to the *Electronic Communications Act 2000* (UK) to identify what would qualify as an electronic signature and what constitutes a certification.[21] The Explanatory Memorandum to the Act states that the provisions do 'not disapply the formal statutory or common law requirements relating to deeds and documents but deems compliance with them'.

Electronic contracts

Contract formality does not alter merely because an electronic medium was used. Simple contracts requiring no formality are entered into daily. A contract which can be entered into orally can of course be entered into by use of email and other forms of electronic communication. This is not new. The first electronic contract was entered into, in all likelihood, in the mid 1800s, when the telegraph was first in commercial use. The courts have had little difficulty recognising contract formation by electronic means.[22] However, the perception of the 1990s was that commercial parties were uncertain of the use of electronic media where formal requirements such as writing, signature, production and retention were concerned.

The title to the nine Australian legislative instruments and the New Zealand Act includes the words 'electronic transactions'. The word 'contract' does not appear in the Commonwealth or New Zealand legislation, and in all states and territories it only appears in the expanded meaning of 'transactions', as including 'any transaction in the nature of a contract, agreement or other arrangement, and also . . . any transaction of a non-commercial nature'. Nevertheless, 'transactions' implicitly includes contractual communications and documents.

20 Section 91(10) provides: 'In this section, references to an electronic signature and to the certification of such a signature are to be read in accordance with section 7(2) and (3) of the *Electronic Communications Act 2000* (c. 7).'

21 See *Signatures*, below.

22 For example, see *Brinkibon Ltd v Stahag Stahl und Stahlwarenhandelsgesellschaft mbH* [1983] 2 AC 34, *Reese Bros Plastics Ltd v Hamon-Sobelco Aust. Pty Ltd* (1988) 5 BPR 11,106 and *Databank Systems Ltd v Commissioner of Inland Revenue* [1990] 3 NZLR 385.

Notwithstanding the general absence of references to contracts, the Electronic Transactions Acts do affect key elements of contracts, such as the timing and place of communications, records and documents, and the basic constructs of offer and acceptance.

Similarly, the UNCITRAL Model Law deals with general concepts applying to 'any kind of information in the form of a data message used in the context of commercial activities'[23] rather than contract issues specifically, but is to be interpreted broadly 'so as to cover matters arising from all relationships of a commercial nature, whether contractual or not'.[24] Its provisions are intended to affect contract formation and performance. Article 5 provides:

> Information shall not be denied legal effect, validity or enforceability solely on the grounds that it is in the form of a data message.

'Data message' is defined as: 'information generated, sent, received or stored by electronic, optical or similar means including, but not limited to, electronic data interchange (EDI), electronic mail, telegram, telex or telecopy'.[25]

The expression 'data message' is used as a technology-neutral term. In the Australian and New Zealand legislation the expression used is 'electronic communication'.

Common law

In their zeal, legislators seem to have assumed that the common law would not recognise functional equivalence nor the validity of electronic communications and electronic documents. The necessity for the Electronic Transactions Acts is questionable. Historically the courts have proved most flexible in such circumstances.

Two questions arise from this assumption. First, what would the courts decide in circumstances where the Electronic Transactions Acts do not apply? Second, what is the effect of exemptions listed within the Act or its Regulations?

Application of the common law

Despite the fact that for more than a century the common law has recognised electronic communications – by telegraph, radio and telephone, and more recently by telex and facsimile – the emergence of the internet, EDI (Electronic Data

23 Model Law, Art. 1.
24 Footnote to Art. 1. The note expands the scope, stating, 'Relationships of a commercial nature include, but are not limited to, the following transactions: any trade transaction for the supply or exchange of goods or services; distribution agreement; commercial representation or agency; factoring; leasing; construction of works; consulting; engineering; licensing; investment; financing; banking; insurance; exploitation agreement or concession; joint venture and other forms of industrial or business cooperation; carriage of goods or passengers by air, sea, rail or road.'
25 Model Law, Art. 2.

Interchange) and email has led governments to legislate new rules. A handful of cases show that the common law courts would have managed.

In *McGuren v Simpson*,[26] McGuren claimed Simpson was barred by the *Limitation Act 1969* (NSW) from commencing legal proceedings for debt. Section 54 provides that the limitation period recommences where a confirmation or acknowledgment is made before expiry of the period, provided such acknowledgment is in writing and signed by the maker. The plaintiff's cause of action arose on 26 November 1993, the disputed email acknowledgment was sent and received on 29 September 1999 and proceedings commenced on 26 August 2002. The limitation period was six years. Initially, in holding that the limitation period had been extended, Lulham LCM incorrectly applied the *Electronic Transactions Act 2000* (NSW). On appeal to the NSW Supreme Court all parties agreed that as the Act commenced in 2001 it did not apply to the 1999 email. However, Lulham LCM stated that his decision was also based on the common law: 'If the name of the party to be charged is printed or written on a document intended to be a memorandum of the contract, either by himself of [sic] his authorised agent, it is his signature whether it is at the beginning or middle or foot of the document.'[27]

The issue before the NSW Supreme Court was whether at common law the email amounted to writing and signature for the purposes of section 54 of the *Limitation Act 1969* (NSW). The court concluded: 'It is [this court's] view that . . . s54 of the Act ought to be read to accommodate technological change and that, accordingly, the email sent by the plaintiff constitutes a written document.'

The court regarded the words of the email 'yes I spent the money and I shouldn't have', together with the name appearing in the email, as a written and signed confirmation of McGuren's obligation to Simpson.

In 2007, in *Hume Computers Pty Ltd v Exact International BV*,[28] Jacobson J applied 'domestic law' because, for international jurisdictional reasons, *lex fori* applied. His Honour rejected the submission 'because an email is an electronic representation of writing and therefore does not constitute writing itself as required by [the Distributorship agreement between the parties]'.[29] He stated:

> The requirement of written notice is to be construed in light of the fact that this is a commercial agreement made between two companies engaged in the computer software business. I would be blinding myself to commercial and technological realities to find that an email communication in the present circumstances was not written notice.[30]

26 [2004] NSWSC 35.

27 Lulham LCM cited a passage from Cheshire and Fifoot's *Law of Contract* (7th edn) and the case *Darryl v Evans* (1962) H&C 174 at 191. See also the conflicting US decisions of *Ballas v Tedesco* 41 F Supp 2d 531, at 541 and *Graham Technology Solutions Inc. v Thinking Pictures Inc.* 949 F Supp. 1427 (ND Cal 1997) and *Lockheed-Arabia v Owen* [1993] 3 WLR 468.

28 [2007] FCA 478.

29 [2007] FCA 478, para 48.

30 [2007] FCA 478, para 49.

In *Wilkens v Iowa Insurance Commissioner*,[31] the Iowa Court of Appeals held that a requirement to keep a written record of an insurance contract was satisfied by an insurer keeping written records on its computer system. The Court of Appeals applied the common law as the case preceded the US legislation.

The Law Commission for England and Wales, in its paper 'Electronic commerce: Formal requirements in commercial transactions – advice from the Law Commission', reached a view consistent with that expressed in *Wilkens*: a document which can be printed and stored is 'in writing'.[32]

The Electronic Transactions Acts were premature. The Acts at best complicate commercial transactions and contracts unnecessarily and at worst are redundant and superfluous.

Exemptions

Each of the nine Australian jurisdictions permits exemptions from the application of their respective Electronic Transactions Acts.[33] All except Queensland permit exemptions by regulations. Queensland's exemptions appear in the Schedule to its Act.[34] There are considerable inconsistencies in the approaches taken among the jurisdictions. Uniform legislation would remedy this defect. The Commonwealth's jurisdiction is inherently different from that of the states and territories and its Act's regulations list more than 150 specific exemptions.[35]

Only two states exempt dispositions of land.[36] Some exempt testamentary dispositions specifically.[37] Others use a variation of the words 'a requirement or permission for a document to be attested, authenticated, verified or witnessed by a person other than the author of the document'.[38] This expression is a minefield, applying to many transactions, contracts, affidavits, statutory declarations, notices, deeds and much more. Curiously, the order of words varies. In New South Wales and Western Australia it is 'verified, authenticated, attested or witnessed'; in Queensland 'attested, authenticated, verified or witnessed' and in South Australia 'witnessed, attested, verified or authenticated'.

31 (1990) 457 NW 2d.

32 Paragraphs 3.5 to 3.23; Law Commission for England and Wales paper, available at: www.lawcom.gov.uk/docs/e-commerce.pdf.

33 The ACT is the only jurisdiction not to enact exemptions.

34 This is in accordance with the Queensland Parliament's principle that legislation should not contain Henry VIII clauses. See *Legislative Standards Act 1992* (Qld) s4(5).

35 Electronic Transactions Regulations 2000 (Cth) Schedule 1. For example, bills of exchange and cheques.

36 Electronic Transactions Regulations 2007 (NSW) regulations 4, 7; Electronic Transactions Regulations 2002 (SA) regulation 4.

37 Electronic Transactions (Northern Territory) Regulations regulation 2; Electronic Transactions Regulations 2001 (Tas) regulation 4(a); Electronic Transactions (Victoria) Regulations 2000 regulation 5; Electronic Transactions Regulations 2003 (WA) regulations 3, 5.

38 *Electronic Transactions (Queensland) Act 2001* (Qld) cl6, Schedule 1; Electronic Transactions Regulations 2007 (NSW) regulation 5(f) ; Electronic Transactions Regulations 2002 (SA) regulation 5; Electronic Transactions Regulations 2003 (WA) regulations 3, 5. Note Western Australia, the last to enact its legislation, chose to exempt testamentary dispositions specifically and include the general expression.

One might wonder why the drafters chose different phrases and whether courts might interpret the phrases differently. Other exemptions include filing and producing documents for judicial proceedings, personal service and powers of attorney.

The rationale for these exemptions is unstated. However, it seems clear that there is a perception by the drafters that electronic communications and documents ought not to be regarded as functionally equivalent in certain circumstances. The Guide to the UNCITRAL Model Law explains that the matter of specifying exclusions should be left to enacting states, to take better account of differences in national circumstances. However, it warned that the objectives of the Model Law 'would not be achieved' if legislators used 'blanket exceptions'.[39] The Commonwealth, with its numerous exemptions, and the four states which use the 'attested, authenticated, verified or witnessed' document exemption, risk falling into this blanket exemption category. More serious is the effect on the principle of functional equivalence. Underlying these exemptions is an apparent mistrust of electronic communications and documents. The 'attested, authenticated, verified or witnessed' document exemption may arise from the perceived impracticality of witnessing electronic documents and a lack of understanding as to how such witnessing can be achieved.[40]

Exemption does not equate to a paper requirement

Many of the Commonwealth exceptions and the state exemptions may be based on the assumption that traditional paper or hard copy will be required in exempted situations. Such an assumption is fundamentally flawed. An exemption does not equate to a paper or hardcopy requirement. It merely means that the Act does not apply in that situation. The situation then falls to the common law: the courts will in these circumstances independently determine that a given electronic communication or document will or will not suffice for a particular requirement. Indeed there are a handful of cases which have already so held (discussed above).[41]

New Zealand

The *Electronic Transactions Act 2002* (NZ) provides for exemptions to be listed in the Schedule to the Act (section 14). The Schedule lists more than 70 specific and general exemptions, including: notices that are required to be given to the

39 *Official Guide to Enactment of the UNCITRAL Model Law on Electronic Commerce* paras 52 and 69.
40 See A McCullagh, W Caelli and P Little, 'Signature stripping: A digital dilemma', (2001) 1 *Journal of Information, Law and Technology*.
41 See *Common law*, above; *McGuren v Simpson* [2004] NSWSC 35; *Hume Computers Pty Ltd v Exact International BV* [2007] FCA 478; *Wilkens v Iowa Insurance Commissioner* (1990) 457 NW 2d 1.

public; information that is required to be given in writing either in person or by registered post; affidavits, statutory declarations, or other documents given on oath or affirmation; powers of attorney; testamentary instruments; negotiable instruments; and documents to files or produced in connection with judicial proceedings. The list is so broad as to risk falling into the UNCITRAL Guide's 'blanket exemption' and to thwart functional equivalence. Parties entering into contracts using electronic means under New Zealand law must be cognisant of the length and breadth of these exemptions, which substantially reverse the Act's effect.

Validity of electronic transactions

The Australian Electronic Transactions Acts provide that:

> a transaction is not invalid because it took place wholly or partly by means of one or more electronic communications.[42]

The New Zealand *Electronic Transactions Act 2002* states:

> To avoid doubt, information is not denied legal effect solely because it is-
> (a) in electronic form or is in an electronic communication;
> (b) referred to in an electronic communication that is intended to give rise to that legal effect.[43]

The immediate distinction is that the Australian provision is predicated on a transaction, while the New Zealand provision is predicated on information. Both are variations on the Model Law, which provides that:

> Information shall not be denied legal effect, validity or enforceability solely on the grounds that it is in the form of a data message.[44]

This general rule is subject to other provisions that deal with the validity of transactions, specific requirements of such elements as writing and signature, and the specific exemptions. Common law courts have never questioned the validity of electronic contracts.[45] Should an electronic communication of a document be challenged, it is more likely that one of the specific provisions of the Act would be utilised rather than this general provision. Nevertheless, this provision establishes the general rule and context for validity, and the form of functional equivalence underlying the legislation.

42 Cth – s8; NSW – s7; Qld – s8; SA – s7; Tas – s5; Vic – s7; WA – s7; ACT – s7; NT – s7. Queensland added the word 'merely' before the word 'because', perhaps giving rise to an argument that it may be a factor.
43 NZ – s8.
44 UNCITRAL Model Law, Art. 5. See also Art. 5bis (added in 1998) and Art. 11, specifically dealing with the formation and validity of contracts.
45 See note 22.

Writing

The writing provisions of the Electronic Transactions Acts are designed to provide functional equivalence of electronic writing, on conditions, where writing is otherwise required or permitted by law.

Many pieces of legislation require contracts to be in writing or evidenced in writing. Such legislation is typically based on the *Statute of Frauds 1677* (Imp), the aim of which was to help protect people and their property against fraud and sharp practice by legislating that certain types of contract could not be enforced unless there was written evidence of its existence and of its terms.[46]

The decision by UNCITRAL to formulate the template legislation was taken because in 'a number of countries the existing legislation governing communication and storage of information is inadequate or outdated because it does not contemplate the use of electronic commerce'.[47] In most common law jurisdictions, the *Statute of Frauds* has been re-enacted in several pieces of legislation. Section 4 of the original statute[48] applies to charges on, among other things, agreements upon consideration of marriage or upon sale of lands, tenements or hereditaments. It also states that a person is not able to sue upon such contracts unless 'some memorandum or note thereof shall be in writing, and signed by the party to be charged therewith, or some other person thereunto by him lawfully authorised'. Section 17 has similar provisions on the purchase of goods for £10 or over:

> . . . *no action shall be brought* whereby to charge any executor or administrator upon any special promise, to answer damages out of his own estate; or *whereby to charge the defendant upon any special promise to answer for the debt, default or miscarriages of another person;* . . . or upon any contract or sale of lands, tenements or hereditaments, or any interest in or concerning them; or upon any agreement that is not to be performed within the space of one year from the making thereof, *unless the agreement upon which such action shall be brought, or some memorandum or note thereof shall be in writing, and signed by the party to be charged therewith, or some other person thereunto by him lawfully authorized.*[49]

Legislation based on the Model Law does not change the impact of Statute of Frauds legislation. Instead it dictates circumstances in which electronic writing is to be regarded as equivalent. Australian legislation derived from the *Statute of Frauds* typically permits transactions of certain interests subject to stated formalities (such as the requirement for writing).[50]

46 See Sharon Christensen and Rouhshi Low, 'Moving the *Statute of Frauds* to the digital age', (2003) 77 *Australian Law Journal* 416 and Alan Davidson, 'Electronic transactions and contracts', (2001) 21 *Proctor* 6, 38.

47 *Official Guide to Enactment of the UNCITRAL Model Law on Electronic Commerce* para 3.

48 *Statute of Frauds* 29 Car 2. c. 5.3.4.

49 Emphasis added.

50 For example in relation to dispositions of land; see *Conveyancing Act 1919* (NSW) s54A and *Property Law Act 1974* (Qld) ss10, 59. Also bills of exchange, *Bills of Exchange Act 1909* (Cth) s8; promissory notes, *Bills of Exchange Act 1909* (Cth), s89; cheques, *Cheques Act 1986* (Cth) s10; assignments of copyright, *Copyright*

Australian provisions

> If, under a law of this jurisdiction, a person is required to give information in writing, that requirement is taken to have been met if the person gives the information by means of an electronic communication, where:
> (a) at the time the information was given, it was reasonable to expect that the information would be readily accessible so as to be useable for subsequent reference, and
> (b) the person to whom the information is required to be given consents to the information being given by means of an electronic communication.[51]

In all Australian Electronic Transactions Acts this provision is repeated with the word 'permitted' replacing 'required'.[52] Hence the provisions apply to any requirement or permission to give information in writing. Information is defined as 'information in the form of data, text, images or sound'.[53] Let us examine these provisions in more detail.

'required to give information in writing'

The expression 'required to give information in writing' is a clear reference to the *Statute of Frauds* requirements. The provision has no application to simple contracts which can be formed orally.

'requirement is taken to have been met'

The expression deems electronic writing to be equivalent.

'electronic communication'

The expression 'electronic communication' is defined as:

> (a) a communication of information in the form of data, text or images by means of guided or unguided electromagnetic energy, or both, or
> (b) a communication of information in the form of sound by means of guided or unguided electromagnetic energy, or both, where the sound is processed at its destination by an automated voice recognition system.[54]

This media-neutral and technology-neutral approach allows the provision to apply to as broad a range of circumstances as possible.

'at the time the information was given, it was reasonable to expect'

In the phrase 'at the time the information was given, it was reasonable to expect' the use of the expression 'reasonable' is both a strength and a weakness. The term

Act 1968 (Cth) s196(3); marine insurance contracts, *Marine Insurance Act 1909* (Cth) ss27, 28. Some sale of goods legislation has similar restrictions: for example, see *Sale of Goods Act 1895* (WA) s4.

51 Electronic Transactions Acts: NSW – s8(1); Qld – s11; SA – s8(1); Tas – s6(1); Vic – s8(1); ACT – s8(1) and NT – s8(1). Western Australia has substantially similar provisions – s8(1). The Commonwealth has substantially similar provisions in s9(1), but also makes special provision for Commonwealth entities.

52 Cth – s9(2); NSW – s8(2); Qld – s12; SA – s8(2); Tas – s6(2); Vic – s8(2); WA – s8(2); ACT – s8(2) and NT – s8(2).

53 Except in the Commonwealth Act, where 'speech' was used instead of 'sound'. The state and territory legislators chose the wider term with hindsight.

54 The Commonwealth Act again limits application to 'speech' instead of 'sound'.

gives those who write legislation (the drafters) and those who interpret the law (in the end the courts) great flexibility. There exist myriad factual circumstances in which the provisions could apply, including the use of future technology. This gives the provision strength to be applied where sound judgment and reason so dictate. But the same provision is also a weakness. One of the fundamental objects of the legislation is to provide certainty, but here a party to a transaction may be unable to predict with precision what a court would regard as reasonable. The lawyer in court will argue various meanings and permutations; first what is reasonable and second, whether it was reasonable 'at the time the information was given'. The uncertainty cancels out the benefit of the provision. The provision is no better than leaving the issue to the courts, which in any event would have found functional equivalence where it was reasonable to do so.

'the information would be readily accessible so as to be usable for subsequent reference' 'reasonable to expect'

The expression 'the information would be readily accessible so as to be useable for subsequent reference' is predicated on the principle of functional equivalence. A feature of ink and paper writing is that it is typically accessible for subsequent reference. Again, interpreters of the provision may question the distinction between 'accessible' and the use in this provision of 'readily accessible'. Is this broader or narrower than for paper, or the same? Paper has different durability. The drafters of wills and other important documents typically use a high grade paper, aware that the document may be required many years in the future. Ink can and does fade. Some government departments have the requirement that documents must only be in black ink. This has been based on the traditional concern that ink fades and the modern concern that light-coloured ink does not photocopy faithfully. Paper oxidises and can disintegrate. In these circumstances it is appropriate to read the expression in conjunction with the qualifying expression 'reasonable to expect'. For example, it would be reasonable to expect that electronic mail which is stored when created and by the recipient on receipt, would be readily accessible for subsequent reference. However, the typical chat room uses electronic exchanges without the expectation that the communications will be stored; such a use would fall foul of the provision.

With instant messaging the exchange appears on each party's computer screens and disappears after a short period of time or when the session is ended. Hence it is unavailable for subsequent reference. However, it is possible, for example, to copy and paste the exchange to a regular computer file, log the conversations, or use a screenshot or screen capture. Should the parties agree to record the exchange in this fashion, then it can be reasonable to expect that the information would be 'readily accessible so as to be useable for subsequent reference'.

Western Australia, being the last Australian state to enact an Electronic Transactions Act, and perhaps with greater time to reflect, replaced the phrase 'the information would be readily accessible so as to be useable for subsequent reference' with 'the information would be readily accessible then and for subsequent reference'. This contemplates that the parties must know at the time of creation that it was reasonable that it would be accessible for subsequent reference. That is, if it is not known at the time and only becomes reasonable to expect at a later date, the provision would not apply.

'giving information'

For the purposes of the section 'giving information' is defined in each of the Australian Acts as including: making an application, making or lodging a claim, giving, sending or serving a notification, lodging a return, making a request, making a declaration, lodging or issuing a certificate, making, varying or cancelling an election, lodging an objection and giving a statement of reasons.

Paragraph (b) Consent

The consent precondition – that the person the information is required to be given to must consent to the information being given by means of an electronic communication – has had incongruous and unintentional consequences. The consent provision will be considered separately.[55]

New Zealand

The New Zealand provision avoids the use of the terms 'requirement' and 'permission', removes the condition 'reasonable to expect' and uses the expression 'electronic form'. The provision is uncomplicated yet positive:

> A legal requirement that information be in writing is met by information that is in electronic form if the information is readily accessible so as to be usable for subsequent reference.[56]

Signatures

Legislation which requires writing typically also requires a signature by the maker.[57] A person who under a law is required to give a signature may, pursuant to and subject to conditions under the Electronic Transactions Acts, use an alternative method of authenticating their identity in relation to an electronic communication of information.[58]

55 See *Consent*, below.

56 *Electronic Transactions Act 2002* (NZ) s18.

57 For example, legislation based on the *Statute of Frauds* see *Conveyancing Act 1919* (NSW) s54A; *Law of Property Act 1936* (SA) s26; *Property Law Act 1974* (Qld) s59; *Conveyancing Law of Property Act 1884* (Tas) s36; *Instruments Act 1958* (Vic) s126; *Statute of Frauds 1677* (Imp.) (WA) s2; *Civil Law (Property) Act* 2006 (ACT) s201; *Law of Property Act 2000* (NT) s62.

58 Cth – s10; NSW – s9; Qld – ss14–15; SA – s8; Tas – s7; Vic – s9; WA – s9; ACT – s9; NT – s9.

Legislation should not be based on the various functions that a signature may have in a paper-based environment. Such an approach places undue emphasis upon concepts peculiar to paper, but also risks tying the legislation to a particular state of technical development. The UNCITRAL Model Law concentrates upon two basic functions of a signature: first, to identify the author of a document and second, to confirm that the author approved the content of that document. It does not deal specifically with the integrity of the document itself.[59]

Australian provisions

Each of the nine Australian Electronic Transactions Acts includes a provision deeming electronic signatures on certain conditions to meet the requirements under the law of signatures:

> If, under a law of this jurisdiction, the signature of a person is required, that requirement is taken to have been met in relation to an electronic communication if:
> (a) a method is used to identify the person and to indicate the person's approval of the information communicated, and
> (b) having regard to all the relevant circumstances at the time the method was used, the method was as reliable as was appropriate for the purposes for which the information was communicated, and
> (c) the person to whom the signature is required to be given consents to that requirement being met by way of the use of the method mentioned in paragraph (a).[60]

The Australian signature provision is substantially similar to the 'writing' provision.[61] Curiously, however, whereas the provision regarding writing applies where there is a requirement for writing or where writing is permitted, the provision for signatures only applies where the signature is required. This provision does not deem functional equivalence to signatures where they are not required, but are simply permitted. For example, where a signature is used as corroborating evidence proving the existence of a contract or transaction where the signature was not required, the Electronic Transactions Acts have no application. The acceptability and equivalence of an electronic signature in such circumstances should be a matter for the common law. The limited cases and commentaries have to date ignored this point. Again, let us now examine the provision in detail.

'signature of a person'

'Signature of a person' is a surprisingly thorny concept.[62] The meaning, and the use and custom, of a signature have been many and varied. Person can

59 See *Official Guide to Enactment of the UNCITRAL Model Law on Electronic Commerce*.
60 NSW – s9(1); Qld – s14; SA – s8(1); Tas – s7(1); Vic – s9(1); WA – s9(1); ACT – s9(1); NT – s9(1). The Commonwealth has substantially similar provisions in s10(1), but also makes special provision for Commonwealth entities.
61 See above.
62 See Chapter 5.

mean an individual or corporation. The signature of a corporation can be its corporate seal accompanied by one or more officers' personal signatures. The seal may be a rubber stamp, impressed indentation into paper or even a wax seal.

Traditional signatures most typically are physically written, printed or impressed. They can be readily reproduced by the maker and recognised and identified by other parties and experts. To varying degrees they are difficult to forge. Traditional signatures have acquired such 'fungibility' that they have been the standard for identification and for the execution of legal instruments, and used in financial institutions, for credit cards, accounts, cheques and so forth. Additionally, alteration and removal are difficult. When faced with a new form of signature the common law courts looked to the accepted functions and purposes of signatures and to the signatory's intention to determine if the electronic signature offered the same.

'signature of the person . . . in relation to an electronic communication'

The type of signature envisaged by the provision is an 'electronic signature', although that expression is not used. Instead the provision applies to the broader expression, 'signature of the person . . . in relation to an electronic communication'. There is no definition of electronic signature in the Australian legislation. The common law courts would most likely look to the pertinent attributes of a traditional signature to determine whether a given electronic signature meets legal requirements.

The term 'electronic signature' should not be confused with 'digital signature'. The latter refers to a specific attachment which uses an asymmetric cryptosystem, a hash function and public and private 'keys' for authentication and verification. An 'electronic signature' is any means of electronic authentication of the identity of a person and of the intent of that person associated with an electronic record. The term has no universally accepted meaning and internationally is variously defined.

Paragraph (a) Method used

Paragraph (a) establishes the principle that, in an electronic environment, the basic legal functions of a signature are performed by way of a method that identifies the originator of a data message and confirms that the originator approved the content of that message.

In 2006 in *J Pereira Fernandes SA v Mehta*,[63] the English courts had cause to consider the nature of an electronic signature. Mehta emailed Fernandes' solicitors offering to provide a personal guarantee in favour of Fernandes and to make a payment of £5000 on certain terms. The email did not show Mehta's name at the foot of the message, but was described in the header as having come from

63 [2006] EWHC 813.

'Nelmehta@aol.com'. On receiving the email, a clerk employed by the solicitors telephoned Mehta and agreed to the proposal. The solicitor then sent a written guarantee to Mehta. However, Mehta did not sign or return it.

Judge Pelling QC held that the email would be a note or memorandum to which section 4 of the *Statute of Frauds* applied. However, his Honour also considered whether or not it was sufficiently signed:

> a party can sign a document for the purposes of Section 4 by using his full name or his last name prefixed by some or all of his initials or using his initials, and possibly by using a pseudonym or a combination of letters and numbers (as can happen for example with a Lloyds slip scratch), providing always that whatever was used was inserted into the document in order to give, and with the intention of giving, authenticity to it.[64]

In determining what would amount to a signature Judge Pelling QC examined the underlying purpose. His Honour stated that the purpose of the *Statute of Frauds*:

> is to protect people from being held liable on informal communications because they may be made without sufficient consideration or expressed ambiguously or because such a communication might be fraudulently alleged against the party to be charged.[65]

In relation to an electronic document his Honour commented that:

> if a party creates and sends an electronically created document then he will be treated as having signed it to the same extent that he would in law be treated as having signed a hard copy of the same document. The fact that the document is created electronically as opposed to as a hard copy can make no difference.[66]

The issue before the court was whether the automatic insertion of a person's email address after the document has been transmitted constitutes a signature for the purposes of *Statute of Frauds*. His Honour's conclusion was an email address insertion was 'incidental'. The header is 'divorced from the main body of the text of the message'. If there is no further evidence in relation to the maker's intention, 'it is not possible to hold that the automatic insertion of an email address is . . . intended for a signature'. To conclude otherwise would, in his Honour's view, undermine or potentially undermine the purpose of the *Statute of Frauds*.

'electronic communication'

The expression 'electronic communication' is defined as:

(a) a communication of information in the form of data, text or images by means of guided or unguided electromagnetic energy, or both, or

64 [2006] EWHC 813, para 26.
65 [2006] EWHC 813, para 16.
66 [2006] EWHC 813, para 28.

(b) a communication of information in the form of sound by means of guided or unguided electromagnetic energy, or both, where the sound is processed at its destination by an automated voice recognition system.[67]

This gives a media neutral and technology neutral approach to the provision to apply to as broad a range of circumstances as possible.

'a method used to identify the person and to indicate the person's approval'

This expression is used in place of the simpler concept of 'electronic signature'. It ensures technology neutrality. The 'method used' could be as simple as a sender typing a name at the end of an email,[68] or an unseen digital signature logically associated and integrated with an electronic file authenticating the identity of the sender to a mathematical certainty and ensuring the integrity of the message.[69] The former is insecure. Anyone can type such a signature. Some email senders may place a digitised image of their signature at the end. This is equally insecure, as anyone can cut and paste the image. The 'method used' could be a simple code. Two parties may by agreement determine that, for example, the number '37' should be placed at the end with the intention that it will signify both the identity and approval of the sender. Others may improve on this rudimentary approach and require the code to be dynamic, yet simple; such as today's date multiplied by five plus the number of the month multiplied by eight. The resulting number appearing at the end of each email appears random, and yet the parties accept by agreement with some deal of certainty and security the authenticity and integrity of the message.[70] Digital signatures that use an asymmetric cryptosystem, or public and private 'keys' and a hash function, to ensure authentication and verification of the message and authorship are very secure.

Paragraph (b) Reliability of the method used

Paragraph (b) provides a flexible approach to the level of security to be achieved by the method of identification used under paragraph (a). In determining whether or not the method used is appropriate, legal, technical and commercial factors should be taken into account. These might include, for example:

- the sophistication of the equipment used by each of the parties;
- the nature of their trade activity;
- the frequency with which commercial transactions take place between the parties;
- the kind and size of the transaction;
- the function of signature requirements in the given statutory and regulatory environment;
- the capability of communication systems;

67 The Commonwealth Act again limits application to 'speech' instead of 'sound'.
68 As approved by Judge Pelling QC in *J Pereira Fernandes SA v Mehta* [2006] EWHC 813, para 8.
69 See Chapter 5.
70 The example for 5 November would be 113.

- compliance with authentication procedures set forth by intermediaries;
- the range of authentication procedures made available by the intermediary;
- compliance with trade customs and practice;
- existence of insurance coverage against unauthorised messages;
- the importance and value of the information contained in the data message;
- the availability of alternative methods of authentication and the cost of their use; and
- the degree of acceptance or non-acceptance of the method of identification in the relevant industry or field both at the time the method was agreed upon and at the time the data message was communicated.[71]

Although this is unstated, the provision envisages the 'method used' including electronic signatures. The Singapore *Electronic Transactions Act 1998* contains a useful definition of 'electronic signature':

> any letters, characters, numbers or other symbols in digital form attached to or logically associated with an electronic record, and executed or adopted with the intention of authenticating or approving the electronic record.[72]

This would cover the simple email signature, a code attached or appended to an email or an electronic document or the use of a digital signature.[73]

In *Faulks v Cameron*[74] the NT Supreme Court considered emails that ended with the type-written words 'Regards Angus' and 'Regards Angus Cameron'. The court had to determine whether the emails were 'signed' as a consequence of the *Electronic Transactions (Northern Territory) Act 2000* (NT). With surprisingly little analysis, it was held that the emails had been signed. Young AM was satisfied that:

> the printed signature on the defendant's emails identifies him and indicates his approval of the information communicated, that the method was reliable as was appropriate and that the plaintiff consented to the method. I am satisfied that the agreement is 'signed'.[75]

'of the information communicated'

These words require a connection between the information and the signature. Hence the method used must indicate the signatory's intention that the signature be attached to or logically associated with the information.

In *McGuren v Simpson*[76] the NSW Supreme Court regarded the words of an email 'yes I spent the money and I shouldn't have', together with the name

71 *Official Guide to Enactment of the UNCITRAL Model Law on Electronic Commerce.*
72 *Electronic Transactions Act 1998* (Singapore), s2.
73 See Chapter 5.
74 [2004] NTSC para 61; (2004) 32 Fam LR 417.
75 [2004] NTSC para 64.
76 [2004] NSWSC 35, see *Writing*, above.

appearing in the email, as a written and signed confirmation of McGuren's obligation to Simpson for the purposes of the *Limitation Act 1969* (NSW). The court approved the passage in *Halsbury's Laws of Australia*:

> Where the name of the party to be charged appears on the alleged note or memorandum, for example, because it has been typed in by the other party, the so-called 'authenticated signature fiction' will apply where the party to be charged expressly or impliedly acknowledges the writing as an authenticated expression of the contract so that the typed words will be deemed to be his or her signature.[77]

The court held that the email was recognisable as a signed note of a concluded agreement: 'the method must be as reliable as appropriate for the purposes for which the information was communicated' . . . 'having regard to all the relevant circumstances at the time the signature method was used'.

Paragraph (b) envisages that technological advances may result in signature technology becoming unsuitable even though it may have been suitable for a particular transaction at an earlier time. The legislature's intention was to link this requirement to the time that the signature method was used to ensure that a signature method that was appropriate at the time it was used is not later rendered invalid. This flexible technology-neutral standard enables signature methods to meet appropriate subjective standards at the time they are used.[78]

Paragraph (c) Consent

Paragraph (c) duplicates the consent provision for 'writing', and raises the same issues, comments and concerns. The consent provision is considered separately.[79]

New Zealand provisions

> 22(1) Subject to subsection (2), a legal requirement for a signature other than a witness' signature is met by means of an electronic signature if the electronic signature—
> (a) adequately identifies the signatory and adequately indicates the signatory's approval of the information to which the signature relates; and
> (b) is as reliable as is appropriate given the purpose for which, and the circumstances in which, the signature is required.
> (2) A legal requirement for a signature that relates to information legally required to be given to a person is met by means of an electronic signature only if that person consents to receiving the electronic signature.[80]

The New Zealand provisions for signatures are based on the UNCITRAL Model Law. Unlike Australia, however, the New Zealand provisions refer specifically to

77 110 *Contract* at [110–1030].
78 Electronic Transactions Bill 1999 (Cth) Explanatory Memorandum.
79 See *Consent*, below.
80 *Electronic Transactions Act 2002* (NZ) s22.

the use of 'electronic signatures' and to their equivalence in set circumstances. 'Electronic signature' in relation to information in electronic form is defined as 'a method used to identify a person and to indicate that person's approval of that information'.[81]

The provision parallels the Australian provision in relation to being 'as reliable as is appropriate given the purpose for which, and the circumstances in which, the signature is required'.

Section 22 duplicates the consent provision for 'writing', and raises the same issues, comments and concerns. The precondition that the person receiving the signature is required to be given consents to that requirement being met by way of the use of the method mentioned is considered separately.[82]

The New Zealand Electronic Transactions Act includes a specific provision for witnessing a document by way of an electronic signature. The requirements parallel the provision in section 22. An electronic signature used to witness a document must comply with section 22 and:

> 23(1) (b) in the case of the witnessing of a signature or a seal, the electronic signature of the witness—
> (i) adequately identifies the witness and adequately indicates that the signature or seal has been witnessed; and
> (ii) is as reliable as is appropriate given the purpose for which, and the circumstances in which, the witness' signature is required.[83]

The consent provision for section 22 is repeated, with the same impact.[84]

Section 24 contains a presumption that certain electronic signatures are reliable for the purposes of the Act:

> 24(1) For the purposes of sections 22 and 23, it is presumed that an electronic signature is as reliable as is appropriate if—
> (a) the means of creating the electronic signature is linked to the signatory and to no other person; and
> (b) the means of creating the electronic signature was under the control of the signatory and of no other person; and
> (c) any alteration to the electronic signature made after the time of signing is detectable; and
> (d) where the purpose of the legal requirement for a signature is to provide assurance as to the integrity of the information to which it relates, any alteration made to that information after the time of signing is detectable.

This section describes the operation of a digital signature but does not preclude other secure forms of electronic signatures.[85]

81 *Electronic Transactions Act 2002* (NZ) s5.
82 See *Consent*, below.
83 *Electronic Transactions Act 2002* (NZ) s23(1)(b) .
84 See *Consent*, below.
85 See Chapter 5.

Production of documents

A person who is required or permitted by law to produce a document that is in the form of paper, an article or other material may instead produce the document in electronic form:

> If, under a law of this jurisdiction, a person is required to produce a document that is in the form of paper, an article or other material, that requirement is taken to have been met if the person produces, by means of an electronic communication, an electronic form of the document, where:
> (a) having regard to all the relevant circumstances at the time the communication was sent, the method of generating the electronic form of the document provided a reliable means of assuring the maintenance of the integrity of the information contained in the document, and
> (b) at the time the communication was sent, it was reasonable to expect that the information contained in the electronic form of the document would be readily accessible so as to be useable for subsequent reference, and
> (c) the person to whom the document is required to be produced consents to the production, by means of an electronic communication, of an electronic form of the document.[86]

The underlying assumption of both the Model Law and the Australian legislation has been that original documents are paper based; the provisions deal with an electronic equivalent. The New Zealand Electronic Transactions Act includes a provision not foreshadowed in either of these: section 29 deals with the position where the electronic form is the original. The section aims to give a paper copy functional equivalence if certain criteria are met:

> A legal requirement to provide or produce information that is in electronic form is met by providing or producing the information—
> (a) in paper or other non-electronic form; but, if the maintenance of the integrity of the information cannot be assured, the person who must provide or produce the information must—
> (i) notify every person to whom the information is required to be provided or produced of that fact; and
> (ii) if requested to do so, provide or produce the information in electronic form in accordance with paragraph (b); or
> (b) in electronic form, whether by means of an electronic communication or otherwise, if—
> (i) the form and means of the provision or production of the information reliably assures the maintenance of the integrity of the information, given the purpose for which, and the circumstances in which, the information is required to be provided or produced; and
> (ii) the information is readily accessible so as to be usable for subsequent reference; and

86 NSW – s10; Qld – ss16–18; SA – s10; Tas – s8; Vic – s10; WA – s10; ACT – s10; NT – s10; New Zealand – s28. The Commonwealth has substantially similar provisions in s11, but also makes special provision for Commonwealth entities.

(iii) the person to whom the information is required to be provided or produced consents to the provision or production of the information in an electronic form and, if applicable, by means of an electronic communication.[87]

Consent

The Commonwealth Parliament introduced consent as an additional precondition for functional equivalence of electronic writing, electronic signatures and electronic production.[88] The person to whom the electronic writing or signature is required to be given, or documents produced, must consent to that requirement being met electronically. The Commonwealth Parliament made certain changes to the Model Law template, which were followed by the states, territories and New Zealand. This is not unusual. Most countries have modified the language and rules to suit their particular culture and domestic position. In New Zealand, however, consent is required for retention and access as well as for writing, signatures and the production of documents.

Although the underlying principle of the legislation is functional equivalence,[89] the notion of consent was considered significant enough to override that precept. It has been suggested that the consent provision thwarts functional equivalence.[90] The reasoning for its inclusion is ill-conceived and it creates an unwarranted impediment.

The consent provision is absent from the UNCITRAL Model Law. The Model Law is based on 'the recognition that legal requirements prescribing the use of traditional paper-based documentation constitute the main obstacle to the development of modern means of communication'.[91] The UNCITRAL drafters based the Model Law on the 'functional equivalent approach', stating that it is 'based on an analysis of the purposes and functions of the traditional paper-based requirement with a view to determining how those purposes or functions could be fulfilled through electronic-commerce techniques'.[92]

They noted that:

> Among the functions served by a paper document are the following: to provide that a document would be legible by all; to provide that a document would remain unaltered over time; to allow for the reproduction of a document so that each party would hold a copy of the same data; to allow for the authentication of data by means of a signature; and to provide that a document would be in a form acceptable to public authorities and courts.[93]

87 *Electronic Transactions Act 2002* (NZ) s29.
88 See *Electronic Transactions Act 1999* (Cth) ss9(1)(d) , 9(2)(d) , 10(1)(d) , 11(1)(e) and 11(2)(e) and the state and territory equivalents.
89 See *Writing*, above.
90 Alan Davidson, 'The Electronic Transaction Acts – in action or inaction?', (2007) 27 *Proctor* 7, 47; Alan Davidson, 'A matter of consent', (2004) 24 *Proctor* 11, 25. See also CC Nicoll, 'Consent – Luddite's lifeline', (2000) 9(3) *Information & Communications Technology Law* 195.
91 *Official Guide and Comments to the UNCITRAL Model Law of Electronic Commerce*.
92 Ibid., para 16.
93 Ibid.

The UNCITRAL drafters also believed that:

> electronic records can provide the same level of security as paper and, in most cases, a much higher degree of reliability and speed, especially with respect to the identification of the source and content of the data, provided that a number of technical and legal requirements are met.[94]

The only stated reservation was that the adoption of the functional equivalent approach 'should not result in imposing on users of electronic commerce more stringent standards of security (and the related costs) than in a paper-based environment'.[95]

When adopting the functional equivalent approach, the UNCITRAL drafters gave due attention to the existing hierarchy of form requirements, which provides distinct levels of reliability, traceability and unalterability with respect to paper-based documents. The Model Law does not attempt to define a computer-based equivalent to any kind of paper document. Instead, it sets out basic functions of paper-based form requirements. Once these requirements have been met, all transactions should 'enjoy the same level of legal recognition as corresponding paper documents performing the same function'.[96]

The Explanatory Memorandum to the Commonwealth's Electronic Transactions Act states that the inclusion of the consent provision was based on the government's:

> general policy that a person should not be compelled to use an electronic communication to conduct a transaction in order to satisfy requirements or permissions to give information in writing under Commonwealth law.[97]

The definition of 'consent' in the Australian Acts includes consent that can reasonably be inferred from the conduct of the person concerned, but does not include consent given subject to conditions unless the conditions are complied with. The definition is intended to ensure that express consent is not required prior to every electronic communication.[98]

The Explanatory Memorandum states that the power only applies where a person is receiving an electronic communication and that it is not necessary to state whether or not that person consents beforehand because 'the provisions are clearly drafted to provide a person with the ability to *choose* whether or not to satisfy their legal obligations by using an electronic communication' (emphasis added). However, this statement is questionable where consideration is given to contract formation. For a contract to be formed utilising the deeming features of the Electronic Transactions Acts, consent must be given by the recipient, and known to be given, in advance. It is insufficient for the recipient to consent after receipt.

94 Ibid., para 16.x.
95 Ibid.
96 Ibid., para 17. For example, see Art. 11 – Formation and validity of contracts.
97 *Electronic Transactions Act 1999* (Cth) Explanatory Memorandum, 49.
98 See *Ilich v Baystar Corp. Pty Ltd* [2004] WASTR 25.

Example

If the law requires writing or a signature, and a person emails an acceptance, what is the effect of the consent provision on the timing of the formation of the contract? To simplify the problem, let us ignore the postal acceptance rule and assume that the contract is formed on communication. Assume an offer is made by letter and contains the requisite material terms and the signature of the offeror. The offeree sends an email acceptance including an otherwise acceptable electronic signature. If it is known that the offeror will consent to an electronic communication of acceptance in advance, whether expressly, impliedly or by prior conduct, the Act applies and the requirement of writing and signature 'is taken to have been met'.

If the offeror has not consented in advance, two possibilities arise. First, on receipt of the acceptance the offeror considers his or her position, weighing up whether to consent to the 'electronic' writing and signature, and after some deliberation – minutes or days – the offeror decides to consent and all appears to be well. This seems to be the position taken by the Explanatory Memorandum. What is not addressed by the legislation is the timing of the formation of the contract. Is the contract formed at the point of communication by the offeree or when the offeror actually consents to the use of electronic writing? What if the offeror obtains legal advice regarding the electronic writing and so takes one or two weeks to consider whether or not to consent? What would be the effect should a time is of the essence provision operate during the deliberation period? To suggest that a contract is entered into not on communication but at a later time would be to create a new concept: deferred formation of contract.

These issues were encountered in *Ilich v Baystar Corp. Pty Ltd*.[99] The WA Strata Titles Referee determined that the language of the WA writing provision[100] requires consent to be known in advance. Kronberger R stated:

> both the 'giving' of the information and the 'consent' are expressed in the present tense, which may well indicate a legislative intention that, at the time the information was given, consent must exist. In other words, subsequent words or conduct are, strictly, irrelevant.[101]

The only result can be that where consent is not known in advance, the requirements of writing and signature are not 'taken to have been met' for the application of the Electronic Transactions Acts. This is contrary to express statements in the Explanatory Memorandum, although the Explanatory Memorandum was not specifically referring to contract formation. This leads to two unusual and perhaps surprising consequences. Where consent cannot be determined in advance, the offeror cannot enforce a contract on receiving an email acceptance. Second, the offeree sending the email knows that the acceptance cannot be enforced

99 [2004] WASTR 25.
100 *Electronic Transactions Act 2003* (WA) s8.
101 [2004] WASTR 25, para 11.

against him or her, to the same extent that the offeree knows that on oral acceptance cannot be enforced. Of course additional conduct may give the offeror a remedy using doctrines of estoppel or part performance, but in the absence of any other factor, a contract has not been formed. So the Australian consent provisions remove functional equivalence, one of the stated aims of the Model Law, and lead to incongruous unintentional results. A recipient of paper-based communications has no choice but to accept them (in terms of contract formation). Treating electronic communications differently erodes the functional equivalence principle and downgrades electronic commerce.

There remains one significant caveat. These comments are made for the purposes of the application of the Electronic Transactions Acts only. The relevant provisions merely state that equivalence requirements are not met if certain preconditions apply. The Act does not state that if the preconditions are not met the electronic writing shall not be equivalent. The next step is to determine whether the common law would regard the electronic writing and signature as equivalent.[102]

Other countries' provisions

Several others countries have included consent provisions. The US *Uniform Electronic Transactions Act 1999* contains a general provision for the application of the entire Act:

> This [Act] applies only to transactions between parties each of which has agreed to conduct transactions by electronic means. Whether the parties agree to conduct a transaction by electronic means is determined from the context and surrounding circumstances, including the parties' conduct.[103]

The Official Comments to this Act state that 'the paradigm of this Act is two willing parties doing transactions electronically'. The drafters considered this consent provision to demonstrate that the application of the Act is to be voluntary and to allow 'the greatest possible party autonomy to refuse electronic transactions'.[104]

The US *Electronic Signatures in Global and National Commerce Act* includes a similar consent provision to protect consumers. The Federal Trade Commission argues that the provision discourages deception and fraud by those who might fail to provide consumers with information the law requires that they receive.[105]

The Electronic Transactions Act of Ireland contains a consent provision substantially similar to that in the Australian Acts. In a section headed 'Use not mandatory', the Canadian *Electronic Transactions Act 2001* states: 'Nothing in this Act requires a person to provide, receive or retain information or a record in electronic form without the person's consent.'[106] The Official Comment explains

102 See *Common law*, above.
103 Section 5(b).
104 *Uniform Electronic Transactions Act 1999* (US) s5, Comment cl2.
105 Federal Trade Commission, 'Report to Congress on the Electronic Signatures in Global and National Commerce Act: The consumer consent provision in section 101(c)(1)(C)(ii)', (2001) www.ftc.gov/os/2001/06/esign7.htm.
106 *Electronic Transactions Act 2001* (Canada) s4.

that the section ensures that the Act will not be used to compel people to use electronic documents against their will, that some people are still 'uncomfortable' with such documents and do not have the 'capacity' to use them. The Act, it explains, is intended to provide 'certainty, not compulsion'.

The Canadian Uniform Electronic Commerce Act provides that 'Nothing in this Act requires a person to use or accept information in electronic form, but a person's consent to do so may be inferred from the person's conduct.' The Official Comment to the Act explains that the section ensures that the Act will not used to compel people to use electronic documents 'against their will'. It too states that many people 'are still uncomfortable with such documents', and do not yet have the capacity to use them.

Comment

Two final comments are warranted to this section. The first point, that the Australian consent provision thwarts functional equivalence, has been made. Under normal circumstances parties cannot refuse traditional paper communications and documents. Instead of providing certainty by recognising equality where the electronic writing, signature or production is functionally equivalent, the consent provision adds an unnecessary precondition creating less certainty, with consequences that are yet to be tested and analysed in court, in the end leaving such matters to the common law.

Second, the consent provision is futile. Where there is no applicable consent, the relevant provision does not state that the electronic writing, signature or production shall not be legally and functionally equivalent. The consequence is that this provision of the Act does not apply in any circumstances. Where the Act does not apply the courts must turn to the common law. The common has, in the few cases to date, held equivalence in any event.[107]

Retention of information and documents

The requirement to record information in writing may be met by recording the information in electronic form.[108] To be acceptable it must be reasonable to expect that the information will continue to be accessible for future reference and the method for storing the information must comply with any requirements of the regulations under the Act as to the kind of data storage device on which the information is to be stored.

107 See *Common law*, above; *McGuren v Simpson* [2004] NSWSC 35; *Hume Computers Pty Ltd v Exact International BV* [2007] FCA 478; *Wilkens v Iowa Insurance Commissioner* (1990) 457 NW 2d 1; *SM Integrated Transware v Schenker Singapore* Ltd [2005] 2 SLR 651.
108 Cth – s12(1); NSW – s11(1); Qld – s19; SA – s11(1); Tas – s9(1); Vic – s11(1); WA – s11(1); ACT – s11(1); NT – s11(1); New Zealand – s25.

Retention in paper form

The requirement to retain a document in the form of paper, an article or other material for a particular period[109] may be met by recording or retaining the information in electronic form.[110] The requirement is taken to have been met if the person retains, or causes another person to retain, an electronic form of the document throughout that period, where:

(a) having regard to all the relevant circumstances at the time of the generation of the electronic form of the document, the method of generating the electronic form of the document provided a reliable means of assuring the maintenance of the integrity of the information contained in the document, and
(b) at the time of the generation of the electronic form of the document, it was reasonable to expect that the information contained in the electronic form of the document would be readily accessible so as to be useable for subsequent reference, and
(c) if the regulations require that the electronic form of the document be retained on a particular kind of data storage device–that requirement has been met throughout that period.

The integrity of information contained in a document is maintained 'if, and only if'[111] the information has remained complete and unaltered, apart from (a) the addition of any endorsement, or (b) any immaterial change, which arises in the normal course of communication, storage or display.[112]

Retention in electronic form

The requirement to retain information the subject of an electronic communication for a particular time may be met by recording or retaining the information in electronic form.[113] To be acceptable, that requirement is taken to have been met if the person retains, or causes another person to retain, in electronic form, the information throughout that period, provided that:

(a) at the time of commencement of the retention of the information, it was reasonable to expect that the information would be readily accessible so as to be useable for subsequent reference, and
(b) having regard to all the relevant circumstances at the time of commencement of the retention of the information, the method of retaining the information in electronic form provided a reliable means of assuring the maintenance of the integrity of the information contained in the electronic communication, and
(c) throughout that period, the first person also retains, or causes the other person to retain, in electronic form, such additional information obtained by the

109 The New Zealand provisions refer to 'paper or other non-electronic form', and do not refer to any particular period.
110 Cth – s12(2); NSW – s11(2) ; Qld – s20; SA – s11(2); Tas – s9(2); Vic – s11(2); WA – s11(2); ACT – s11(2); NT – s11(2); New Zealand – s25.
111 The Northern Territory altered this to 'only if'.
112 This additional provision is absent from the New Zealand provision.
113 Cth – s12(4); NSW – s11(4); Qld – ss19–21; SA – s11(4); Tas – s9(4); Vic – s11(4); WA – s11(4); ACT – s11(4); NT – s11(4); New Zealand – ss26–27.

first person as is sufficient to enable the identification of the following:
   (i) the origin of the electronic communication,
   (ii) the destination of the electronic communication,
   (iii) the time when the electronic communication was sent,
   (iv) the time when the electronic communication was received, and

(d) at the time of commencement of the retention of the additional information covered by paragraph (c), it was reasonable to expect that the additional information would be readily accessible so as to be useable for subsequent reference, and
(e) if the regulations require that the information be retained on a particular kind of data storage device — that requirement has been met throughout that period.

The requirements for origin, destination and time are aimed at electronic mail. The application to other technologies such as SMS messaging, instant messaging and chat rooms where communication is often effectively instantaneous may prove problematic.

As with the previous requirement to retain in a paper form, the integrity of information contained in a document is maintained 'if, and only if'[114] the information has 'remained complete and unaltered, apart from the addition of any endorsement, or any immaterial change, which arises in the normal course of communication, storage or display'.[115]

Time and place of dispatch and receipt of electronic communications

The time and place of the dispatch and receipt of electronic communications can have significant impact in commerce, particularly in the law of contract. Strict rules have developed in relation to the time and place of offer and acceptance in contract law. This may have a bearing on the applicable jurisdiction or contract formation. The Electronic Transactions Acts of Australia and New Zealand include provisions based on the UNCITRAL Model Law that attempt to resolve uncertainties in this area. However, the application of the provisions raises serious questions. The time of dispatch and time of receipt must be considered in conjunction with common law rules of offer and acceptance and the timing of contract formation. One must also ask whether the postal acceptance rule can or should apply to electronic mail and other forms of electronic communications and, if it does, how it interacts with the legislation.

114 The Northern Territory altered this to 'only if'.
115 Cth – s12(3); NSW – s11(3); Qld – s 20(3); SA – s11(3); Tas – s9(3); Vic – s11(3); WA – s11(3); ACT – s11(3); NT – s11(3); this additional provision is absent from the New Zealand provision.

Time of dispatch

(1) For the purposes of a law of this jurisdiction, if an electronic communication enters a single information system outside the control of the originator, then, unless otherwise agreed between the originator and the addressee of the electronic communication, the dispatch of the electronic communication occurs when it enters that information system.

(2) For the purposes of a law of this jurisdiction, if an electronic communication enters successively 2 or more information systems outside the control of the originator, then, unless otherwise agreed between the originator and the addressee of the electronic communication, the dispatch of the electronic communication occurs when it enters the first of those information systems.[116]

The dispatch of the electronic communication occurs when it enters an information system outside the control of the originator. In the most typical case this would be when the originator clicks the 'Send' button of an email. The broad term 'electronic communication' applies to emails, SMS messages, present and future technologies. The provision for time of dispatch is a default rule. The parties may have 'otherwise agreed' to alternative arrangements.

In *Szaeg v Minister for Immigration*,[117] the court applied the Commonwealth Electronic Transactions Act to Australia Post's facsimile service, finding that the sender's document was dispatched when 'handed' to an employee of Australia Post, because that system was outside the control of the originator.

Problems arise with regard to email and communication systems not envisaged when this provision was drafted by UNCITRAL in 1996. Today, some systems permit the originator to recall an email before it has been flagged as read. However, it may be possible that the recipient could 'read' the name of the sender and the subject line without formally opening the email. If we assume the subject line contained an acceptance to an offer with words such as, 'Your offer accepted unconditionally', a paradox arises. The email is 'received' according to the Act; and the acceptance has been communicated as required by the rules of contract at common law; and yet according to the Electronic Transactions Acts it has not been dispatched. The courts are yet to consider this. In this situation, the provision is not triggered. This is a result not anticipated by the legislators, but one which can, in any event, be resolved by applying common law principles. This is another area in which the legislation is redundant and somewhat ineffectual.

The New Zealand provision in relation to the time of dispatch is much less verbose than the Australian equivalents, but contains the same notions and principles:

116 NSW – s12; other Australian jurisdictions have identical or substantially identical provisions; Cth – s14; Qld – s23; SA – s13; Tas – s11; Vic – s12; WA – s13; ACT – s13; NT – s13.

117 [2003] FMCA 258.

An electronic communication is taken to be dispatched at the time the electronic communication first enters an information system outside the control of the originator.[118]

Time of receipt

(3) For the purposes of a law of this jurisdiction, if the addressee of an electronic communication has designated an information system for the purpose of receiving electronic communications, then, unless otherwise agreed between the originator and the addressee of the electronic communication, the time of receipt of the electronic communication is the time when the electronic communication enters that information system.

(4) For the purposes of a law of this jurisdiction, if the addressee of an electronic communication has not designated an information system for the purpose of receiving electronic communications, then, unless otherwise agreed between the originator and the addressee of the electronic communication, the time of receipt of the electronic communication is the time when the electronic communication comes to the attention of the addressee.[119]

'Designated information system' is not defined in the Australian or New Zealand legislation nor in the Model Law, but 'information system' is defined as:

a system for generating, sending, receiving, storing or otherwise processing electronic communications.[120]

The system would be designated when an email address or mobile phone number has been communicated by the addressee to the originator. However, the path of an electronic communication may be quite complex. For example, with a typical email, once sent, the email enters the internet, going from router to router. It then enters the addressee's Internet Service Provider (ISP), if it is operational. At a later time the addressee's computer will contact the ISP to check for updates. This may be done when the addressee turns on the computer, or automatically at regular intervals. In any event the email will be passed onto the addressee's computer. Debate arises whether the legislature intended the internet, the ISP or the addressee's computer to be the designated information system.[121] Most likely it is the ISP, which relates to that part of an email address to the right of the '@' sign. In *Szaeg v Minister for Immigration*,[122] a facsimile application was handed to an Australia Post employee. The court found that the facsimile 'was sent to some address unknown [and] it was not received by the [addressee] . . . the facsimile transmission . . . did not enter the [addressee's] designated information system for the purposes of s14(3) of the *Electronic Transactions Act 1999* (Cth)'.[123]

118 *Electronic Transactions Act 2002* (NZ) s10(1).
119 NSW – s12; other Australian jurisdictions have identical or substantially identical provisions; Cth – s14; Qld – s24; SA – s13; Tas – s11; Vic – s12; WA – s13; ACT – s13; NT – s13.
120 Cth – s5; NSW – s5; Qld – schedule 2; SA – s5; Tas – s3; Vic – s3; WA – s5; ACT – dictionary; NT – s5.
121 For example see Patrick Quirk and Jay Forder, *Electronic Commerce and the Law*, 2nd edn, Wiley Press, Brisbane, 2003, p. 67.
122 [2003] FMCA 258.
123 [2003] FMCA 258, para 25.

In itself the provision can form a quasi postal acceptance rule, as the email is deemed received before communication. The problem is complicated by rules relating to acceptances and, in common law countries, by the possible application of the postal acceptance rule.

Where an information system is not designated an email is deemed received when it 'comes to the attention of the addressee'. It remains unclear whether the words 'comes to the attention of the addressee' mean when the addressee has read the communication or when the addressee has received a notice, such as, 'you have mail'. Potentially, a recipient could note that an email has been received from the offeree, and deliberately avoid reading it until the offer lapses. This issue must await a court's deliberation.

As with the time of dispatch, the New Zealand provision, in relation to the time of receipt, is much less verbose than the Australian equivalents, but contains the same notions and principles. For New Zealanders, the time of receipt is:

> (a) in the case of an addressee who has designated an information system for the purpose of receiving electronic communications, at the time the electronic communication enters that information system; or
> (b) in any other case, at the time the electronic communication comes to the attention of the addressee.[124]

Acceptance by electronic communication and the postal acceptance rule

Contracts are formed only when there is a meeting of the minds. There is an offer and a corresponding acceptance. The general common law principle is that acceptance must be communicated to the offeror, at which point a legally enforceable contract comes into existence. However, from the 19th century cases of *Henthorn v Fraser*,[125] based on *Adams v Lindsell*,[126] an exception known as the postal acceptance rule arose. As propounded by Lord Herschell in the former case:

> Where the circumstances are such that it must have been within the contemplation of the parties that . . . the post might be used as a means of communicating the acceptance of an offer, the acceptance is complete as soon as it is posted.[127]

This exception applies whether the letter is delivered, delayed or even lost. It has been regarded by the House of Lords as a rule of 'commercial expediency' and as a 'foundation of convenience'.[128] Without the rule the acceptor may be placed in an uncertain position, not knowing when or whether the acceptance had been duly received and communicated. The rule avoids the necessity of a confirmation and of a confirmation of the confirmation *ad infinitum*.

124 *Electronic Transactions Act 2002* (NZ) s11.
125 [1892] 2 Ch 27.
126 (1818) B & Ald 681.
127 [1892] 2 Ch 27, 33.
128 Lord Wilberforce, in *Brinkibon Ltd v Stahag Stahl* [1983] 2 AC 34.

While the rule has been held to apply to telegrams, the courts have been reluctant to extend the principle. They have consistently held that it does not apply to instantaneous forms of communication such as the telephone and (in some cases) telexes.

In *Entores Ltd v Miles Far East Corp.*[129] Lord Denning discussed various scenarios, such as acceptance shouted across a river which was not heard due to a passing plane, the telephone going dead, and where a recipient informs a sender of the failure of the teleprinter during the transmission of a telex. Initially Lord Denning commented: 'In all the instances I have taken so far, the man who sends the message of acceptance knows that it has not been received or he has reason to know it. So he must repeat it.' However, he then considered the position where the sender does not know the acceptance had not been communicated. Perhaps 'the listener on the telephone does not catch the words of acceptance' and does not ask that the words be repeated; or 'the ink on the teleprinter fails at the receiving end' and the clerk does not ask for the message to be repeated. As a result 'the man who sends an acceptance reasonably believes that his message has been received . . . The offeror in such circumstances is clearly bound because he will be estopped from saying that he did not receive the message of acceptance. It is his own fault . . .'[130] However, his Lordship continued, if the offeror 'without any fault on his part does not receive the message of acceptance – yet the sender of it reasonably believes it has got home when it has not – then I think there is no contract'.[131] Parker LJ in the same case notes that although the operation of a telex 'is not completely instantaneous, the parties are to all intents and purposes in each other's presence, just as if they were in telephonic communication'.[132]

The same cannot be said of all electronic communications. The parties often regard the communications as akin to mail. There may be no expectation of being in each other's presence.

In *Brinkibon Ltd v Stahag Stahl und Stahlwarenhandelsgesellschaft mbH*[133] the House of Lords confirmed on the facts before them that the telex should be treated as instantaneous communication. However, Lord Wilberforce qualified the principle. Although it was still too early to consider electronic mail his Lordship set out a number of guidelines, reminiscent of electronic mail use, which could be used to assess whether a form of communication might have the postal acceptance rule applied to it:

> The message may not reach, or be intended to reach, the designated recipient immediately: Messages may be sent out of office hours, or at night, with the intention, or upon the assumption, that they will be read at a later time. There may be some error or default at the recipient's end which prevents receipt at the time contemplated and believed in by the sender. The message may have been sent and/or received through

129 [1955] 2 QB 327.
130 [1955] 2 QB 327, 333.
131 [1955] 2 QB 327, 333.
132 [1955] 2 QB 327, 337.
133 [1983] 2 AC 34.

> machines operated by third persons. And many other variations may occur. No universal rule can cover all such cases: they must be resolved by reference to the intention of the parties, by sound business practice and in some cases by a judgment where the risks should lie.[134]

There is no blanket rule: as Lord Wilberforce noted, 'no universal rule can cover all such cases'. The mistake made by many commentators has been to examine only the technology and to categorise it as instantaneous or not. This has led to misdirection. The correct approach should be to examine the use of the technology, as described by Lord Denning in Entores' case and Lord Wilberforce above. There have been misgivings about the blanket categorisation of the telex and facsimile as instantaneous. When employed via third parties (indeed on occasions through the post office) similar delays, if not identical delays, occur as occur with standard post. If the use anticipates such delay, there is no reason in principle that the postal acceptance rule should be regarded as inapplicable.

Messenger systems and chat rooms provide users with effectively real-time communications that should be categorised as instantaneous. Acceptance of offers made using such technology should follow the rule of acceptance on communication. Acceptances may be by numerous means – voicemail, VOIP, SMS messaging, bulletin boards, electronic mail, to name a few. To classify the application of the 19th century rule based purely on whether or not it can be instantaneous is fraught with danger.

Some technologies may be used in more than one way: like the post and in an instantaneous manner. Email is sometimes used as a chat room or instant messaging. Two users may send and receive numerous emails within a matter of minutes. It may be conversational, with questions and answers clarifying each other's position before a formal offer and acceptance occurs. This use of communications should be regarded as instantaneous, in which case the general 'communication acceptance rule' ought to apply. Similar use of other technology should be treated likewise.

However, some users utilise electronic mail like standard mail. A given office's protocol may be to access email once per day and then carefully draft a reply for sending later in the day. Such an office may require a supervisor to check the contents of all mail before it is sent, whether it is standard or electronic mail. Once approved, the electronic mail may be sent in one batch, for example at a time similar to the dispatch of regular mail. Clearly this use would be more akin to standard mail.

The focus of the application of the rule should be on the use, bearing in mind Wilberforce's threefold test of intention, sound business practice and risk.

Four scenarios may be considered in determining whether and how the postal acceptance rule should apply to electronic communications:

134 [1983] 2 AC 34, 42.

1 The postal acceptance rule does not apply because the Electronic Transactions Acts override the common law and specifically provide that 'receipt' occurs when the email enters the recipient's 'designated information system'. Hence it is the intention of the legislature that a quasi postal acceptance rule applies to electronic communications.
2 The postal acceptance rule does not apply to electronic communications because such communications are to be regarded as instantaneous communications.
3 The postal acceptance rule applies to electronic communications independently of the Electronic Transaction Acts.
4 The postal acceptance rule applies in circumstances involving commercial expediency, depending upon intention of the parties and use.

The first approach is unsustainable because the underlying stated rationales of the Electronic Transactions Acts do not deal directly with contracts or the postal acceptance rule.[135] The Acts are designed to deal with a broad range of electronic transactions. The second and third approaches presume that the postal acceptance rule is technology based. This is a misconception perpetuated by scant reading of the cases. The cases refer to use, intention, sound businesses practice and risk. For example Lord Wilberforce in Brinkibon's case[136] envisaged circumstances where facsimiles may nevertheless be subject to the postal acceptance rule. The fourth approach looks to the ramifications of usage, such as the inconvenience of delay in the context of intention, business practice and commerce. If the intention of the parties cannot be determined, the 'rule of commercial expediency' should apply.

Today, commercial parties have at their disposal numerous methods of communication virtually beyond comprehension at the time of Lord Denning in Entores' case and Lord Wilberforce in Brinkibon's case. Prophetically, Lord Brandon stated, in Brinkibon's case:

> The reason for the exception is commercial expediency . . . That reason of commercial expediency applies to cases where there is bound to be a substantial interval between the time when the acceptance is sent and the time when it is received. In such cases the exception to the general rule is more convenient, and makes on the whole for greater fairness, than the general rule itself would do.[137]

Where the parties' intention is that communications shall be electronic and there is expected to be, and is, a substantial interval between the time when the acceptance is sent and the time when it is received, the postal acceptance rule ought to apply. The rationale remains commercial expediency, sound business practice and an appropriate assessment of risk.

135 See Electronic Transactions Bill 1999 (Cth) Explanatory Memorandum.
136 [1983] 2 AC 34.
137 [1983] 2 AC 34, 48.

Place of dispatch and place of receipt

(5) For the purposes of a law of this jurisdiction, unless otherwise agreed between the originator and the addressee of an electronic communication:
  (a) the electronic communication is taken to have been dispatched from the originator's place of business, and
  (b) the electronic communication is taken to have been received at the addressee's place of business.

(6) For the purposes of the application of subsection (5) to an electronic communication:
  (a) if the originator or addressee has more than one place of business, and one of those places has a closer relationship to the underlying transaction–it is to be assumed that that place of business is the originator's or addressee's only place of business, and
  (b) if the originator or addressee has more than one place of business, but paragraph (a) does not apply–it is to be assumed that the originator's or addressee's principal place of business is the originator's or addressee's only place of business, and
  (c) if the originator or addressee does not have a place of business–it is to be assumed that the originator's or addressee's place of business is the place where the originator or addressee ordinarily resides.[138]

These rules are important in determining the law to be applied, or the forum, in the event of a dispute. It is not proposed to deal with the various rules of private international law and domestic conflict of laws rules here. However, circumstances may arise where the place of contract formation determines which laws or which forum may be applicable.

Again the term 'electronic communication' applies to emails, instant and SMS messages, present and future technologies. The provision is a default rule. The parties may have 'otherwise agreed' to alternative arrangements.

Attribution of electronic communication

Traditional forms of paper-based communications can be typically attributed to the author by such factors as handwriting, signature and letterhead. Electronic communications can be generic. The electronic text can be indistinguishable regardless of the source. However, to whom do we attribute the message? For example using one of the popular public email services, such as Hotmail or Yahoo, an impersonator could register a person's name and send an email as johndoe@hotmail.com.[139] A second example is where the impersonator accesses another's computer and, using the actual email system on the computer, sends a

138 NSW – s12; other Australian jurisdictions have identical or substantially identical provisions; Cth – s13; Qld – s25; SA – s13; Tas – s11; Vic – s12; WA – s13; ACT – s13; NT – s13; New Zealand – ss12–13. The New Zealand provisions contain the same principle, but are much less verbose.

139 Hotmail and other such email services are so popular that a given name may not be available, but registrants typically a variation, for example a number, *johndoe3120@hotmail.com*. Nevertheless the impersonation remains.

message. The recipient of such a message has every indication that the message is genuine.

The Australian Electronic Transactions Acts resolve this conundrum in a manner that is functionally equivalent to that of the paper world. A person is not bound by an electronic communication unless the communication was sent by, or with the authority of, that person:

> unless otherwise agreed between the purported originator and the addressee of an electronic communication, the purported originator of the electronic communication is bound by that communication only if the communication was sent by the purported originator or with the authority of the purported originator.[140]

In the paper world a person is not bound by forgeries or communications sent without authority, unless there is some culpability. Such authority may be given ostensibly or impliedly in accordance with agency principles. Importantly, the parties may agree to exclude this provision in advance. Authority includes delegation, agency and the employer–employee relationship. Culpability may arise through estoppel or negligence.

This Australian provision differs from the UNCITRAL Model Law. The advantage of the Australian approach is its functional equivalence. The disadvantage is that a person receiving an apparently genuine message may be faced with a subsequent argument or defence that it was sent without authority. In addition, the Model Law provides for a presumption that the purported originator is in fact the originator. The recipient may act on the electronic communication where, in order to ascertain whether the communication is of the originator, the recipient properly applied a procedure previously agreed to for that purpose, or the communication as received resulted from the actions of a person whose relationship with the originator, or with any agent of the originator, enabled that person to gain access to a method used by the originator to identify the electronic communication as his or her own.[141]

The Australian Acts preserve the law of agency.[142] The decision to reject the UNCITRAL model follows a recommendation made by the Electronic Commerce Expert Group (ECEG). ECEG argued that the UNCITRAL proposal favoured electronic commerce over paper-based communication. It noted that the use of signatures on paper for commerce at a distance (by mail or facsimile) involves the risk of forged or unauthorised signatures but there is no general legislative rule that entitles the addressee to presume that the signature is genuine. If the UNCITRAL proposal was accepted, addressees of electronically signed data messages would be better placed than those who received manually signed paper-based messages.

140 Cth – s15(1); NSW – s14(1) ; Qld – s26(1); SA – s14(1); Tas – s12(1); Vic – s14(1); WA – s14(1); ACT – s14(1); NT – s14(1).
141 UNCITRAL *Model Law of Electronic Commerce*, Art. 13, which is quoted in full in Appendix B.
142 Cth – s15(2); NSW – s14(2); Qld – s26(2); SA – s14(2); Tas – s12(2); Vic – s14(2); WA – s14(2); ACT – s14(2); NT – s14(2).

The New Zealand Electronic Transactions Act does not include an attribution provision. The result should be that the courts apply the same standards for non-electronic communications.

Originals

The New Zealand Electronic Transactions Act includes a provision dealing with originals:[143]

> A legal requirement to compare a document with an original document may be met by comparing that document with an electronic form of the original document if the electronic form reliably assures the maintenance of the integrity of the document.[144]

The section is predicated on the assumption that the original document is a hardcopy. No provision exists where the original is the electronic document. The UNCITRAL Model Law makes the same narrow assumption.[145]

Electronic Case Management System

Part 2A of the *Electronic Transactions Act 2000* (NSW) includes specific provisions for electronic documents to be used in a court which has adopted the Electronic Case Management System. These provisions override the general provisions of the Act. Part 2A permits the NSW Attorney General to establish an electronic case management system to enable documents with respect to legal proceedings to be created, issued, used, served and communicated in electronic form.

Comment

The impact of the legislation will be broad, extending beyond internet transactions. Careful consideration will need to be given to transactions to be exempted by the regulations and legislation. Within Australia the differences between the states and territories' legislation must be noted. There are several stylistic changes, particularly in the Queensland legislation, which a court may regard as deliberate and significant in a given factual situation. Though the states and territories generally resisted divergences in an attempt to maintain national uniformity, stability and certainty, completely uniform legislation would be preferable.

143 The Australian legislators choose not to include a provision dealing with originals.
144 *Electronic Transactions Act 2002* (NZ) s30.
145 Art. 8.

The major advantage of legislation based on the UNCITRAL Model Law is that it provides an even starting platform for national legislatures. An international approach provides a level of uniformity that is a desirable if not necessary factor for international commerce.

Although the Electronic Transactions Acts create a regulatory regime for the use of electronic communications in transactions they do not remove any legal obligations that may be imposed upon a person by other laws. The major purpose of the Acts is to enable people to use electronic communications in the course of business operations and to satisfy their legal obligations.

However, the legislation is largely unnecessary and often redundant. It was introduced before the courts had the opportunity to adequately apply common law precepts to the new technologies and circumstances. The limited cases to date support this view.[146] The analysis of the provisions of the legislation, the misconceived consent requirement and the advent of new and changing technologies not considered by the legislators all lead to the conclusion that the legislation should be reviewed, with consideration being given to their repeal. The Law Commission for England and Wales, in its paper 'Electronic Commerce: Formal requirements in Commercial Transactions – Advice from the Law Commission', recommended that there was 'no need' for legislation based on the UNCITRAL Model Law for most purposes:

> We conclude that in most contexts e-mails (and attachments) and website trading (but not EDI) are already capable of satisfying the statutory form requirements existing in English law in the areas considered in this Advice. To that extent we suggest that there is no need to consider adoption in this country of the UNCITRAL Model Laws.[147]

The consent provisions are ill-considered and unsound. The writing, signature, production and retention requirements are superfluous. The exemptions are inconsistent across jurisdictions, giving rise to confusion and inconsistencies. The exemptions under the Commonwealth legislation are so broad as to frustrate the Act's operation. A major joint state and territory review is recommended, with a view to curtailing unwarranted and undesirable provisions and enacting streamlined uniform legislation.

Further reading

Sharon Christensen, William Duncan and Rouhshi Low, 'The requirements of writing for electronic land contracts – the Queensland experience compared with other jurisdictions', (2003) 10(3) *E Law – Murdoch University Electronic Journal of Law (MurUEJL)*, www.murdoch.edu.au/elaw/issues/v10n3/christensen103.html.

146 See *Common law*, above; *McGuren v Simpson* [2004] NSWSC 35, *Hume Computers Pty Ltd v Exact International BV* [2007] FCA 478 and *SM Integrated Transware v Schenker Singapore Ltd* [2005] 2 SLR 651.
147 Paragraph 2.15, Electronic Commerce: Formal requirements in Commercial Transactions – Advice from the Law Commission, available at: www.lawcom.gov.uk/docs/e-commerce.pdf.

Henry Gabriel, 'The new United States Uniform Electronic Transactions Act: Substantive provisions, drafting history and comparison to the UNCITRAL Model Law on Electronic Commerce'; www.unidroit.org/english/publications/review/articles/2000–4.htm. The author was a Commissioner on the National Conference of Commissioners on Uniform State Laws at the time of the drafting.

Minyan Wang, 'The impact of information technology development on the legal concept – a particular examination on the legal concept of "signatures"', (2006) 15(3) *International Journal of Law and Information Technology*.

D Witte, 'Comment: Avoiding the un-real estate deal: Has the Uniform Electronic Transactions Act gone too far?', (2002) 35 *John Marshall Law Review* 311 at 321. Also see *Uniform Electronic Transactions Act 1999* (US) s5(b) Comment para 4.

4

Shrinkwrap, clickwrap and browsewrap contracts

Online commerce, local and global, has become commonplace in the past two decades. Numerous traders have embraced the new opportunities by creating global meeting places and auction houses such as eBay and amazon. The sale of software, cars, travel, books, music and videos are just some of the many commercial transactions carried out in cyberspace. More than a trillion dollars in trade is transacted online annually. The online contracts are typically entered into without paper, without the use of a pen and without the use of the spoken word. This chapter examines cases known as the shrinkwrap, clickwrap and browsewrap cases, which have explored how contracts may be entered into and how conditions may be incorporated into the transaction. This chapter also mentions contracts entered into by electronic agents.

Accepting terms and conditions by conduct, notice or click of the mouse has raised questions reminiscent of the contract cases known as the ticket cases.[1] The typical ticket case scenario arises where a person is handed a ticket – for example, for train travel, parking or dry cleaning. The ticket may state that the holder is subject to terms which the holder may have had little or no chance to read, negotiate or agree. The courts have been required to determine the point of time at which these contracts are entered into and their terms and conditions . The starting point has been determining which terms are to be incorporated into the contract. In *Thornton v Shoe Lane Parking Station*[2] Lord Denning, referring to an extremely onerous exclusion of liability clause on a parking ticket, stated:

1 For example, *Parker v The South Eastern Railway Co.* (1877) 2 CPD 416; *Olley v Marlborough Court* [1949] 1 KB 532; *Sydney City Council v West* (1965) 114 CLR 481; *Thornton v Shoe Lane Parking Station* [1971] 2 QB 163.
2 [1971] 2 QB 163.

> It [the exclusion clause] is so wide and so destructive of rights that the court should not hold any man bound by it unless it is drawn to his attention in the most explicit way... In order to give sufficient notice, it would need to be printed in red ink with a red hand pointing to it – or something equally startling.[3]

Similarly, in *Interfoto Picture Library Ltd v Stiletto Visual Programmes Ltd*[4] the English Court of Appeal held that to incorporate onerous terms into a contract, reasonable notice must be given.

Such an approach is prophetic for contracts in cyberspace. The principles of the ticket cases, such as the opportunity to access terms and conditions and the extent to which such clauses will be binding, have been explored and paralleled progressively in the shrinkwrap, clickwrap and browsewrap cases.

Shrinkwrap

Shrinkwrap contracts derive their name from the clear plastic wrapping that encloses the goods (such as software packages). The software packages typically include a notice saying that by opening the shrinkwrap, the purchaser agrees to the terms and conditions enclosed. Such contracts typically include provisions such as an arbitration clause, a choice of law and forum clause, disclaimers, limitations of warranties and limitations of remedies. The major criticism of such contracts is that the consumer will be bound by terms and conditions unknown at the time the contract is entered into. In a Dilbert cartoon, Scott Adams parodied shrinkwrap contracts thus:

> Dilbert: I didn't read all of the shrinkwrap license on my new software until after I opened it. Apparently I agreed to spend the rest of my life as a towel boy in Bill Gates' new mansion.
> Dogbert: Call your lawyer.
> Dilbert: Too late. He opened software yesterday. Now he's Bill's laundry boy.[5]

The initial response was to limit the application of such clauses. In *Step-Saver Data Sys Inc. v Wyse Tech*[6] the court held that the terms of the shrinkwrap licence were not enforceable because Step-Saver had not assented to them. However, later, in *ProCD Inc. v Zeidenberg*,[7] the appeal court noted that it would be impossible to print the entire contract on the exterior of the box. The court found that:

3 [1971] 2 QB 163, 170.
4 [1989] 1 QB 433.
5 Scott Adams, *Dilbert*, United Feature Syndicate Inc., 14 January 1997. It is interesting that as early as 1997 the problem was perceived to be broad enough to warrant public parody.
6 939 F. 2d 91 (3d Cir 1991).
7 86 F. 3d 1447 (7th Cir 1996).

> notice on the outside, terms on the inside, and a right to return the software for refund if the terms are unacceptable . . . may be a means of doing business valuable to buyers and sellers alike.[8]

The implications have extended to mass marketing, the distribution of software online, which is most significant, and to the purchase of tickets for concerts, air travel and sporting events where the purchaser pays in advance and receives the tickets with the terms included. In *Hill v Gatway 2000 Inc.*[9] Easterbrook J commented that cashiers and sales personnel cannot be expected to read the terms of each contract to the customer before ringing up the sale. In that case his Honour explicated the cost benefits to the customer where 'telephonic recitation' was avoided and use was made of a 'simple approve-or-return device'. His Honour concluded, 'Competent adults are bound by such documents, read or unread.'[10]

Clickwrap

Clickwrap agreements are a method of including terms and conditions in an online contract. A clickwrap contract is formed on the internet. The name is derived from 'shrinkwrap'. However, with a clickwrap contact, the user, for example, assents to a list of terms by clicking an onscreen button marked, for example, 'Agree' or 'I accept'. The clickwrap contract has the advantage of allowing the user to read the terms and conditions before assenting. The vendor can dictate the number of steps for the user to pass through before reaching the assent stage.

In the typical situation the vendor uses an interactive web page which may ultimately require personal and credit card details. The contract needs no paper or signature. Assent is given in one of two principal methods. The first is 'type and click', where the user must type, for example, 'I agree' or 'Yes' and then click a send button. The second uses clicks alone to indicate assent. The trader must take reasonable steps to bring the terms and conditions to the attention of the buyer at or before the time the contract is formed.[11]

The first decision on the enforceability of clickwrap contracts was *Hotmail Corp. v Van$ Money Pie Inc.*, in 1998.[12] At the time, Hotmail provided free email services to more than 10 million customers.[13] An email account would be allocated to the user only after the user clicked an 'I agree' button which listed hotmail's specific terms and conditions. These included the prohibition on transmitting unsolicited commercial email – that is, spam. Assent occurred by clicking the appropriate button. The defendant sent spam mail with pornographic

8 See G Evans and B Fitzgerald, 'Information Transactions under UCC Article 2B: The Ascendancy of Freedom of Contract in the Digital Millennium?', (1998) *UNSWLJ* 21, 416.
9 105 F. 3d 1147 (7th Cir 1997).
10 105 F. 3d 1147 (7th Cir 1997), 1149.
11 *Parker v The South Eastern Railway Company* (1877) 2 CPD 416.
12 47 USPQ 2d 1020 (1998). There are no Australian or New Zealand cases to date.
13 By 2008 the number had grown to more than 400 million.

materials, and altered the return address to make it appear that it came from a different source. Hotmail applied for an injunction to restrain both the spamming and the false source. The defendant claimed, among other things, that the terms could not form part of any agreement as he had not agreed to the terms. The court had little difficulty in holding that the defendant was bound by the terms and conditions by clicking the button 'I agree' and that there was an enforceable agreement. The key is to determine whether or not the user had given consent to the terms and conditions by a positive and unambiguous act.

In *Steven J Caspi v Microsoft Network*[14] the terms of the agreement appeared in a scrollable window next to two blocks containing the words 'I agree' and 'I disagree'. The user could not use Microsoft's network without selecting the affirmative choice. The appellate court held that this created an enforceable contract:[15]

> To conclude that plaintiffs are not bound by [the forum] clause would be equivalent to holding that they were bound by no other clause either, since all provisions were identically presented. Plaintiffs must be taken to have known that they were entering into a contract; and no good purpose, consonant with the dictates of reasonable reliability in commerce, would be served by permitting them to disavow particular provisions or the contracts as a whole.

In *Register.com Inc. v Verio Inc.*[16] the court described the key to a clickwrap agreement being that the decision to accept or reject must occur before access is given. Clickwrap agreements must expressly and unambiguously manifest assent prior to the user being given access to the product.

In *AV et al. v iParadigms LLC*[17] a US court applied the clickwrap principle to minor high school students:[18]

> The Court finds that the parties entered into a valid contractual agreement when Plaintiffs clicked 'I Agree' to acknowledge their acceptance of the terms of the Clickwrap Agreement. The first line of the Clickwrap Agreement, which appears directly above the 'I Agree' link, states: 'Turnitin and its services . . . are offered to you, the user, conditioned on your acceptance without modification of the terms, conditions and notices contained herein.' Also the Clickwrap Agreement provides that iParadigms will not be liable for any damages 'arising out of the use of this web site.' By clicking 'I Agree' to create a Turnitin profile and enter the Turnitin website, Plaintiffs accepted iParadigm's offer and a contract was formed based on the terms of the Clickwrap Agreement.[19]

14 323 NJ Super 118 (1999).
15 See also *I Lan Systems Inc. v Netscout Service Level Corp.* (D Mass 2002).
16 356 F. 3d 393 (2nd Cir 2004).
17 Company Civ Act. No. 07–0293 (ED Va 2008).
18 In relation to the applicability to minors, the court quoted from *Williston on Contracts*, Section 9: 14, 4th edn, 2007: 'if an infant enters into any contract subject to conditions or stipulations, he cannot take the benefit of the contract without the burden of the conditions or stipulations'. The court concluded: 'Plaintiffs received benefits from entering into the agreement with iParadigms. They received a grade from their teachers, allowing them the opportunity to maintain good standing in the classes in which they were enrolled . . . Plaintiffs cannot use the infancy defense to void their contractual obligations while retaining the benefits of the contract. Thus, plaintiffs' infancy defense fails' (p. 10).
19 Company Civ Act. No. 07–0293 (ED Va 2008) at 8, emphasis in original.

Browsewrap

The term 'browsewrap' refers to the situation where a user enters into an agreement without giving unambiguous consent to the terms and conditions. For example, a vendor of software to be downloaded may give the user the opportunity to browse the terms but has not made access to the product conditional on reading the available terms and conditions. The vendor may merely place a link to the 'terms of download' on the download page, leaving it optional to view those terms.

In *Specht v Netscape Communications Corp.*,[20] Specht downloaded Netscape's SmartDownload software. Specht claimed that as a result, private information was wrongfully and surreptitiously transmitted to Netscape. Netscape sought to compel arbitration arguing the applicability of the online agreement, which included an arbitration clause. To download the software, users click the button marked 'Download'. The only reference to the terms and conditions appeared only if the user scrolled down the page. The user would then see the words 'please review and agree to the terms of the Netscape SmartDownload license agreement before downloading and using the software'. Next to this was a link which, if clicked, opened a web page containing the terms, including the arbitration clause. The court held that Specht was not bound by the terms. The fact that users were not required to give a positive assent before proceeding was critical to the court's reasoning. The court doubted whether such browsewrap agreements were enforceable:

> Promises become binding when there is a meeting of the minds and consideration is exchanged. So it was at King's Bench in common law England; so it was under the common law in the American colonies; so it was through more than two centuries of jurisprudence in this country; and so it is today. Assent may be registered by a signature, a handshake, or a click of a computer mouse transmitted across the invisible ether of the Internet. Formality is not a requisite; any sign, symbol or action, or even wilful inaction, as long as it is unequivocally referable to the promise, may create a contract.
>
> Where click-wrap license agreements and the shrink-wrap agreement at issue in ProCD require users to perform an affirmative action unambiguously expressing assent before they may use the software, that affirmative action is equivalent to an express declaration stating, 'I assent to the terms and conditions of the license agreement' or something similar. For example, Netscape's Navigator will not function without a prior clicking of a box constituting assent. Netscape's SmartDownload, in contrast, allows a user to download and use the software without taking any action that plainly manifests assent to the terms of the associated license or indicates an understanding that a contract is being formed.[21]

Actual or presumptive knowledge of the browsewrap terms and conditions, for example by prior access, will suffice. The reasoning in *Ticketmaster Corp. v Tickets.com Inc.*[22] was consistent with Specht's case, but without deciding the point

20 150 F. Supp. 2d 585 (SD NY 2001).
21 150 F. Supp. 2d 585 (SD NY 2001), 587.
22 54 USPQ 2d 1344 (CD Cal 2000).

the court left open the possibility that prior use of a website, together with knowledge of the terms and conditions, could create a subsequent binding contract. The plaintiff offered tickets to entertainment events for sale online. The website included terms and conditions purporting to govern use of the site. These were located at the bottom of the website's front page. Users were not required to confirm assent to the terms and conditions or to indicate whether or not the terms and conditions had been read. The defendant sold similar tickets online but also provided information for other sites which included deep links to the plaintiff's site. These deep links bypassed the plaintiff's front page, but included the statement: 'These tickets are sold by another ticketing company. Although we can't sell them to you, the link above will take you directly to the other company's web site where you can purchase them.' The plaintiffs sued for breached the contract, infringement of copyright, unfair competition, unjust enrichment and interference with business advantage. The court left open the possibility that use of a website together with knowledge of the terms and conditions could create a binding contract, but stated:

> [The terms and conditions provide] that anyone going beyond the home page agrees to the terms and conditions set forth, which include that the information is for personal use only, may not be used for commercial purposes, and no deep linking to the site is permitted. In defending this claim, Ticketmaster makes reference to the 'shrink-wrap license' cases, where the packing on the outside of the CD stated that opening the package constitutes adherence to the license agreement (restricting republication) contained therein. This has been held to be enforceable. That is not the same as this case because the 'shrink-wrap license agreement' is open and obvious and in fact hard to miss. Many web sites make you click on 'agree' to the terms and conditions before going on, but Ticketmaster does not. Further, the terms and conditions are set forth so that the customer needs to scroll down the home page to find and read them. Many customers instead are likely to proceed to the event page of interest rather than reading the 'small print'. It cannot be said that merely putting the terms and conditions in this fashion necessarily creates a contract with anyone using the web site.[23]

In relation to deep linking, the court added that:

> hyperlinking does not itself involve a violation of the Copyright Act . . . since no copying is involved . . . the customer is automatically transferred to the particular genuine web page of the original author. There is no deception in what is happening. This is analogous to using a library's card index to get reference to particular items, albeit faster and more efficiently . . . deep linking by itself (i.e. without confusion of source) does not necessarily involve unfair competition.[24]

In *Net2Phone Inc. v Los Angeles Superior Court*[25] the court stated that knowledge or even presumptive knowledge of the existence of the terms may be a sufficient to form a contract. The court preferred a firm assent by a 'clickthrough' process,

23 54 USPQ 2d 1344 (CD Cal 2000).
24 54 USPQ 2d 1344 (CD Cal 2000).
25 109 Cal App 4th 583 (Cal Ct App 2003).

but noted that cases involving cruise tickets and parking tickets have established that such assent is not necessary for formation.[26]

In *Register.com Inc. v Verio Inc.*[27] the court recognised that contract offers on the internet often required the offeree to click on an 'I agree' icon. However, the court stated that this was not necessary 'in all circumstances':

> While new commerce on the Internet has exposed courts to many new situations, it has not fundamentally changed the principles of contract. It is standard contract doctrine that when a benefit is offered subject to stated conditions, and the offeree makes a decision to take the benefit with knowledge of the terms of the offer, the taking constitutes an acceptance of the terms, which accordingly become binding on the offeree.[28]

The approach in Australia and New Zealand is yet to be firmly determined, though the US cases are a good guide. The approach will most likely be to use or modify dicta such as Lord Denning's 'red hand' passage in *Thornton v Shoe Lane Parking Station Ltd*, where the clause did not exempt the defendants from liability as they had not taken reasonable steps to bring it to the attention of Thornton.

In the absence of cases in Australia and New Zealand some attempts have been made to consider the application of the unconscionability provisions of the *Trade Practices Act 1974* (Cth), the *Contracts Review Act 1980* (NSW), selected parts of the Fair Trading Acts, or to enact specific legislation to protect electronic contracts.[29] However, internationally the courts have provided an adequate response in the past decade and there is no reason to assume that Antipodean courts will not respond in the same way. At the very least, legislation which protects consumers generally applies equally to electronic contracts.

Electronic agents

It is possible for computer users to instruct the computer to carry out transactions automatically. For example, in today's supermarket the computer updates its inventory as items are scanned for sale. When the stock of an item falls to a predetermined level the computer is programmed, without human intervention, to contact the computer of the supplier and place an order for replacement stock. The supplier's computer, without human intervention, accepts the order and the next morning automatically prints out worksheets and delivery sheets for the supply and transport staff.

26 See also *Comb v PayPal Inc.* 218 F. Supp. 2d 1165 (ND Cal 2002).
27 356 F. 3d 393 (2d Cir NY 2004).
28 356 F. 3d 393 (2d Cir NY 2004), 403.
29 See D Clapperton and S Corones, 'Unfair terms in "clickwrap" and other electronic contracts', (2007) 35 *Australian Business Law Review* 152.

These electronic agents are programmed by and with the authority of the buyer and supplier. The legal status of electronic agents has not been clarified by the courts, but the most common view is that like any other piece of equipment under the control of the owner, the owner accepts responsibility. The computer is a tool programmed by or with a person's authority to implement their intention to make or accept contractual offers.[30]

Further reading

Dale M Clapperton and Stephen G Corones, 'Unfair terms in "clickwrap" and other electronic contracts', (2007) 35 *Australian Business Law Review* 152.

Alan Davidson, 'Shrinkwrap, clickwrap and browsewrap contracts', (2003) 23 *Proctor* 9, 41.

Debora Halbert, 'The open source alternative: Shrink-wrap, open source and copyright', (2003) 10 *E-Law Journal* 4.

30 For an example of a legislative provision see *Uniform Electronic Transactions Act* (US) s14.

5

Electronic signatures

The signature has been the prime method a person uses as a proof of identity, and as a material expression of intent and execution of documents. A signature on a document indicates the provenance of the document and the intention of the signatory with regard to that document. With the advent of the electronic era, a form of signature is adopted for electronic documents. This chapter examines the regulation, use and security of electronic signatures and the types of electronic signatures, such as digital signatures and SSL technology, being used in electronic commerce.

Traditional signatures

Understanding an electronic signature involves understanding the purpose and use of the traditional signature. The status of traditional signatures has been taken for granted or assumed. The law has developed for centuries with notions of deeds and documents being signed, sealed and delivered, witnessed, notarised and so forth. In some situations a signatory is bound on signing, and in others the signatory is not bound until there is an affirming act. The underlying intention of the signatory may be to be bound only by the subsequent act of delivery. There may be many signatures on a contract: one person may sign intending to be bound by the terms of the contract, while another is merely a witness, with no legal interest in the terms. Their intention is paramount. Even if a signature appears on a document, the signor may not be not bound because he or she lacked the requisite intention – there could be duress, undue influence, *non est factum*, unconscionability or some other vitiating factor. The signature of an illiterate may be an 'X'. An incapacitated person such as a quadriplegic may use another person to place the signature on the document. In all circumstances it is the underlying intention of the signatory that is determinative.

The cases assist in determining the characteristics of a signature. The inclusion of a mark in some form intended as the signatory's authorisation, approval or execution of the contents of the document is paramount.

In 1682 in *Lemayne v Stanley*,[1] the full court regarded a will, handwritten by the testator, with his name appearing in the will, as 'sufficient signing'.[2] The majority also held that the testator's seal was a sufficient signing, stating, 'for *signum* is no more than a *mark*, and sealing is a sufficient *mark* that this is his *will*'.[3] In 1745 in *Ellis v Smith*[4] the court held that 'signing' for legal purposes means the affixing of a personal signature. In 1794 in *Knight v Crockford*,[5] a letter beginning 'I James Crockford, agree to sell' was held to be sufficiently signed under s4 of the *Statute of Frauds*.[6] The case of *Lobb v Stanley*[7] in 1844 considered the effect of a signature in the body of writing. Lord Denman CJ stated:

> it is a signature of a party when he authenticates the instrument by writing his own name in the body. Here it is true, the whole name is not written, but only 'Mr Stanley'. I think more is not necessary.[8]

Wightman J stated 'if a party insert (sic) his name, either at the beginning or in the body of the document, for the purpose of authenticating it, that is enough, and no other signature is wanted'.[9] In 1892 in *Evans v Hoare*,[10] the defendant prepared a document which showed the plaintiff's address at the top, and the defendant's name and address in the body. Cave J stressed the importance of the 'intention for a signature' whether the purported signature is at the beginning, in the body or at the end.

These cases regarded the writing of one's name within the document to be equivalent to a signature. However, the case of *Tourret v Cripps*[11] in 1879 dealt with the proposed terms of a lease in the defendant's own handwriting on a sheet of paper on which was printed at the head of the page 'From Richd L. Cripps', with the defendant's address. The court held that the document had been signed. The essential element of intention was rising to the fore. In *Caton v Caton*[12] in 1867 Lord Westbury said that what is alleged to constitute the signature must:

1 (1682) 3 Lev 1; 83 ER 545.
2 (1682) 3 Lev 1, 1 per North, Wyndham, Charlton and Levinz JJ; 83 ER 545, 546, emphasis in original.
3 83 ER 545, 546.
4 (1754) 1 Ves Jun 11 at 12; 30 ER 205. As late as 1954 Denning LJ, in a dissenting judgment, stated that 'the virtue of a signature lies in the fact that no two persons write exactly alike, and so it carries on the face of it a guarantee that the person who signs has given his personal attention to the document': *Goodman v J Eban Ltd* [1954] 1 QB 550 at 561.
5 (1794) 1 Esp N P C 190; 170 ER 324.
6 For the text of *Statute of Frauds* s4 and its implications see Chapter 3.
7 (1844) 5 QB 574; 114 ER 1366.
8 (1844) 5 QB 574, 581; 114 ER 1366, 1369.
9 (1844) 5 QB 574, 583; 114 ER 1366, 1369.
10 [1892] 1 QB 593, see Cave J at 597. See also *Schneider v Norris* (1814) 2 M & S 286.
11 (1879) 48 LJ Ch 567; 27 WR 706.
12 (1867) LR 2 HL 127.

> be so placed as to show that it was intended to relate and refer to, and that in fact it does relate and refer to, every part of the instrument . . . It must govern every part of the instrument. It must shew that every part of the instrument emanates from the individual so signing, and that the signature was intended to have that effect. It follows that if a signature be found in an instrument incidentally only, or having relation and reference only to a portion of the instrument, the signature cannot have legal effect and force which it must have in order to comply with the statute, and to give authenticity to the whole of the memorandum.

The case of *R v Moore; Ex Parte Myers*[13] in 1884 dealt with a pawnbroker's pledge ticket that was not signed in accordance with the relevant legislation but was signed by an authorised agent. The name of the pawnbroker was printed on the ticket. Higginbotham J stated:

> A signature is only a mark, and where the Statute merely requires a document shall be signed, the Statue is satisfied by proof of the making of the mark upon the document by or by the authority of the signatory . . . where the Statute does not require that the signature shall be an autograph, the printed name of the party who is required to sign the document is enough . . . or the signature may be impressed upon the document by a stamp engraved with a facsimile of the ordinary signature of the person signing . . . proof in these cases must be given that the name printed on the stamp was affixed by the person signing, or that such signature has been recognised and brought home to him as having been done by his authority so as to appropriate it to the particular instrument.[14]

Binding oneself to the contents of the document therefore does not require the physical act of putting pen to paper, but can be achieved by an agent or through the use of some mechanical means, such as an impress stamp bearing a facsimile of the person's signature. *R v Moore* determines that the object of a signature affixed to a document is to authenticate the genuineness of the document.[15] Further, the case holds that a person, in order to be bound, must put his or her mind to the act of signing the document, as opposed to simply providing an autograph.

In *Lazarus Estates Ltd v Beasley*,[16] Denning LJ (as he then was) made the following comment in relation to a company stamp:

> The statutory forms require the documents to be 'signed' by the landlord, but the only signature on these documents (if such it can be called) was a rubber stamp 'Lazarus Estates Ltd', without anything to verify it. There was no signature of a secretary or of any person at all on behalf of the company. There was nothing to indicate who affixed the rubber stamp. It has been held in this court that a private person can sign a

13 (1884) 10 VLR 322.
14 (1884) 10 VLR 322, 324. See also *Smith v Greenville County* 188 SC 349; 199 SE 416, 419 (1938), where it was held: 'A signature may be written by hand, printed, stamped, typewritten, engraved, photographed or cut from one instrument and attached to another, and a signature lithographed on an instrument by a party is sufficient for the purpose of signing it: it being immaterial by what kind of instrument a signature is made.'
15 Higginbotham J added, 'It was observed by Patterson J in *Lobb v Stanley*, that the object of all Statutes which require a particular document to be signed by a particular person is to authenticate the genuineness of the document.'
16 [1956] 1 All ER 341.

document by impressing a rubber stamp with his own facsimile signature on it . . . but it has not been held that a company can sign by its printed name affixed with a rubber stamp.

Modern signatures

In the United States in 1948, in *Joseph Denunzio Fruit Company v Crane*,[17] the Californian District Court commented that it 'must take a realistic view of modern business practices', suggesting that the court should take judicial notice of the extensive use of the teletype machine. In holding that the point was *res nova*, the court held teletype messages satisfy the *Statute of Frauds* in California.

In relation to facsimiles and telexes it could be argued that each message contains identifying information which could be regarded as a signature in law.[18] However, the identifiers indicate the machine used, not necessarily the particular user or author of the message. Office facsimiles may be used by many parties. It is possible to alter the telex machine or facsimile machine so that it sends a false identification message. These possibilities may weigh heavily against any suggestion that a telex or facsimile should be treated as signed.[19]

Electronic signing

Electronic documents may be signed electronically with the same underlying intent and purpose as traditional documents. Cases requiring a mark with a corresponding intent of authorising or executing the document have equal application to electronic documents with electronic signatures. Thus signing an electronic communication such as an email by simply typing the sender's name at the end would be an acceptable electronic signature if inserted with the requisite intention. Similarly, including the sender's name in the body of the email as envisaged in *Lobb v Stanley*[20] and *Evans v Hoare*[21] will suffice.

In the Queensland Court of Appeal, in *R v Frolchenko*,[22] Williams J recognised that modern communication, such as email, may not bear a personal signature. His Honour stated that such an electronic document could be authenticated by looking at other factors, such as whether the name appears in typescript at the end of the document.

17 79 F. Supp. 117 (DC Cal 1948).

18 See C Reed, 'Authenticating electronic mail messages – some evidential problems, (1989) 52 *Modern Law Review* 649 and Queensland Law Reform Commission Issues Paper, 'The receipt of evidence by Queensland courts: Electronic records', www.qlrc.qld.gov.au/wp52.html.

19 In the United States, some commentators have adopted a pragmatic attitude to the potential difficulties raised by authentication; see B Wright, 'The verdict on plaintext signatures: They're legal', (1994) 10 *The Computer Law and Security Report* 311–12.

20 (1844) 5 QB 574; 114 ER 1366.

21 [1892] 1 QB 593, see Cave J at 597.

22 (1998) QCA 43.

In *Mehta v J Pereira Fernandes SA*,[23] the court considered an email which was not signed at the foot of the message, but was described in the header of the email as having come from Nelmehta@aol.com. The court cited *Evans v Hoare* with approval and determined that whether or not an email address in the header amounts to a signature depends upon the intention of the sender. The defendant, by the email, offered to provide a personal guarantee in favour of the plaintiff in relation to a payment of £5000, on certain terms. The *Statute of Frauds* requires such guarantees to be in writing. Judge Pelling QC held that the email would be a note or memorandum to which section 4 of the *Statute of Frauds* applied. However, his Honour considered whether it was sufficiently signed, stating:

> a party can sign a document for the purposes of Section 4 by using his full name or his last name prefixed by some or all of his initials or using his initials, and possibly by using a pseudonym or a combination of letters and numbers (as can happen for example with a Lloyds slip scratch), providing always that whatever was used was inserted into the document in order to give, and with the intention of giving, authenticity to it.[24]

In determining what would amount to a signature Judge Pelling QC examined the underlying purpose. The purpose of the *Statute of Frauds*:

> is to protect people from being held liable on informal communications because they may be made without sufficient consideration or expressed ambiguously or because such a communication might be fraudulently alleged against the party to be charged.[25]

In relation to an electronic document his Honour commented that:

> if a party creates and sends an electronically created document then he will be treated as having signed it to the same extent that he would in law be treated as having signed a hard copy of the same document. The fact that the document is created electronically as opposed to as a hard copy can make no difference.[26]

However, the issue before the court was whether the automatic insertion of a person's email address after the document has been transmitted constitutes a signature for the purposes of the *Statute of Frauds*. His Honour's conclusion was that an email address insertion was 'incidental', and that the header was 'divorced from the main body of the text of the message'. If there is no further evidence in relation to the maker's intention, 'it is not possible to hold that the automatic insertion of an email address is . . . intended for a signature'. To conclude otherwise would, in his Honour's view, undermine or potentially undermine the purpose of the *Statute of Frauds*. Judge Pelling QC relied on the passage from Lord Westbury in *Caton v Caton*[27] quoted earlier, and stressed the importance of the signatory's intention in placing of inserting the purported signature, stating:

23 [2006] EWHC 813.
24 [2006] EWHC 813, para 26.
25 [2006] EWHC 813, para 16.
26 [2006] EWHC 813, para 28.
27 (1867) LR 2 HL 127.

'it is not possible to hold that the automatic insertion of an email address is, to use Cave J's language, "intended for a signature"'.[28]

In the US case *Doherty v Registry of Motor Vehicles*,[29] Agnes J held that a police report made 'by means of email or some other electronic method' is regarded as signed, subjecting the reporting officer to possible perjury charges.

Acceptance at face value and risk

Parties to traditional hardcopy documents had accepted signatures at face value, accepting the associated practical and legal risks. An email can be made to appear to be from a third party and false addresses and pseudonyms can be employed. However, this is not new. Letters, facsimiles, telexes and so forth have also been faked. Fraudsters utilise all means at their disposal.

The genuineness of a document does not rely solely on the signature. Typically a range of proofs are used for verification purposes. It is quite rare that standard hardcopy documents are adduced into evidence and proven as genuine only by the signature. More often the origin and genuineness are determined from the facts, conduct and acts of the parties and the surrounding circumstances of the case. Parties to contracts generally accept signatures at face value. They have no technical process to prove the genuineness of a signature, handwritten or otherwise, and there is typically no practice requiring additional verification until a dispute arises. Similarly, there is no technical proof of origin of a telegram or telex. Commercial parties have accepted this risk in the past.

The courts and legislators should resist establishing any rule, principle or law that an electronic signature is not valid merely because of the possibility of fraud or tampering, or that such signatures are fallible. Lawmakers should recognise the commercial parties' acceptance of risk for commercial expediency, particularly in like circumstances.

Functions of signatures

The functions of a signature of a person historically do much more than merely authenticate the genuineness of a document. For example, three signatures typically appear on a testamentary disposition such as a will. The intention and function depend on whether the signature is that of the testator or witness. Functions include:

- to identify the signatory;
- to clarify and identify the personal involvement of the signatory in the act of signing;

28 (1867) LR 2 HL 127, para 29.
29 No. 97CV0050 (Mass 1997), www.state.ma.us/itd/legal/case.htm.

- to associate a particular person with the contents of the document;
- to witness another person's signature;
- to approve the contents of the document;
- to indicate authorship of the document by the signatory.[30]

In the absence of vitiating factors such as fraud, *non est factum*, undue influence and unconscionable conduct, the law regards a signature as binding on the signatory in relation to the contents of the document even if the signatory has not read the document.[31]

There appear to be five main functions of signature requirements:

- evidentiary – to ensure the availability of admissible and reliable evidence when applying legislation such as the *Statute of Frauds*;
- cautionary – to act as a warning that this is a serious document with legal consequences;
- reliance – the signature requirement may indicate that the veracity of document or record may be relied on by others later;
- channelling – to act as a demarcation between 'intent to act in a legally significant way and intent to act otherwise'; and
- record-keeping – for compliance with laws such as taxation, money-laundering and statutes of limitation.[32]

Electronic Transactions Acts

The requirement for signature in documents, particularly those that document, evidence or create contracts, was discussed in Chapter 3.

Where the law requires a signature, the Electronic Transactions Acts deem that the requirement is taken to have been met in relation to an electronic communication where certain conditions are met. The underlying rationale is based on functional equivalence. Where the electronic signature is functionally equivalent to the traditional signature, it ought to be treated equally in law. Each of the nine Australian Electronic Transactions Acts includes a provision deeming electronic signatures, on certain conditions, to meet the requirements under law of signatures. The signature provision states:[33]

> If, under a law of this jurisdiction, the signature of a person is required, that requirement is taken to have been met in relation to an electronic communication if:

30 See A McCullagh, P Little and W Caelli, 'Electronic signatures: Understand the past to develop the future, [1998] *UNSWLJ* 56 and A McCullagh, W Caelli and P Little, 'Signature stripping: A digital dilemma', [2001] *JILT* (1).

31 *L'Estrange v Graucob* [1934] 2 KB 394.

32 See Electronic Commerce Expert Group, 'Electronic commerce: Building the legal framework, Report of the Electronic Commerce Expert Group to the Attorney-General', (1998), para 2.7.28.

33 NSW – s9(1); Qld – s14; SA – s8(1); Tas – s7(1); Vic – s9(1); WA – s9(1); ACT – s9(1) and NT – s9(1). The Commonwealth has substantially similar provisions in s10(1), but also makes special provision for Commonwealth entities.

(a) a method is used to identify the person and to indicate the person's approval of the information communicated, and
(b) having regard to all the relevant circumstances at the time the method was used, the method was as reliable as was appropriate for the purposes for which the information was communicated, and
(c) the person to whom the signature is required to be given consents to that requirement being met by way of the use of the method mentioned in paragraph (a).[34]

The requirement that the court must have 'regard to all the relevant circumstances at the time the method was used, the method was as reliable as was appropriate for the purposes for which the information was communicated', is a serious weakness of the signature provision. It is intended to create a standard for functional equivalence. However, the provision will require the court, through counsel, to cover a significant number of factors – legal, technical and commercial – before making the equivalence determination. These might include, for example:

- the sophistication of the equipment used by each of the parties;
- the nature of their trade activity;
- the frequency with which commercial transactions take place between the parties;
- the kind and size of the transaction;
- the function of signature requirements in a given statutory and regulatory environment;
- the capability of communication systems;
- compliance with authentication procedures set forth by intermediaries;
- the range of authentication procedures made available by the intermediary;
- compliance with trade customs and practice;
- existence of insurance coverage mechanisms against unauthorised messages;
- the importance and value of the information contained in the data message;
- the availability of alternative methods of authentication and the cost of implementing them;
- the degree of acceptance or non-acceptance of the method of identification in the relevant industry or field both at the time the method was agreed upon and the time when the data message was communicated; and
- any other relevant factor.[35]

34 The following expressions and phrases from the legislation were considered in detail in Chapter 3: 'signature of person'; 'signature of the person . . . in relation to an electronic communication'; 'method used'; 'requirement is taken to have been met'; 'electronic communication'; 'a method used to identify the person and to indicate the person's approval'; reliability of the method used'; 'of the information communicated'; 'the method must be as reliable as appropriate for the purposes for which the information was communicated' – 'having regard to all the relevant circumstances at the time the signature method was used' and 'consent'.

35 See *UNCITRAL Guide to the Model Law of Electronic Commerce*, paras 52–61, in particular para 58; www.uncitral.org.

'Electronic signature' defined

An 'electronic signature' is any means of electronic authentication of the identity of a person and of the intent of that person to indicate approval or to be associated with an electronic record. The term has no universally accepted meaning and internationally is variously defined in different statutes. An 'electronic signature' may be defined as any electronic data, including any letters, characters, numbers or other symbols, attached to or logically associated with an electronic record, used with the intention of authenticating or approving the electronic record.

The Electronic Transactions Acts in Australia do not define 'electronic signature'. They do not need to. Instead, they logically connect the use of a 'signature' with an 'electronic communication'. By necessity, such a signature must be electronic. The expression 'electronic communication' is defined as:

> (a) a communication of information in the form of data, text or images by means of guided or unguided electromagnetic energy, or both, or
> (b) a communication of information in the form of sound by means of guided or unguided electromagnetic energy, or both, where the sound is processed at its destination by an automated voice recognition system.[36]

The expression 'electronic communication' may be regarded as narrower than 'electronic record' or 'electronic document', but this point remains untested. In all likelihood a court would apply the expression to all forms.

The UNCITRAL Model Law of Electronic Commerce does not use the expression 'electronic signature'. However, the UNCITRAL Model Law of Electronic Signatures defines an 'electronic signature':

> 'Electronic signature' means data in electronic form in, affixed to or logically associated with, a data message, which may be used to identify the signatory in relation to the data message and to indicate the signatory's approval of the information contained in the data message.[37]

The New Zealand *Electronic Transactions Act 2002* provides: '"electronic signature", in relation to information in electronic form, means a method used to identify a person and to indicate that person's approval of that information'.

The expression only appears once in each of the Australian Acts, and then only to provide that 'Certain other laws [are] not affected.' This savings provision is designed to ensure that the general deeming provision of the Electronic Transactions Acts does not affect the operation of a specific law which may call for an electronic signature 'howsoever described'.

The Singapore *Electronic Transactions Act 1998* includes extensive provisions dealing with electronic and digital signatures. The Singapore Act defines 'electronic signature' as:

36 NSW – s5; Qld – s6; SA – s5; Tas – s3; Vic – s3; WA – s5; ACT – s5; and NT – s5. The Commonwealth *Electronic Transactions Act 1999* uses the narrower word 'speech' in place of 'sound'. New Zealand defines 'electronic communications' in s5: '"electronic communication" means a communication by electronic means'.
37 Art. 2.

> any letters, characters, numbers or other symbols in digital form attached to or logically associated with an electronic record, and executed or adopted with the intention of authenticating or approving the electronic record.[38]

An electronic signature may be as simple as typing a name at the end of an email. It may be a complex mathematical transformation designed to provide a level of security to ensure that the electronic message is from the purported sender and is unaltered. Both forms of signatures have data in digital form attached to or logically associated with an electronic document. The level of security to be used is a matter for the parties, depending upon factors such as the commercial and legal risk involved. Parties entering into transactions over the internet frequently need to authenticate the message and verify the identity of the sender.

Uses

An electronic signature may be used to sign any electronic document or record, with the same intention and for the same purpose as a traditional signature on a paper record or document. However, certain types of electronic signatures have a much higher level of security than the traditional signature, and may be used for authentication and verification purposes with a greater degree of certainty.

Security of electronic signature

Traditional signatures can be easily forged. Nevertheless the traditional signature is used extensively in commerce, on contracts, cheques and other negotiable instruments. Electronic signatures have different risks. While the simple electronic signature merely typed at the end of an electronic record may be less secure than the traditional signature, there are many choices available to parties, commercial and otherwise, to raise the level of authentication and verification to well in excess of that of a handwritten signature.

Anyone could type a person's name at the end of a typed electronic communication. Such a signature, while valid in the right circumstances, is most insecure. Two parties could agree upon a simple code. Parties could agree upon a simple number or a string of characters. For example, parties agree on the number '333'. This number may be placed at the end of the message with the agreed meaning that this verified that the communication came from the other party. Such a simple signature identifies the signor by agreement, and indicates authentication or approval. The number may be dynamic; for example the agreed number could be the sum of five times the date plus six times the month: 5 November would thus be '91'. It would be difficult for a third party to work out the authenticating

38 *Electronic Transactions Act 1998* (Singapore) s2.

signature even by intercepting a handful of communications. There are more sophisticated methods available commercially.

Digitised signatures

A digitised signature is a handwritten signature, scanned into a computer and then placed electronically into an electronic document, such as an email, to give the appearance of a traditional signature. Such a signature is an electronic signature, and if placed with the requisite intention operates in the same manner as any other electronic signature. A digitised signature has a low level of security, as anyone intercepting such an email can extract it and use it.

Digital signatures

The digital signature is a subset of the electronic signature. Digital signatures are attached to specific data, such as an email, computer file or web page. Although often used as a signature in the usual sense, the expression 'digital signature' is a misnomer. It may be better described as a 'digital certificate', as it functions as certification of the document and sender (or creator) rather than as a formal signature. A digital signature is:

> an electronic signature consisting of a transformation of an electronic record using an asymmetric cryptosystem and a hash function such that a person having the initial untransformed electronic record and the signer's public key can accurately determine —
> (a) whether the transformation was created using the private key that corresponds to the signer's public key; and
> (b) whether the initial electronic record has been altered since the transformation was made.[39]

A digital signature permits both verification and authentication of data.

The recipient verifies the digital signature by a simple automatic computation involving the data, the purported digital signature and the sender's public key. The computation determines whether or not the correct mathematical relational exists. If it does, it reports that the digital signature is verified. An unverified digital signature may be an indication that there is a hoax sender or that the message has been altered. The recipient should then take appropriate steps, such as determining the *bona fides* of the message, and seek a retransmission.

Standard encryption requires that the sender and the recipient of a message know and use the same secret code. This method is symmetrical: the process to decrypt is the reverse of the process to encrypt. The method is called private encryption. The major difficulty with it is agreeing on the secret code with confidence that no one else finds out. Anyone who overhears or intercepts the key

39 *Electronic Transactions Act 1998* (Singapore) s2. See also the UNCITRAL Model Law of Electronic Signatures.

can potentially read all encrypted messages. Accordingly, private encryption has its own inherent security risks. Also, in modern communications one may have little notice of who one's correspondent may be, and thus of when one may need this level of encryption.

Public encryption was developed in 1976 to resolve private encryption dilemmas. The sender and recipient each possess a pair of keys: the public key and the private key. Each person's public key is freely available to anyone. The private key is kept secret. There is no need for both parties to share information about the private key. All communications involve only the public key; no private key is ever transmitted or shared. There is a mathematical relationship between the public key and the private key, but the private key cannot be determined from the public key. One or more public keys may be created from the private key. The program makes complex mathematical calculations to do this.

The holder of a private key uses that key to place a digital signature on the relevant electronic document. Any person holding the corresponding public key is able to verify that the message has not been altered and authenticate that it could only come from the holder of the private key. The 'hash function' means applying an algorithm mapping or translating one sequence of the electronic record into another, generally smaller, set (the 'hash result') such that: a record yields the same hash result every time the algorithm is executed using the same record as input; it is computationally infeasible that a record can be derived or reconstituted from the hash result produced by the algorithm; and it is computationally infeasible that two records can be found that produce the same hash result using the algorithm. The level of security available is such that it is regarded as impossible to duplicate.

A verified result of a digital signature is regarded as certain in the industry so far as the computing aspects are concerned. The courts should accept such an authentication.

Authentication of electronic data messages will become increasingly important for lawyers and commercial parties for evidentiary purposes. The Australian Evidence Acts do not address all aspects of email communications. Some state Acts make presumptions regarding the sending and receipt of postal articles, telexes, lettergrams and telegrams. However, there is no similar presumption regarding email. Nonetheless, the Commonwealth and NSW Evidence Acts state:

> The hearsay rule does not apply to a representation contained in a document recording a message that has been transmitted by electronic mail or by a fax, telegram, lettergram or telex so far as the representation is a representation as to:
> (a) the identity of the person from whom or on whose behalf the message was sent, or
> (b) the date on which or the time at which the message was sent, or
> (c) the message's destination or the identity of the person to whom the message was addressed.[40]

40 *Evidence Act 1995* (Cth) s71 and *Evidence Act 1995* (NSW) s71.

Nevertheless, courts will need to be satisfied regarding the authenticity of transmissions.

The enactment of legislation dealing with digital and electronic signatures needs to be considered with caution. Only recently have international models been available for national legislatures. In July 2001 UNCITRAL released the Model Law of Electronic Signatures.[41] This Model Law is intended to bring greater legal certainty to the use of electronic signatures. It establishes the presumption that electronic signatures are to be treated as equivalent to handwritten signatures where certain criteria of technical reliability are met. The Model Law uses technology-neutral language and establishes rules of conduct for assessing responsibilities and liabilities of the signatory, the relying party and trusted third parties that might intervene in the signature process. In a similar vein, the European Union passed a Directive on a Community Framework for Electronic Signatures.[42] It establishes a legal framework for electronic signatures and certain certification services.

Given the pace of technological development, it is more appropriate for the market to determine practice issues, such as the levels of security and reliability required for electronic signatures. Legislation should deal simply with the legal effect of electronic signatures.

While the Commonwealth *Electronic Transactions Act 1999* is based on the UNCITRAL Model Law of Electronic Commerce, the federal Attorney-General's view is that a legislative regime concerning digital and electronic signatures is not required. The debate in relation to the legal issues raised by electronic commerce is often clouded by the discussion of digital and electronic signatures. 'Electronic signature' is a term used to refer to a range of technologies intended to ensure the security and certainty of electronic commerce.

Australian Business Number Digital Signature Certificates

To aid electronic commerce, the federal government developed the Australian Business Number Digital Signature Certificate (ABN-DSC), which is a digital certificate linked to an entity's ABN. It facilitates online service delivery and is intended to foster the use of digital certificates and electronic commerce in Australia. Since March 2001 Commonwealth agencies have accepted the certificates to identify businesses for online transactions. Only Certification Authorities accredited under the Commonwealth's Gatekeeper Public Key Infrastructure framework are able to issue ABN-DSCs.

41 The text of the UNCITRAL Model Law on Electronic Signatures was adopted on 5 July 2001. It is available at www.uncitral.org/english/texts/electcom/ml-elecsig-e.pdf. Australia has not enacted any legislation based on this Model Law.

42 Directive 1999/93/Ec of the European Parliament and of the Council of 13 December 1999, on a Community Framework for Eelectronic Signatures, available at www.europa.eu.int/comm/internal_market/en/media/sign/Dir99–93-ecEN.pdf.

In 1997, the federal government established the Online Council following agreement by states, territories and local government to cooperate on online issues. In March 2002, all states and territories agreed to urgently consider accredited and cross-recognised certificates which are ABN-DSC compliant. This will help meet the e-commerce objectives of governments by making digital certificates widely available to businesses. This development has direct implications for B2G (business to government) electronic commerce and facilitates B2B (business to business) electronic commerce by providing authentication, confidentiality, integrity and non-repudiation.[43]

Secure Socket Layer – Transport Layer Security

Secure socket level (SSL) technology refers to a sequence of processes that ensure that information stored in electronic form or transmitted over networks, such as the internet, is not accessible to any person not authorised to view that information.[44]

Transport Layer Security (TLS) is a security protocol that ensures privacy between communicating applications and their users on the internet. When a server and client communicate, TLS ensures that no third party may interfere with or read any message. TLS is the successor to SSL.

Both SSL and TLS use a form of digital signature technology. For evidentiary purposes, communications would be regarded as secure, authenticated and verified. SSL and TLS provide endpoint authentication and communications privacy over the internet using cryptography. In typical use, the server is authenticated and the client remains unauthenticated; mutual authentication requires deploying public key infrastructure to clients. The protocols allow server applications to communicate in a way designed to prevent eavesdropping, tampering and message forgery. This allows users to securely provide personal details – such as name, address and credit card numbers – to sites employing the protocol.

The SSL and TLS protocols exchange records. Each record can be compressed, encrypted and packed with a message authentication code. Each record has a content type field that specifies the type of protocol being used. When the connection is initiated, the record level encapsulates another protocol, the handshake protocol. Client and server negotiate a common secret called 'master secret'. Other key data is calculated from the 'master secret', which is passed through a 'Pseudo Random Function'.

43 See generally the website of the Australian Government Information Management Office: www.agimo.gov.au/archive/publications_noie/2003/08/framework/glossary.

44 ALRC, Discussion Paper 72, para 6.5. For an example of an SSL application, see SSL-Explorer Enterprise Edition at www.optimati.com/index.php?option=com_content&task=view&id=47&Itemid=63.

Applications

The SSL and TLS protocols are commonly found with hypertext transfer protocol (HTTP), which is the most common protocol used on the World Wide Web. When used, the protocol becomes HTTPS. HTTPS is used to secure World Wide Web pages for applications such as electronic commerce. It uses public key certificates to verify the identity of endpoints.

Further reading

Sharon Christensen, William Duncan and Rouhshi Low, '*The Statute of Frauds* in the digital age – maintaining the integrity of signatures', (2003) *MurUEJL* 44.
Alan Davidson, 'Electronic signatures', (2002) 22 *Proctor* 8, 31.
Alan Davidson, 'Signatures on electronic documents', (2004) 24 *Proctor* 7, 29.
Yee Fen Lim, 'Digital signature, certification authorities and the law', (2002) *MurUEJL* 29.
A McCullagh, W Caelli and P Little, 'Signature stripping: A digital dilemma', (2001) *JILT* (1).

6

Copyright issues in electronic commerce

The internet facilitates the swift copying and exchange of digital material. Hyperlinking and framing can result in new and novel possibilities for infringement of copyright. Works protected by copyright, such as music and videos, can be easily transferred in peer-to-peer dealings. Such transfers have become the target of copyright owners. This chapter explores these issues and developments and the contiguity between copyright and electronic commerce. It is not intended to state the law relating to copyright, other than a brief overview. The advent of electronic commerce and the internet have necessitated a rethink of intellectual property issues by the World Intellectual Property Organisation (WIPO), the courts and the legislature. The proliferation of material on the internet – written, aural and graphic – has posed new questions and resulted in the creation of new rights internationally. The next chapter addresses the interplay between electronic commerce and intellectual property rights in trade marks, patents and circuit layouts.

The nature of copyright

Intellectual property comprises state-sanctioned rights entitling the holder of such rights to a limited monopoly on exploiting and controlling the property for a predetermined period of time. The state gains the benefit of, for example, literary and musical works, new inventions, medicines, research, designs and innovation. Inventors, writers, composers and designers have the incentive of potential rewards for their efforts. Their works, patents and designs will form part of the public domain when the predetermined period expires.

Copyright arose only after the invention of the printing press. Previously copies were made by hand and were relatively expensive and time consuming. Charles II of England encouraged the passing of the *Licensing Act of 1662*. However, the *Statute of Anne* in 1702[1] is regarded as the first copyright legislation. It gave publishers rights for fixed periods. From copyright's initial application to books and maps it has now extended to such things as sound, films, choreography and software. The Berne Convention of 1886 established international recognition of copyright among member nations. The Berne Convention pioneered the concept that copyright automatically arose at the time of creation. That is, registration of a declaration of copyright is not necessary. Copyright arises once a work is written down, painted or drawn, filmed or taped. It protects the expression of an idea, but not the idea itself; it protects only against the copying of a work, not against independent creation. An idea is not protected until it is documented in writing or recorded in some way. Copyright is intended to reward authors economically by giving them control of the use of their work.

Works must be original.[2] Originality is assessed only according to whether or not the creator produced the work using his or her own skill and effort and did not copy another's work. If two authors express the same idea and their respective works are created independently, they both have copyright protection.

The regulations of the Berne Convention are incorporated into the World Trade Organization's TRIPS agreement (1995), giving the Berne Convention near global application. The 2002 World Intellectual Property Organisation (WIPO) Copyright Treaty enacted greater restrictions on the use of technology to copy works in the nations that ratified it.

Copyright, at its basic level, is the right to copy. Copyright arises only by the operation of statute: in Australia, the *Copyright Act 1968* (Cth).[3] The right usually vests in the creator or author of a work at the time when the work is first produced in material form. For example if one is said to own the copyright in a musical work, no other person has the right to make a copy of that work without permission.[4]

Copyright covers two categories of material: original and secondary. Original 'works', as defined by the Copyright Act, may be literary, dramatic, musical or artistic. Secondary copyright refers to 'subject matter other than works', such as sound recordings, cinematograph films (this covers computer-generated or interactive computer games), television and sound broadcasts and published editions of works.[5] The former is dealt with by Part III of the Copyright Act and

1 *8 Anne c 19.* Long title: 'An Act for the Encouragement of Learning, by vesting the Copies of Printed Books in the Authors or purchasers of such Copies, during the Times therein mentioned'.

2 See *Macmillan v Cooper* (1923) 93 LJPC 113; *Cramp & Sons Ltd v Frank Smythson Ltd* [1944] AC 328; *Feist Publications Inc. v Rural Telephone Service Co. Inc.* 737 F. Supp. 610, 622 (Kan 1990). See also *Lott v JBW & Friends PL* [2000] SASC 3.

3 Copyright is specifically a federal power pursuant to Australian Constitution, s51(xviii) Copyrights, patents of inventions and designs, and trade marks. The New Zealand legislation is the *Copyright Act 1994*.

4 See *APRA v Canterbury-Bankstown League Club Ltd* [1964] NSWR 138.

5 See *APRA v Commonwealth Bank of Australia* (1993) 25 IPR 157; *Telstra Corp. Ltd v APRA* (1997) 38 IPR 294.

the latter by Part IV. The copyright in a secondary work exists independently of the copyright in the original work. From a copyright perspective, the internet is a mixture of original works and secondary works. Web pages may include graphics and images, text, compilations, and music and video clips, each of which is likely to be protected by copyright.

Generally copyright in a literary, dramatic, musical or artistic work lasts for 70 years after the end of the calendar year in which the author of the work died.[6] If, before the death of the author of a literary work (other than a computer program) or a dramatic or musical work, the work had not been published, performed in public or broadcast, and records of the work had not been offered or exposed for sale to the public, copyright lasts for 70 years after the end of the calendar year in which the work is first published, performed in public, or broadcast, or records of the work are first offered or exposed for sale to the public, whichever is the earliest of those events.[7] A reference to the doing of an act in relation to a work includes a reference to the doing of that act in relation to an adaptation of the work.[8] If the first publication of a literary, dramatic, musical or artistic work is anonymous or pseudonymous, copyright lasts for 70 years after the end of the calendar year in which the work was first published. However, if, at any time before the end of the period the identity of the author of the work is generally known or can be ascertained by reasonable inquiry, the period is calculated from the death of the author.[9] Where the work is the product of joint authorship, the period of protection runs from the death of the last surviving joint author.

Exclusive rights

The Copyright Act grants certain exclusive rights to the holder of the copyright depending upon the type of work:[10]

> 31(1) For the purposes of this Act, unless the contrary intention appears, copyright, in relation to a work, is the exclusive right:
> (a) in the case of a literary, dramatic or musical work, to do all or any of the following acts:
> (i) to reproduce the work in a material form;
> (ii) to publish the work;
> (iii) to perform the work in public;
> (iv) to communicate the work to the public;
> (v) to make an adaptation of the work;
> (vi) to do, in relation to a work that is an adaptation of the first-mentioned work, any of the acts specified in relation to the first mentioned work in subparagraphs (i) to (iv), inclusive; and

6 *Copyright Act* (Cth) s33(2); 50 years in New Zealand, *Copyright Act 1994* (NZ) s22.
7 *Copyright Act* (Cth) s33(3).
8 *Copyright Act* (Cth) s33(4).
9 *Copyright Act* (Cth) s24.
10 For New Zealand, see *Copyright Act 1994* (NZ) s16.

(b) in the case of an artistic work, to do all or any of the following acts:
  (i) to reproduce the work in a material form;
  (ii) to publish the work;
  (iii) to communicate the work to the public; and
(c) in the case of a literary work (other than a computer program) or a musical or dramatic work, to enter into a commercial rental arrangement in respect of the work reproduced in a sound recording; and
(d) in the case of a computer program, to enter into a commercial rental arrangement in respect of the program.

The relief that a court may grant in an action for an infringement of copyright includes 'an injunction (subject to such terms, if any, as the court thinks fit) and either damages or an account of profits'.[11] There may be additional damages for flagrant breaches or as punishment, to deter the infringer (and, presumably, others). The aggrieved party may also claim damages for conversion and delivery up of infringing copies or plates used for making such copies.[12]

Where a proved infringement on a commercial scale involves a communication of a work to the public, and it is likely that there were other infringements of the copyright by the defendant that the plaintiff did not prove in the action, the court is directed by the Copyright Act to have regard to the likelihood of other infringements, as well as the proved infringement. In determining whether the proved infringement and the likely infringements were on a commercial scale, the court must take into account the volume and value of any articles that are infringing copies and any other relevant matter. For the purposes of this consideration, 'article' includes a reproduction or copy of a work in electronic form.[13]

Infringement

Infringement may be direct or indirect. Direct infringement is where a person exercises an exclusive right of the copyright owner without permission.[14] The downloading of music from the internet, for example, involves direct copying. Indirect infringement includes the importing of infringing articles, selling or conducting other trade dealings with infringing articles and permitting a place of public entertainment to be used for a public performance of a work.[15]

Generally the infringement occurs where a 'substantial part' of the work is copied.[16] In addition, generally there needs to be proof of a sufficient degree of objective similarity between the work and the copy, and a causal connection between the two works.

11 *Copyright Act 1968* (Cth) s115(2).
12 *Copyright Act 1968* (Cth) s116, and ss119–125.
13 *Copyright Act 1968* (Cth) s115(5), (6), (7) and (8).
14 *Coogi Australia Pty Ltd v Hysport International Pty Ltd* [1998] FCA 1059.
15 See *Copyright Act 1968* (Cth) s39(1). See *Peer-to-peer file sharing*, below.
16 *Copyright Act 1968* (Cth) s14.

Substantial part

Copyright is infringed where the work or a substantial part of the work has been copied.[17] What constitutes a substantial part is a question of fact, and is determined on a case by case basis. In 1853 Harriet Beecher Stowe sued to stop an unauthorised German translation of her novel *Uncle Tom's Cabin* – and lost.[18] Notwithstanding that the entire work was involved, Judge Robert Grier held that copyright applies only to the 'precise words'. Calling a translation 'a copy of the original', his Honour described as ridiculous the proposition that the translation was a breach of the copyright. In 1870 the US Congress remedied this position, as have all modern jurisdictions. The copyright holder now has the right to make an adaptation (this includes translations) of the work and to do all the things with that adaptation that are allowed to be done with the work itself.[19]

Prior to the internet and computer programs, the cases established that the most important factor in determining whether or not the part of the work copied was a substantial part was the quality of what was copied rather than the quantity.[20] Copying a few notes of music of a popular song could constitute a substantial part if those notes constitute the main theme of the song.[21]

The question of whether or not something is a substantial part of a computer program created difficulty in *Autodesk Inc. v Dyason No. 1*[22] and *Autodesk Inc. v Dyason No. 2* (Autodesk cases).[23] The question centred on a 127-bit series. The 127 bits were tiny, typically representing a mere 16 characters in a typed document. But as used in the program they were an important construct. Dawson J said that the 127-bit series was 'a substantial, indeed *essential*, part of that program . . . [it] processes the information which it receives'.[24] Brennan J said that the series was 'but a minute fraction of the bytes in the whole . . . program. Nevertheless, the series . . . is both *original and critical*'.[25] Gaudron J said that the series was 'the *linchpin* of the program . . . It was *critical* . . . there is, in my view, simply no basis for an argument that the [series] was not a substantial part'.[26]

17 See *Copyright Act 1968* (Cth) ss14, 31(1)(a)(i) and (b)(i).
18 *Stowe v Thomas* 23 F. Cas. 201 (CCED Pa 1853).
19 *Copyright Act 1968* (Cth) s31(1)(a)(vi) and (vii).
20 *Ladbroke (Football) Ltd v William Hill (Football) Ltd* [1964] 1 WLR 273, House of Lords.
21 *Hawkes & Son (London) Ltd v Paramount Film Service Ltd* [1934] 1 Ch 593. See also *Blackie & Sons Ltd v Lothian Book Publishing Co. Pty Ltd* (1921) 29 CLR 396; *Warner Bros Pictures v Majestic Pictures Corp.* (1934) 70 F. 2d 310, titles of works; *Exxon Corp. v Exxon Ins. Consultations* [1982] Ch 119; *Arica Institute Inc. v Palmer* (1992) 970 F. 2d 1067, short phrases and slogans.
22 (1992) 173 CLR 330.
23 (1993) 176 CLR 300.
24 (1992) 173 CLR 330 at 346, emphasis added.
25 (1992) 176 CLR 300 at 311, emphasis added.
26 (1992) 176 CLR 300 at 330, emphasis added.

In *Data Access Corporation v Powerflex Services Pty Ltd*[27] (Data Access) the High Court considered the approaches in the Autodesk cases to come close to a 'but for' analysis: 'but for' the 127-bit series, the program would not work. The reasoning in the Autodesk cases did not find favour with the High Court, although Gaudron J published a separate judgment explaining her Honour's position. The High Court noted that in general, a computer program may not work, or may not work as planned, if even one character is altered and so it could be argued that each character is 'essential'. In *Cantor Fitzgerald International v Tradition (UK) Ltd*,[28] in the English Patents Court, Pumfrey J observed of *Autodesk (No. 1)* that it 'would result in any part of any computer program being substantial since without any part the program would not work, or at best [would] not work as desired'.[29] In *Autodesk (No. 2)* Mason CJ had dissented, stating that 'substantial part':

> refers to the *quality* of what is taken rather than the *quantity* . . . it is important to inquire into the importance which the taken portion bears in relation to the work as a whole: is it an 'essential' or 'material' part of the work?[30]

The High Court in Data Access was unable to agree with the approach taken by the majority in the Autodesk cases. The High Court felt that there was 'great force' in the criticism that the 'but for' essentiality test was not practicable. Mason CJ's dissenting opinion was preferred. Accordingly, a person who does no more than reproduce those parts of a program which are data or related information and which are irrelevant to its structure (that is, that are not choice of commands and combination and sequencing of commands), will be 'unlikely' to have reproduced a substantial part of the computer program. The High Court used the term 'unlikely' in preference to 'impossible' because it conceived that data, considered alone, could be sufficiently original to be a substantial part of the computer program in the right circumstances.

In Data Access the High Court also considered whether a 'Huffman Compression' table was a literary work. The Huffman compression is a method of reducing the amount of memory space consumed by data files. In evidence Dr Bennett, the third respondent, explained the method, stating, for example that the letter 'e' was normally encoded as the bit string '01100101', but as a common character it might be encoded instead as the bit string '101', with a space (and thus also efficiency) saving of 62.5 per cent.[31] This process could be used for many letters, strings and phrases. There was no allegation that Dr Bennett copied the source code of the Huffman algorithm from the original program. Dr Bennett wanted the plaintiff's original program to be able to compress and decompress his files,

27 [1999] HCA 49. Gleeson CJ, McHugh, Gummow and Hayne delivered a joint judgment with which Gaudron J agreed effectively in full, only making a separate comment regarding her Honour's earlier judgment in *Autodesk v Dyason No. 2* (1993) 176 CLR 300.
28 [2000] RPC 95.
29 [2000] RPC 95, 131.
30 (1993) 176 CLR 300, 305, emphasis added.
31 [2000] RPC 95, 114.

and for his program to compress and decompress other files in an identical way so that the programs could work together.[32] Consequently, he needed to be able to replicate precisely the default Huffman compression table. Dr Bennett did not have access to the original file. His evidence was that he refrained from 'decompiling or looking inside the Dataflex runtime', and instead carried out a process to recreate and deduce the original table. This 'reverse engineering method' was highly ingenious. Nevertheless, the High Court held that the original Huffman table fell 'squarely within the statutory definition of a literary work'. Both the High Court and the Full Federal Court considered that the process undertaken by Dr Bennett constituted a 'reproduction' of the original Huffman table:

> The fact that Dr Bennett used an ingenious method of determining the bit string assigned to each character does not make the output of such a process any less a 'reproduction' than if Dr Bennett had sat down with a print-out of the table and copy-typed it.[33]

Objective similarity and causal connection

The requisite degree of similarity depends on the type of work. An exact duplicate will be objectively similar. Objective similarity may arise after transformation into another media, such as a book into a film. Two pieces of computer software may be objectively similar if one has been translated into another computer language; the deliberate copying of computer data may also produce something that is objectively similar.

Gibbs CJ, in *SW Hart & Co. Pty Ltd v Edwards Hot Water Systems*,[34] described the test in the following terms:

> The notion of reproduction, for the purposes of copyright law, involves two elements – resemblance to, and actual use of, the copyright work, or, to adopt the words which appear in the judgment of Willmer LJ in *Francis Day & Hunter Ltd v Bron* (1963) Ch 587, at p 614, 'a sufficient degree of objective similarity between the two works' and 'some causal connection between the plaintiffs' and the defendants' work'.

Lord Reid said, in *Ladbroke (Football) Ltd v William Hill (Football) Ltd*:[35]

> Broadly, reproduction means copying, and does not include cases where an author or compiler produces a substantially similar result by independent work without copying. And, if he does copy, the question whether he has copied a substantial part depends much more on the quality than on the quantity of what he has taken.

Causal connection may be proved directly or indirectly. The copying may be unconscious, such as the reproduction of music heard years before.

32 See *Software*, below.
33 [2000] RPC 95, 124.
34 (1985) 159 CLR 466, 472.
35 (1964) 1 WLR 273, 276. See also *Francis Day & Hunter Ltd v Bron* [1963] Ch 587.

Software

In the past quarter-century the courts and the legislature have grappled with copyright issues of software and computer programs. In 1986, in *Computer Edge v Apple*,[36] the High Court of Australia held that the Copyright Act did not apply to software. The High Court was unable to classify computer code as a literary work, or any other type of work. Computer code may be as basic as a series of unintelligible (to the human reader) ones and zeros, in machine language, or one of dozens of other computer languages with varying degrees of readability. It was not denied that great care and skill were involved in creating the code, but it did not fit in with the then definition of literary work. This resulted in an amendment to the Copyright Act expanding 'literary works' to include a computer program, defined as:

> a set of statements or instructions to be used directly or indirectly in a computer in order to bring about a certain result.[37]

The new definition was tested by the High Court in *Data Access Corporation v Powerflex Services Pty Ltd*[38] (Data Access). The court's finding that the respondents infringed the appellant's copyright in the computer compression table was described as having 'considerable practical consequences'.[39] There are significant ramifications for anyone who seeks to produce a computer program that is compatible with a program produced by others. The court, concerned about the impact of its decision, stated, 'matters . . . can be resolved only by the legislature reconsidering and, if it thinks it necessary or desirable, rewriting the whole of the provisions that deal with copyright in computer programs'.[40]

Dr Bennett became familiar with the applicant's 'Dataflex' system and decided to create and market an application development system which would be compatible with the Dataflex database file structure so that persons who were familiar with the Dataflex system would be able to use his new product. By a process of reverse engineering and study of both the documentation and operation of the Dataflex system, Dr Bennett created computer programs compatible with the Dataflex system. He 'duplicated' and used certain commands and 'Reserved Words' (the precise words used by Dataflex to operate the program). The underlying source code, however, was quite different from the original. He was aware that copying the original source code would be a copyright breach, so he avoided doing so directly. The Full Court of the Federal Court of Australia held that commands in a computer program were not original literary works within the

36 (1986) 161 CLR 171.
37 *Copyright Act 1968* (Cth) s10.
38 [1999] HCA 49. Gleeson CJ, McHugh, Gummow and Hayne delivered a joint judgment with which Gaudron J agreed effectively in full, only making a separate comment regarding her Honour's earlier judgment in *Autodesk v Dyason No. 2* (1993) 176 CLR 300.
39 [1999] HCA 49, para 125.
40 [1999] HCA 49, para 125.

meaning of the definition in section 10 of the Copyright Act, and so copyright did not subsist in them.

The High Court asked whether each of the Reserved Words constituted a 'computer program' within the meaning of section 10(1) of the Act. The High Court held that although each Reserved Word was 'undoubtedly' in 'code or notation', each one was merely a single word. None of the Reserved Words met the criteria set out in section 10. The majority went to some lengths to explain the nature and characteristics of computer code and even set out three short examples of such code.[41] The High Court held that it is not appropriate to relate the Reserved Word back to the underlying computer code to which it refers. It is the particular selection, ordering, combination and arrangement of instructions within a computer program which provide its expression. Similarly, a set of Reserved Words is not a 'computer program'. The simple listing does not cause a computer to perform any identifiable function. The definition of a 'computer program' requires that the set of instructions be intended to cause the computer to perform a particular function.

Right of communication

In 2000 the centrepiece of amendments to the Copyright Act was a new broadly based, technology-neutral right of communication to the public. The right is an exclusive right (for the copyright holder) in literary, musical, artistic and dramatic works, sound recordings, films and broadcasts.[42]

'Communicate' is defined as: 'make available online or electronically transmit (whether over a path, or a combination of paths . . .)'.[43] The right replaced and extended the pre-existing technology-specific broadcasting right. 'Broadcast' is defined as: 'a communication *to the public* delivered by a broadcasting service within the meaning of the *Broadcasting Services Act 1992* (Cth)'.[44] However, a broadcasting service does not include 'a service, such as a teletext service, that provides no more than data or no more than text (with or without associated images); or a service that makes programs available on demand on a point-to-point basis, including a dial-up service'.[45] The communication right includes cable transmissions and encompasses the uploading of material onto an internet server. The technology-neutral approach can apply to an electronic transmission by copper wires, optic fibre cables and microwaves. The right does not apply to the physical distribution of copyright material in a tangible form (such as the distribution of hard copies of books). The expression 'to the public' includes the public within or outside Australia, so it permits Australian copyright holders to control communication directed to overseas audiences.

41 Computer programmers will observe curious use of computer code in their Honours' judgment.
42 *Copyright Act 1968* (Cth) s31(1)(a)(iv) and (b)(iii). The new right does not apply to published editions.
43 *Copyright Act 1968* (Cth) s10.
44 *Copyright Act 1968* (Cth) s10, emphasis added.
45 *Broadcasting Services Act 1992* (Cth) s6.

Exemptions

The Copyright Act exempts dealing with copyright material without the permission of or payment to copyright owners where such use is for the purposes of 'research and study, criticism and review, parody or satire, reporting the news, and professional legal advice'. These are referred to as the 'fair dealing' exemptions.[46] Specific provisions which deal with electronic works are now included as well: there is a fair dealing for research and study related to a published literary work in electronic form (except a computer program or an electronic compilation, such as a database), a published dramatic work in electronic form or an adaptation published in electronic form of such a literary or dramatic work. The fair dealing under these circumstances is either 10 per cent of the number of words in the work or adaptation, or if the work or adaptation is divided into chapters, a single chapter.[47] There are several other exemptions.

Libraries and archives

Libraries and archives are permitted to make reproductions of copyright material for library users for the purposes of research and study, and for other libraries for certain purposes. This access must be provided in a way which will not unreasonably prejudice the interests of copyright owners. The Copyright Act includes such permission for electronic reproduction and communication of copyright material. However, certain restrictions apply: for example, a library may only request an article or a portion of a work in electronic form from another library if that portion or article is not available 'within a reasonable time at an ordinary commercial price'. This is designed to minimise unreasonable conflict with the emerging markets of copyright owners.[48]

Infringing copies made on machines installed in libraries and archives

Following concern over the possibility of legal action against libraries when photocopiers were first introduced, the Copyright Act was amended to remove liability from the library where conspicuous warning notices are in place.[49] More recently this was extended to copies by computers:

> Where:
> (a) a person makes an infringing copy of, or of part of, a work on a machine (including a computer), being a machine installed by or with the approval of the body administering a library or archives on the premises of the library or archives, or outside those premises for the convenience of persons using the library or archives; and
> (b) there is affixed to, or in close proximity to, the machine, in a place readily visible to persons using the machine, a notice of the prescribed dimensions and in accordance with the prescribed form;

46 See *Copyright Act 1968* (Cth) ss40–42 and *Copyright Act 1994* (NZ) ss42, 43, 176.
47 *Copyright Act 1968* (Cth) s40.
48 *Copyright Act 1968* (Cth) Div 5.
49 See *University of New South Wales v Moorhouse* (1975) 133 CLR 1.

neither the body administering the library or archives nor the officer in charge of the library or archives shall be taken to have authorized the making of the infringing copy by reason only that the copy was made on that machine.[50]

Educational statutory licences

Part VB of the Copyright Act permits educational institutions with a statutory licence to make copies of works subject to the payment of equitable remuneration. The Part applies to the reproduction and communication of works in hardcopy form or in electronic form.

Temporary reproductions

When a web page is accessed, the user's web browser makes a copy of the text and graphics on the website and places both in a temporary directory or cache. Depending on the extent of use of the web browser and the size reserved for such files, the text and graphics may remain in place indefinitely. Early questions arose about unauthorised copies of material subject to copyright. One early opinion was that there was an implied licence.[51] That is, if the holder of the copyright placed material on the internet it could be implied that normal use of the internet, including the making of such temporary files, must be implied. However, should the user make any other copy, such as cutting and pasting or some other use of the temporary files, such an implication could not be made.

The Copyright Act provides that temporary or incidental reproductions made in the course of the technical processes of exercising the communication right are exempt, giving statutory effect to the implied licence concept. The Act covers both temporary reproductions made in the course of communication and temporary reproductions of works as part of a technical process of use.[52]

Enforcement measures

The Copyright Act helps copyright owners enforce their rights in the digital environment by providing enforcement regimes for circumvention devices and services, rights management information and broadcast decoding devices. The enforcement measures are designed to combat piracy.

The Act includes civil remedies and criminal sanctions against the manufacture, commercial dealing, importation, making available online, advertising, marketing and supply of a circumvention device or service used to circumvent technological protection measures such as program locks.[53] A 'circumvention device' is defined as:

50 *Copyright Act 1968* (Cth) s39A. The provision also applies to archives.
51 Alan Davidson, 'Digital agenda amendments to the Copyright Act', (2001) 21–3 *Proctor*, 30.
52 *Copyright Act 1968* (Cth) ss43A, 43B respectively.
53 *Copyright Act 1968* (Cth) ss116AK–116AQ.

a device, component or product (including a computer program), that:
(a) is promoted, advertised or marketed as having the purpose or use of circumventing the technological protection measure; or
(b) has only a limited commercially significant purpose or use, or no such purpose or use, other than the circumvention of the technological protection measure; or
(c) is primarily or solely designed or produced to enable or facilitate the circumvention of the technological protection measure'.[54]

This definition includes software tools and is intended to exclude general purpose electrical equipment, such as DVD recorders and computers. A 'circumvention service' has a corresponding meaning.[55]

A 'technological protection measure' generally means:

(a) an access control technological protection measure; or
(b) a device, product, technology or component (including a computer program) that:
    (i) is used in Australia or a qualifying country by, with the permission of, or on behalf of, the owner or exclusive licensee of the copyright in a work or other subject-matter; and
    (ii) in the normal course of its operation, prevents, inhibits or restricts the doing of an act comprised in the copyright;

  but does not include such a device, product, technology or component to the extent that it:
    (iii) if the work or other subject-matter is a cinematograph film or computer program (including a computer game) – controls geographic market segmentation by preventing the playback in Australia of a non-infringing copy of the work or other subject-matter acquired outside Australia; or
    (iv) if the work is a computer program that is embodied in a machine or device– restricts the use of goods (other than the work) or services in relation to the machine or device.[56]

Many products now incorporate Rights Management Information (RMI). The Copyright Act includes criminal offences and civil remedies regarding the intentional removal and alteration of RMI, or the commercial dealing with copyright material where the RMI has been removed.[57] 'Electronic rights management information' is defined as:

> information that is attached to or embodied in a copy of the work or subject-matter, or which appears in connection with a communication, or the making available, of the work or subject-matter, which identifies the work or subject-matter, and its author or copyright owner or identifies or terms and conditions on which the work or subject-matter may be used.[58]

This includes digital watermarks.

54 *Copyright Act 1968* (Cth) s10.
55 *Copyright Act 1968* (Cth) s10.
56 *Copyright Act 1968* (Cth) s10.
57 *Copyright Act 1968* (Cth) ss116B–116D.
58 *Copyright Act 1968* (Cth) s10.

A person will only be directly liable for copyright infringements involved in a communication where they have determined the content of that communication. This provides an exemption for the carrier or the Internet Service Provider (ISP), unless it can be proven that either had authorised the infringement by a user.[59]

Time-shifting, format-shifting and space-shifting

Time-shifting of television and radio programs for later private use is permitted. The recording of television or radio programs solely for private and domestic use by watching or listening to the material broadcast at a time more convenient does not infringe copyright.[60] The exception does not generally apply to podcasts or webcasts, and does not apply if the recording is later sold, rented or distributed (other than to a member of the person's family or household), or played or shown in public or broadcast. There is no express obligation to destroy the recording at any time, although it may be implied that the exception is not meant to permit permanent copies or repeated use.

Space-shifting means that a person who owns a copy of a sound recording, such as a CD, is allowed to make a copy of that recording, for private and domestic use, to play on another device owned by that person. The device could be a CD player, a computer, a car CD player, or a portable device such as an MP3 player or iPod.[61] The provision does not apply if the initial copy is an infringing copy (such as a pirated CD or an unauthorised digital download). The provision also does not apply if the second copy is later sold, rented or distributed (other than to a member of the person's family or household), or played or shown in public or broadcast. The provision may apply to recorded music from a digital download, but does not apply to podcasts.

Format-shifting involves copying from one 'format' to another for private use. An individual may copy books (including novels, children's books, reference books), newspapers and other periodical publications (such as magazines and journals), photographs, and films on videotape but not DVDs, provided the copy is made from a non-infringing copy owned by the individual. For example, print material may be scanned for viewing on a computer screen. A photograph may be converted from a print to an electronic image, or conversely an electronic image may be printed. A video may be copied into a digital format. The provision does not apply to sales, rent or distribution other than to a member of the person's family or household.[62]

59 See *Copyright Act 1968* (Cth) ss39B, 116AA–116AJ, 195AW, 195AXI.
60 *Copyright Act 1968* (Cth) s111.
61 *Copyright Act 1968* (Cth) s109A.
62 See generally Australian Copyright Council, (2008) Information Sheet GO97v01; www.copyright.org.au/pdf/acc/infosheets_pdf/g097.pdf.

Piracy and enforcement

Conduct regarded as 'substantial infringement on a commercial scale' attracts special attention; it includes likely infringements as well as proved infringements when infringement takes place online on a commercial scale. Colloquially this is referred to as piracy. There are also offences relating to unauthorised access to encoded broadcasts, such as pay television.[63]

Peer-to-peer file sharing

The downloading of music, video and other files online is prolific, if not epidemic. Transfers of copyright material – material that would have cost several billion dollars in total to buy – are a source of considerable concern for the music industry. The fact that a substantial number of such files are subject to copyright, and that the right to copy is one of the many exclusive rights provided by the Copyright Act, is typically disregarded by internet users.[64] Peer-to-peer (P2P) transfers have facilitated the majority of the downloads. The transfer of files without copyright protection is uncontroversial. There have been numerous cases brought before the courts aimed at stopping or at least discouraging these downloads. Yet sites providing access, typically indirectly on a P2P basis, continue to flourish. This is a question of supply attempting to meet demand notwithstanding legal and perhaps moral concerns.

The US decision of *A & M Records Inc. v Napster Inc.*[65] was the first significant case challenging P2P transfers for copyright material online. Copyright law in the United States is governed by the *United States Copyright Act 1976*. The Ninth Circuit Court rejected Napster's claim that the online service and alleged infringements were exempted by the 'fair use' defence. (This claim was based on the fact that international copyright law is generally more lenient on personal file sharing where fair use is involved.) The transfer of files without copyright protection is uncontroversial – there is thus considerable legitimate benefit to be gained by using P2P when copyright is not infringed. Australia and New Zealand do not have a general fair use defence; instead they have a relatively narrow range of 'fair dealing' defences for specific circumstances. Napster argued that each transfer was in fact a personal transfer within the meaning of the fair use principle. 'Fair use' is defined in the US copyright statute thus:

> the fair use of a copyrighted work, including such use by reproduction in copies or phonorecords or by any other means specified by that section, for purposes such as criticism, comment, news reporting, teaching (including multiple copies for classroom

63 See *Copyright Act 1968* (Cth) ss115(5)–(8), 132AC, Part VAA.
64 *Copyright Act 1968* (Cth) s31.
65 239 F. 3d 1004 (9th Cir 2001).

> use), scholarship, or research, is not an infringement of copyright. In determining whether the use made of a work in any particular case is a fair use the factors to be considered shall include—
> 1. the purpose and character of the use, including whether such use is of a commercial nature or is for nonprofit educational purposes;
> 2. the nature of the copyrighted work;
> 3. the amount and substantiality of the portion used in relation to the copyrighted work as a whole; and
> 4. the effect of the use upon the potential market for or value of the copyrighted work.[66]

The US courts, applying these four determining factors, have permitted the sharing of material with family or friends. Posting material to a website can be in breach even if the host's intention is not commercial. In *LA Times v Free Republic*[67] the court found that the posting of material with no commercial motive or intent nevertheless allowed others to avoid paying the customary price charged for the works, and was thus an infringement.

In the Napster case, the court determined that Napster had played an integral role in the transfer process and could have blocked access to infringing material had it chosen to do so. The court rejected the defence that the fair use of the end users protected Napster. It questioned whether sampling and space-shifting were protected in any event.

Subsequent online file-sharing programs took note of the Napster case and attempted to circumvent the court's concerns by decentralising their databases and removing their direct ability to control or track transfers.

Attempts have been made by file-sharing proprietors to rely on the principle enunciated in *Sony Corp. v Universal City Studios.*[68] This pre-internet case held that making recordings of television programs to be viewed later, known as time-shifting, was fair use. Universal Studios and other joint applicants argued that the manufacturers of Betamax and VCR devices and of the corresponding blank tapes authorised infringing copies by the purchasers of the tapes or, alternatively, were liable for contributory infringement. The US Supreme Court, by a majority of 5–4, held that there must be a balance between a copyright holder's legitimate demand for effective protection of the statutory monopoly and the rights of others freely to engage in substantially unrelated areas of commerce. The court ruled that:

> the sale of copying equipment, like the sale of other articles of commerce, does not constitute contributory infringement if the product is widely used for legitimate, unobjectionable purposes. Indeed, it need merely be capable of substantial noninfringing uses.[69]

66 *United States Copyright Law 1976*, §107.
67 54 USPQ 2d 1453 (2000).
68 464 US 417 (1984).
69 464 US 417 (1984) at 442.

However, in 2005 in *MGM v Grokster*,[70] in line with the Napster case, the US Supreme Court unanimously held filing-sharing proprietors Grokster and Streamcast liable for authorising others to infringe copyright:

> We hold that one who distributes a device with the object of promoting its use to infringe copyright, as shown by clear expression or other affirmative steps taken to foster infringement, is liable for the resulting acts of infringement by third parties.

Grokster was ordered to pay US$50 million to the music and recording industries. It also announced that it would no longer offer P2P file sharing citing the Supreme Court case as the reason.

The issues of thumbnails, inline linking and fair use were considered in *Kelly v Arriba Soft Corporation*.[71] The defendant used online thumbnail images of copyright material which linked to further information and data. The 9th Circuit Court of Appeals applied the four factors for fair use, and held that creating the thumbnail images as previews was 'transformative' only. The images were not intended to be viewed at high resolution like the original artwork. It was questionable whether the low-resolution images could be regarded as a full replication, and a significant amount of detail by necessity was lacking. The court regarded their use as reasonable and necessary. Finally, the court held that the market for the original photographs was not substantially diminished, and in fact that the thumbnails may increase demand.

Australian copyright law includes fair dealing defences for research and study, criticism and review, parody or satire, reporting the news, and professional legal advice, but does include the four factors found in the US legislation permitting fair use. In the Federal Court of Australia in *Universal Music Australia Pty Ltd v Sharman License Holdings Ltd*[72] Wilcox J considered the P2P file-sharing system known as Kazaa. Despite the fact that the Kazaa website contained warnings against the sharing of copyright files, and an end user licence agreement under which users agreed not to infringe copyright, it had 'long been obvious' that those measures were ineffective to prevent, or even to substantially curtail, copyright infringements by users. It had 'long been known', said Wilcox J, that Kazaa was widely used for the sharing of copyright files. In granting relief, Wilcox J ordered a complex series of arrangements including declarations and an order restraining future violations. Wilcox J stated that he was 'anxious' not to make an order which the respondents would not be able to obey except at the unacceptable cost of preventing the sharing even of files which did not infringe the applicants' copyright. The infringing respondents were restrained from authorising Kazaa users to do any of the infringing acts in relation to any sound recording of which any of the applicants were the copyright owners.

Most significantly, Wilcox J ordered that a comprehensive filtering system and update process be put in place. His Honour stated that Kazaa may continue if

70 545 US 913 (2005).
71 336 F. 3d 811(CA9 2003).
72 [2005] FCA 1242.

a protocol, which included detailed filter technology that was updated regularly were put in place.

In *Cooper v Universal Music Australia Pty Ltd*[73] the Full Federal Court considered the position of the website www.mp3s4free.net and relevant provisions of the *Copyright Act 1968* (Cth). The action included more than 40 parties. The decision of the Full Federal Court, French, Branson and Kenny JJ, particularly considered the meaning of 'authorising' an infringement of copyright in the context of websites, and concluded that website hosts may be liable for authorising copyright infringement, and in certain circumstances, so may ISPs.

Stephen Cooper was the registrant of the domain name mp3s4free.net. The website included links to music files on websites which were independent of Cooper. The links facilitated the transmission of music files, typically in MP3 format, to the user. Several record companies, who held the copyright in the many of the music files, commenced proceedings against Cooper, the ISP hosts and certain named directors for copyright infringement. The Federal Court at first instance[74] found that Cooper, the web hosts and directors of the companies had infringed copyright.

Cooper appealed on the grounds that Tamberlin J erred in finding that, by providing a website with hyperlinks, he authorised the making of copies of the sound recordings and the communication to the public of those recordings without the licence of the record companies. The web hosts and directors also appealed.

The Full Federal Court's decision helped clarify two important issues: first, the meaning of 'authorisation' of copyright infringement in a sound recording, and second, the application of section 112E of the Copyright Act in relation to the protection of the carrier.

Authorisation

Section 13(2) provides, in part:

> the exclusive right to do an act in relation to a work, an adaptation of a work or any other subject-matter includes the exclusive right to authorize a person to do that act in relation to that work, adaptation or other subject-matter.

Section 101 deals with infringement by doing acts comprised in copyright:

> (1) ... a copyright subsisting by virtue of this Part is infringed by a person who, not being the owner of the copyright, and without the licence of the owner of the copyright, does in Australia, or authorizes the doing in Australia of, any act comprised in the copyright.
>
> (1A) In determining ... whether or not a person has authorised the doing in Australia of any act comprised in a copyright ... without the licence of the owner of the copyright, the matters that must be taken into account include the following:

73 [2006] FCAFC 187.
74 *Universal Music Australia Pty Ltd v Cooper* [2005] FCA 1878.

(a) the extent (if any) of the person's power to prevent the doing of the act concerned;
(b) the nature of any relationship existing between the person and the person who did the act concerned;
(c) whether the person took any other reasonable steps to prevent or avoid the doing of the act, including whether the person complied with any relevant industry codes of practice.[75]

Pivotal to the court's reasoning was the application of section 101(1A), which had been inserted by the *Copyright Amendment (Digital Agenda) Act 2000* (Cth). The court considered the position of Cooper, the ISP host and that company's director separately.

The court reasoned that Cooper had the power to prevent the infringement of copyright, as he could determine not to make available this website, which he knew was designed to permit downloading of music files. The hyperlinks could be automatically added to the mp3s4free website. Cooper argued that once the website was established, he did not have any relevant power to control its use or prevent users from accessing and using this site for the purpose of copying music.

In relation to section 101(1A)(a), the court rejected the contention that unless Cooper had power, at the time of the doing of each relevant act comprised in a copyright subsisting by virtue of the Act, to prevent its being done, he had no relevant power within the meaning of paragraph.[76] Cooper was responsible for creating and maintaining the mp3s4free website. The court's conclusion was that:

> a person's power to prevent the doing of an act comprised in a copyright includes the person's power not to facilitate the doing of that act by, for example, making available to the public a technical capacity calculated to lead to the doing of that act.[77]

The 'inexorable inference' is that Cooper made a deliberate choice to establish and maintain this website in a form which did not give him the power immediately to prevent, or immediately to restrict, internet users from using links to copy sound recordings in which copyright subsisted.

The principal content of the website was links to other websites and files on other servers, the overwhelming majority of which were the subject of copyright. It was the intentional choice of Cooper to establish his website in a way which allowed the automatic addition of hyperlinks. The court further concluded that within the meaning of section 101(1A)(a), Cooper did have power to prevent the communication of copyright sound recordings to the public in Australia via the website.[78]

Section 101(1A)(b) deals with the nature of the relationship between, in this case, Cooper and the person who downloaded the music. Cooper submitted that

75 The two different spellings of 'authorise' are faithfully replicated for the provision.
76 [2006] FCAFC 187, para 41.
77 [2006] FCAFC 187, para 41.
78 See generally paras 41–45.

he did not have any relationship with people who made MP3 files generally accessible over the internet or with people who downloaded such files. However, the court considered that an aspect of the relationship was the deliberate attraction of the user-friendly website for downloading. The court considered that Cooper benefited financially from sponsorship and advertisements on the website:

> that is, that the relationship between Mr Cooper and the users of his website had a commercial aspect. Mr Cooper's benefits from advertising and sponsorship may be assumed to have been related to the actual or expected exposure of the website to internet users. As a consequence Mr Cooper had a commercial interest in attracting users to his website for the purpose of copying digital music files.[79]

Section 101(1A)(c) deals with whether the person took any other reasonable steps to prevent or avoid the doing of the act, including complying with relevant industry codes of practice. Cooper submitted that disclaimers on the website amounted to reasonable steps within the meaning of the paragraph. However, the court considered that the disclaimers misstated Australian copyright law in a material way and the inclusion did not constitute a reasonable step to prevent or avoid the infringement of copyright. Tamberlin J at first instance attributed little weight to them as he found their intended purpose was 'merely cosmetic'.[80] The disclaimer stated, in part:

> When you download a song, you take full responsibility for doing so. None of the files on this site are stored on our servers. We are just providing links to remote files . . . This site only provides links to the according sites and no songs are located on our servers . . . We are not responsible for any damage caused by downloading these files, or any content posted on this website or linked websites.[81]

Indeed, Tamberlin J at first instance considered that the disclaimers indicate Cooper's knowledge of the existence of illegal MP3s on the internet and the likelihood that some of the MP3s downloaded constituted infringing copies of copyright music and sound recordings.[82] The Full Federal Court concluded that Cooper did not establish that he took any reasonable steps to prevent or avoid the use of his website for copying copyright sound recordings or for communicating such recordings to the public.

The Full Federal Court considered the High Court decision in *University of New South Wales v Moorhouse*.[83] The High Court held 'authorise' to mean to 'sanction, approve or countenance'.[84] Gibbs J stated that a person who has control of the means by which an infringement of copyright could be committed, in that case a photocopier, and who permitted it to be used for a copyright infringement knowingly, or with reasonable suspicion that it is likely to be used for infringement,

79 Para 48.
80 Para 49.
81 See para 100.
82 *Universal Music Australia Pty Ltd v Cooper* [2005] FCA 972, para 87.
83 (1975) 133 CLR 1.
84 (1975) 133 CLR 1, para 10. The word 'authorise' 'has been held judicially to have its dictionary meaning of "sanction, approve, countenance"': *Falcon v Famous Players Film Co.* (1926) 2 KB 474, at 491; *Adelaide Corporation v Australasian Performing Right Association Ltd* [1928] HCA 10; (1928) 40 CLR 481, at 489, 497.

has authorised the breach, unless reasonable steps were taken to limit such use. The Full Federal Court applied the Moorhouse case by analogy.[85]

Carrier protection

Cooper, the ISP and the directors all claimed protection under section 112E of the Copyright Act, which provides that:

> a carrier . . . who provides facilities for making, or facilitating the making of, a communication is not taken to have authorised any infringement of copyright . . . merely because another person uses the facilities so provided to do something the right to do which is included in the copyright.

The court ruled out Cooper's claim relatively perfunctorily. The director, Mr Bal, was also held liable, as on the facts he was the 'controlling mind' of the company.

In relation to the ISPs the court found that:

> As all of the relevant acts of copyright infringement took place via Mr Cooper's website, I conclude that E-Talk (the internet service provider) had power to prevent the doing of the acts concerned because . . . it had the power to withdraw the hosting of Mr Cooper's website.[86]

Hyperlinking

Hyperlinks permit a copy of data, images and other material to be downloaded onto the user's computer. Trade mark and copyright issues arise.

Fundamental to internet use and navigation are hyperlinks. With a simple click of a computer mouse a massive amount of information and data is available. Some links are internal: that is, the link will take the user to another web page by the same designer and author as the initial page. Most links will cause the web browser to retrieve the web page of another host, anywhere in the world, and display the contents. The target site will have a unique URL (Universal Resource Locator).

The contents of a retrieved web site may be protected by trade mark or copyright regulation.[87] The question of the status and extent of protection available in relation to the 'sign' (for trade marks) or 'copy' (for copyright) displayed on the user's computer has been the subject of speculation. When viewing a website, the content is not a continuous stream, such as with television. Instead a connection is made with the host, typically for a fraction of a second, during which time the contents are copied to a cache or temporary area. The text, graphics files and any other data are copied to the user's computer. The web browser assembles the

85 The Full Federal Court made the distinction between other parties who, unlike Cooper, did not have the ability to take reasonable steps to prevent the copyright infringement.
86 Para 62.
87 In relation to trade marks, see Chapter 7.

copied material into a predetermined meaningful way. The question, whether this amounts to a breach of trade mark right or copyright, has been untested. However, it would be reasonable to assume that the host of the data has granted, at the very least, an *implied licence* for the user to view the content. The host knows, or ought to know that temporary copies are made as part of the operation of the World Wide Web. The host almost always wants or expects the data to be displayed by web users. It is the host that has placed the material into a public area for access on the World Wide Web.

In *Ticketmaster Corporation v Tickets.com Inc.*[88] the court held that:

> hyperlinking does not itself involve a violation of the Copyright Act . . . since no copying is involved . . . the customer is automatically transferred to the particular genuine web page of the original author. There is no deception in what is happening. This is analogous to using a library's card index to get reference to particular items, albeit faster and more efficiently.

In Australia the issue was resolved with the inclusion of section 43A(1) , which provides:

> The copyright in a work, or an adaptation of a work, is not infringed by making a temporary reproduction of the work or adaptation as part of the technical process of making or receiving a communication.

Temporary reproductions made in the course of the technical process of making or receiving electronic communications are excluded from the scope of copyright. However, if any further step is taken to make a copy, such as 'cutting and pasting' text from the screen, or copying the temporary files to another part of the user storage, the protection is inapplicable.

Further reading

Australian Copyright Council: www.copyright.org.au.
Intellectual Property Society of Australia and New Zealand: www.ipsanz.com.au.
IP Australia: www.ipaustralia.gov.au.
World Intellectual Property Organisation (WIPO): www.wipo.org.
WIPO (Copyright): www.wipo.org/about-ip/en/copyright.html.

88 54 USPQ 2d 1344 (CD Cal 2000).

7

Electronic commerce – trade marks, patents and circuit layouts

The digital revolution has necessitated a re-examination of intellectual property issues by intellectual property holders, users and law makers. The nature of the digital age enables data to be easily copied, published and disseminated. Placing data, images, logos and text, for example, on internet websites, is child's play. This means that trade mark holders have a new frontier to battle. Misuse of their trade mark rights, deliberate or incidental, commercial or personal, arises in relation to cybersquatting and domain names, hyperlinking (particularly deep linking), framing in web pages and the use of meta-tags.

This chapter addresses issues relating to the impact of electronic commerce on the specific intellectual property rights of trade marks, patents and circuit layouts. It is intended to be only a brief overview of the law in these areas. The chapter will discuss the use and design of the internet, including hyperlinks, framing and meta-tags. Patents involving software and hardware are considered, and the specific intellectual property type known as circuit layouts is described. Cybersquatting and related issues are discussed in Chapters 8 and 9.

The nature of trade marks

A trade mark is 'a sign used, or intended to be used, to distinguish goods or services dealt with or provided in the course of trade by a person from goods or services so dealt with or provided by any other person'.[1] A trade mark is used in the course of trade to show a connection between a particular business and the goods or services it supplies. Trade marks indicate a standard of quality

1 *Trade Marks Act 1995* (Cth) s17.

associated with a product or service and protect consumers from confusion and deception. Trade marks are protected under common law and under the *Trade Marks Act 1995* (Cth).[2]

Registrants only have the right to use their registered trade mark for the goods and services for which it is specifically registered. In Australia there are 42 classes in which a trade mark can be registered and there is no limit on the number of classes in which an application can be made, as long as the trade mark is actually used or intended to be used with respect to the goods or services nominated. A sign is defined as 'including the following or any combination: any letter, word, name, signature, numeral, device, brand, heading, label, ticket, aspect of packaging, shape, colour, sound or scent'.[3] The inclusion of any aspect of packaging, shape, colour, sound and scent in 1995 significantly broadened the range of features or signs which can be registered.

The system for registering trade marks effectively entitles a person to 'own' a word, logo, phrase or other distinctive sign and to stop others from using that mark or a deceptively similar mark in relation to the same or similar sorts of goods or services. Provided the trade mark is used continually and renewed, typically every 10 years, the trade mark may be retained indefinitely.

Trade mark registration is territorial and each country has its own system of registration. An Australian trade mark registration is effective throughout Australia. Those wishing to protect trade marks (for trade purposes, say) in other countries need to register them in those other countries.

A registered owner of a trade mark has:

> the exclusive rights:
> (i) to use the trade mark; and
> (ii) to authorise other persons to use the trade mark
> in relation to the goods and/or services in respect of which the trade mark is registered and the right to obtain relief under the Act if the trade mark has been infringed.[4]

A registered owner also has 'the right to obtain relief under the Act if the trade mark has been infringed'.[5]

A person may apply for the registration if 'the person claims to be the owner of the trade mark' and 'the person is using or intends to use the trade mark', 'the person has authorised or intends to authorise another person to use the trade mark' or 'the person intends to assign the trade mark to a body corporate that is about to be constituted'.[6]

Business names are not trade marks. The purpose of registering a business name or company name is to identify the owners operating under that name.

2 For the New Zealand equivalent, see the *Trade Marks Act 2002* (NZ).
3 *Trade Marks Act 1995* (Cth) s6.
4 *Trade Marks Act 1995* (Cth) s20.
5 *Trade Marks Act 1995* (Cth) s20.
6 *Trade Marks Act 1995* (Cth) s27.

Registration of a business name or a company name is not, in itself, a defence to an action of infringement of a registered trade mark. Nevertheless, business and company names become valuable identifiers, and they may be protected by provisions of the *Trade Practices Act 1974* (Cth), the states' and the ACT's Fair Trading Acts and the tort of passing off.

Infringement

Section 120 of the Act details circumstances where a registered trade mark is infringed. In essence, infringement occurs where a sign with respect to goods or services is substantially identical with, or deceptively similar to that of the trade mark. Establishing that the use is not likely to deceive or cause confusion constitutes a defence to an infringement charge:

> 120(1) A person infringes a registered trade mark if the person uses as a trade mark a sign that is substantially identical with, or deceptively similar to, the trade mark in relation to goods or services in respect of which the trade mark is registered.
>
> (2) A person infringes a registered trade mark if the person uses as a trade mark a sign that is substantially identical with, or deceptively similar to, the trade mark in relation to:
>
> (a) goods of the same description as that of goods (registered goods) in respect of which the trade mark is registered; or
>
> (b) services that are closely related to registered goods; or
>
> (c) services of the same description as that of services (registered services) in respect of which the trade mark is registered; or
>
> (d) goods that are closely related to registered services.
>
> However, the person is not taken to have infringed the trade mark if the person establishes that using the sign as the person did is not likely to deceive or cause confusion.
>
> (3) A person infringes a registered trade mark if:
>
> (a) the trade mark is well known in Australia; and
>
> (b) the person uses as a trade mark a sign that is substantially identical with, or deceptively similar to, the trade mark in relation to:
>
> (i) goods (unrelated goods) that are not of the same description as that of the goods in respect of which the trade mark is registered (registered goods) or are not closely related to services in respect of which the trade mark is registered (registered services); or
>
> (ii) services (unrelated services) that are not of the same description as that of the registered services or are not closely related to registered goods; and
>
> (c) because the trade mark is well known, the sign would be likely to be taken as indicating a connection between the unrelated goods or services and the registered owner of the trade mark; and
>
> (d) for that reason, the interests of the registered owner are likely to be adversely affected.

(4) In deciding, for the purposes of paragraph (3)(a), whether a trade mark is well known in Australia , one must take account of the extent to which the trade mark is known within the relevant sector of the public, whether as a result of the promotion of the trade mark or for any other reason.

For the purposes of the Act, goods or services are similar to other goods or services, respectively, if they are the same, or if they are of the same description as that of the other goods or services.[7] A trade mark is taken to be deceptively similar to another trade mark if it so nearly resembles that other trade mark that it is likely to deceive or cause confusion.[8]

Sections 120(3) and (4) have extended the infringement action to restrain activities which are likely to adversely affect the interests of the owner of a famous or well-known trade mark by the 'dilution' of its distinctive qualities or of its value to the owner. The courts have held that this 'dilution theory of liability "does not require proof of a likelihood of confusion"; rather, what is protected is "the commercial value or 'selling power' of a mark by prohibiting uses that dilute the distinctiveness of the mark or tarnish the associations evoked by the mark"'.[9]

Hyperlinking

Hyperlinks are fundamental to internet use and navigation.[10] The value of the World Wide Web lay in the process of hyperlinking, which allows users to connect to vast quantities of information, text, graphics and data with a simple mouse click. However, websites can also be designed in a way that may be prejudicial to trade mark rights and copyright of others.[11]

Trade mark infringement can arise where one website links to another with registered trade marks.[12] The linking process can be regarded as a 'use' of such a mark in contravention of section 120(1) and (2). For example, users who link directly to targeted pages within a particular website bypass the home page. This form of hyperlinking is referred to as deep linking. Whatever the intention, the effect may be to reduce the site's advertising revenue, if that is dependent on 'hits' on the home page. Revenue is typically tied to the number of hits or depends upon actual use of the advertising links. Secondly, the use of a deep link may give the impression that there is a connection, affiliation or some relationship between

7 *Trade Marks Act 1995* (Cth) s14.

8 *Trade Marks Act 1995* (Cth) s10. See *Koninklijke Philips Electronics NV v Remington Products Australia Pty Ltd* (2000) 100 FCR 90; *Shell Co. of Australia Ltd v Esso Standard Oil (Aust.) Ltd* (1963) 109 CLR 407; *Berlei Hestia Industries Ltd v Bali Co. Inc.* (1973) 129 CLR 353; *Polaroid Corp. v Sole N Pty Ltd* [1981] 1 NSWLR 491; *Transport Tyre Sales Pty Ltd v Montana Tyres Rims and Tubes Pty Ltd* (1999) 93 FCR 421.

9 *Campomar Sociedad Limitada v Nike International Limited* [2000] HCA 12, paras 42–43, citing *Restatement Third, Unfair Competition*, §25, Comment (a).

10 On hyperlinking generally, see Chapter 6.

11 See Katia Bodard, Bruno de Vuyst and Gunther Meyer, 'Deep linking, framing, inlining and extension of copyrights: Recent cases in common law jurisdictions', (2004) *MurUEJL* 2.

12 Ibid.

the owners of the two sites. This impression may deceive or cause confusion to a user.

The first such case before the courts was the Scottish case *Shetland Times Ltd v Wills and Zetnews Ltd*,[13] before Lord Justice Hamilton in 1996. The defendant ran the newspaper the *Shetland News* and a corresponding website; the applicant[14] ran the *Shetland Times* and a website. The defendant created hyperlinks using the headlines from the applicant's newspaper. The hyperlinks pointed directly to articles on the applicant's website. Because the links bypassed the applicant's home page, his Lordship considered: 'there is a clear prospect of loss of potential advertising revenue in the foreseeable future'. The appearance gave some users the impression that the articles were part of the defendant's website. Additionally his Lordship stated that '[t]here was, in the circumstances, no substance, in my view, in the suggestion that the pursuers were gaining an advantage by their newspaper items being made available more readily through the defenders' website'.[15] The court issued an 'interim interdict' requiring the defendants to remove the links. Unfortunately the judgment was given at a preliminary stage, and the case did not proceed to a full hearing where argument could be made in full. The matter was subsequently settled.

In *Ticketmaster Corporation v Microsoft Corporation*,[16] Microsoft provided a link, on its Seattle Sidewalk entertainment site, to the Ticketmaster home page. Ticketmaster argued that a formal licence agreement would be required before anyone could link to its site. Ticketmaster had been negotiating such a licence with Microsoft, but negotiations had broken down. Instead, Ticketmaster entered into an agreement with CitySearch, a competitor of Microsoft's Sidewalk site. Ticketmaster claimed that Microsoft misused Ticketmaster's name and trade mark, diluting the value of the trade mark and damaging its relationship with sponsors. Microsoft claimed that the use of links is fundamental to the operation of the web and relied on the US constitution's protection of free speech and the fair use doctrine. It further asserted that by placing a website online web hosts impliedly consented to links, including deep links. Ticketmaster claimed that Microsoft was 'feathering its own nest at Ticketmaster's expense . . . committing electronic piracy'. This case was settled out of court, but it demonstrates the commercial concerns involved. The concept of property in a link was unknown to the legal and internet community. Many would argue that freedom of access and use was an inherent feature of the World Wide Web, particularly for websites placed in the public domain and intended to be readily accessed by the public.[17]

Implementing a hyperlink in itself should not be regarded as a trade mark infringement. To succeed in an action, an aggrieved party must show deception or confusion. Since these early cases, such problems have been resolved by

13 [1997] 37 IPR 71; [1997] FSR 604.
14 Described in the judgment as 'the pursuers'.
15 [1997] 37 IPR 71; [1997] FSR 604, para 22.
16 No. 97-3055 DDP (CD Cal 1997).
17 See also *The Washington Post Company v Total News Inc.* 97 Civ. 1190 (PKL) (SD NY).

technology. Web designers can restrict access to 'deep linked' pages, so that the user can only access such pages from the home page. The structure and design of the home page can be such as to reveal the correct authorship, minimise confusion, and of course make available the advertising previously bypassed.[18]

In *Zoekallehuizen.nl v NVM*[19] the respondent operated a specialist search engine that searched for houses for sale and placed links to such information on the websites of local real estate agents. The applicants were the Dutch Association of Real Estate Agents and two local real estate agents. The President of the District Court of Arnhem rejected the claim for a restraining order stating that the websites of the real estate agents did not show substantial investment and were therefore not protected.

In *Bixee v Naukri*,[20] the Delhi High Court in India prohibited Bixee.com from deep linking to Naukri.com. The preliminary injunction was issued on a prima facie finding that Naukri suffered significant financial loss due to the diversion of readers away from the advertisements.

In Europe, in response to concerns that information is inadequately protected by the law, Directive 96/9/EC of the European Parliament and of the Council of 11 March 1996 on the legal protection of databases was passed. The preamble provides that the rationale for the directive is that databases 'are not at present sufficiently protected by Member States' and 'such differences in the legal protection of databases offered by the legislation of the Member States have direct negative effects on the functioning of the internal market as regards databases'. This directive concerns the legal protection of databases in any form.

In Australia, if links are used in a manner which causes deception or confusion, and a trade mark is involved, the aggrieved party may seek redress through section 120 of the *Trade Marks Act* (Cth). In addition, section 121(2) provides that it is a prohibited act to apply or use a trade mark in an altered manner or in relation to registered goods, with use that is likely to injure the reputation of the trade mark.

If no trade mark is involved, remedies may be available under the copyright legislation,[21] the *Trade Practices Act 1974* (Cth) Part V,[22] the corresponding provisions of the state and territory Fair Trading Acts and the tort of passing off.

Framing

Web pages may be displayed in sections called frames. The web designer may create a heading or menu along the top of the page or on the left or right margin. This heading or menu remains static while the body of the web page changes

18 See *Bernstein v JC Penny Inc.* 50 USPQ 2d 1063 (CD Cal 1998).

19 District Court (Arrondissementsrechtbank) Arnhem, 2006, available at www.ivir.nl/files/database/sources/HofZAH.pdf.

20 (2005) IA no. 9733/2005, reported as India's first deep linking case.

21 See Chapter 6.

22 In particular, ss52 and 53.

when links, internal or external, are selected. This permits navigation within a website.

However, hypertext mark-up language (HTML) permits any web page to be displayed within any frame, including within the body of the original site. Problems arise where the static heading or menu continues, but the body of the web page displayed is in fact from another source. Users may well think that the outside contents are part of the framer's web page. Indeed the displayed URL in the web browser may indicate the framed website, not the site that is actually being displayed in the body of the page. Such unauthorised framing can be misleading and deceptive and can amount to infringement of trade mark.

In *Washington Post Co. v Total News*[23] the defendant used framing to present the content of external news sites such as CNN, Times Mirror, Dow Jones and Reuters. The advertisements from the external sites were reduced in size or obscured. The plaintiffs claimed:

> Simply put, Defendants are engaged in the Internet equivalent of pirating copyright material from a variety of famous newspapers, magazines or television news programs; packaging those stories to advertisers as part of a competitive publication or program produced by the Defendants and pocketing the advertising revenue generated by the unauthorized use of that material . . . just as that conduct would not be tolerated in the world of print and broadcasting it is equally unlawful in the world of cyberspace.[24]

In their written preliminary statement to the court, *The Washington Post* had this to say about linking without permission:

> This action arises out of blatant acts of misappropriation, *trademark dilution and infringement*, wilful copyright violations, and other related tortious acts all committed by Defendants in connection with their operation of a parasitic site known as 'total-news.com' on that portion of the Internet known as the World Wide Web.[25]

The dispute was settled on terms that included permitting Total News to continue linking to the plaintiff's website. However, Total News could not use the framing technique.

In *Futuredontics Inc. v Applied Anagramatics Inc.*,[26] another US case, framing was used with respect to referrals to dental websites. The frame was one of several on the defendant's website. Other frames contained information on its business operations. The plaintiff's cause of action was based on copyright infringement and unfair competition. The court refused to grant an interim injunction. However, the reason for this refusal was that the plaintiff failed to prove that any harm had resulted from the conduct of the defendant. This decision was affirmed on appeal. Nevertheless, the case accepted that framing in appropriate circumstances could be the basis for an infringement claim.

23 97 Civ 1190 (SD NY 1997).
24 97 Civ 1190 (SD NY 1997).
25 97 Civ 1190 (SD NY 1997), Preliminary Statement (emphasis added).
26 CV 97–6991 ABC (Manx), 1998 US Dist. Lexis 2265 (9th Cir 1998).

Meta-tags

Meta-tags are pieces of information in the source code of a website. Meta-tags are not displayed on the website, but can be viewed by selecting the 'source' view.[27] There are many types of meta-tags. The 'keyword meta-tag' was designed to assist search engines. Search engines are based on algorithms designed to produce valuable and meaningful results for the user. One important criterion involves the words and phrases used, their frequency and location. A score is given and search results list the results with the highest scores first. One significant factor in allocating a score is the use of words in the keyword meta-tag. The website designer inserts keywords into the meta-tag to indicate the purpose of the website and to assist search engines. In *Playboy Enterprises Inc. v Hie Holdings Pty Ltd*[28] the Registrar of the Australian Trade Marks Office described a meta-tag in these terms:

> [M]eta-tags are words embedded in the code for homepages on the Internet . . . It is common for traders to embed meta-tags in their homepages so that those who are searching on the Internet find those pages via the meta-tags which are located by the search engine that the searcher is using. Thus a cheese-vendor on the Internet might include the various trade marks or generic names of cheeses as meta-tags on his homepage: these words are found by the search engines used by the public. Thus, these tags are invisible to the person trying to locate the cheese but will bring the merchant's homepage up on the search.[29]

Some website authors have misused the meta-tags. The information may be misleading and deceptive. For example, the designer may insert 'best' or 'top ten' before other keywords describing the product or service. Such a tactic increases that website's search engine score for searches that include the word 'best'. Such a manipulation may be a breach of the *Trade Practices Act* 1974 (Cth).[30]

Trade mark infringement becomes a factor where the website designer uses another's trade mark in the meta-tag. The purpose may be to deflect patronage from a competitor, or merely to interfere with the search engine research for another site. Even though the code is hidden from view, such a misuse can amount to an infringement. The 9th Circuit court in the United States described the use in this manner:

> The meta tags are not visible to the websurfer although some search engines rely on these tags to help websurfers find certain websites. Much like the subject index of a card catalog, the meta tags give the websurfer using a search engine a clearer indication of the content of a website.[31]

27 For example the following code may be inserted in the 'head' section of the website's code, for a law firm: <meta name='keywords' content='law legal resource solicitor lawyer securities litigation intellectual property patent trade marks trade secret copyright software computer technology mediation confidential information australia'>.

28 [1999] ATMO 68.

29 In this case the ATMO refused to the application to register 'Playbabe' on the basis of its similarity to the Playboy trade marks.

30 For example ss52 and 53, dealing with deceptive and misleading conduct.

31 *Playboy Enterprises Inc. v Welles* 162 F. 3d 1169 (9th Cir 1998).

In the US case of *Playboy Enterprises Inc. v Calvin Designer Label*[32] the court granted an injunction to restrain the use of Playboy's trade mark in the defendant's domain names (playboyxxx.com and playmatelive.com) in any 'machine-readable code', and in meta-tags:

> The Court finds that Plaintiff PEI is likely to succeed on the merits in proving inter alia trademark infringement, unfair competition, including a false designation of origin and false representation, in Defendants' use of the domain names . . . and the repeated use of the PLAYBOY trademark in machine readable code in Defendants' Internet Web pages, so that the PLAYBOY trademark is accessible to individuals or Internet search engines which attempt to access Plaintiff under Plaintiff's PLAYBOY registered trademark.[33]

In *Playboy Enterprises Inc. v Welles*[34] the defendant had been a Playboy playmate of the year, and used the playboy trade marks on her website, including in her site's meta-tags. The 9th Circuit court stated that the defendant 'used the terms "Playboy" and "Playmate" as meta tags for her site so that those using search engines on the Web can find her website if they were looking for a Playboy Playmate'. The court held that misuse of the trade mark in a meta-tag can be an infringement. However, the defendant's use in this instance was regarded as fair use, as the expressions 'playboy' and 'playmate' may be used and were used appropriately in an editorial fashion.[35]

In *Kailash Center for Personal Development Inc. v Yoga Magik Pty Limited*[36] Jonn Mumford, also known as Swami Anandakapila Saraswati, had developed a reputation with regard to yoga and Eastern spirituality. He entered into an agreement to permit his names and reputation to be associated with the promotion of several websites. After a falling out between the parties Mumford demanded the removal of meta-tags which included his names and the cancellation of the registration of certain domain names. Mumford's name was in the source code, as a meta-tag and on 44 web pages of the defendant. Notwithstanding the prior authorisation and agreement the Federal Court held that such conduct was misleading and deceptive, contravening of section 52 of the *Trade Practices Act 1974* (Cth). Allsop J regarded the use of Mumford's name and pseudonyms as a misappropriation and ordered the discontinuance of such use.

Although not applied in Australian case law to date, section 146 of the *Trade Marks Act 1995* (Cth) is also well suited for a imposing a sanction against the misuse of a trade mark in a meta-tag:

32 985 F. Supp. 2d 1220 (ND Cal 1997).
33 985 F. Supp. 2d 1220 (ND Cal 1997), para 6.
34 162 F. 3d 1169 (9th Cir 1998).
35 See also *Natural Floor Covering Centre Pty Ltd v Monamy (No. 1)* [2006] FCA 518, *Athens v Randwick City Council* [2005] NSWCA 317, *Insituform Technologies Inc. v National Envirotech Group* (1997) Civil Action No. 97–2064 (ED La 1997) and the German Federal Supreme Court judgment May 2006 I ZR 183/03.
36 [2003] FCA 536.

Falsely applying a registered trade mark

146(1) A person is guilty of an offence if the person:

(a) falsely applies a registered trade mark to goods that are being, or are to be, dealt with or provided in the course of trade; or

(b) falsely applies a registered trade mark in relation to goods or services that are being, or are to be, dealt with or provided in the course of trade;

knowing that the trade mark is registered or reckless of whether or not the trade mark is registered.

(2) A person falsely applies a registered trade mark to goods, or in relation to goods or services if the person applies the trade mark or a sign substantially identical with it to the goods or in relation to the goods or services:

(a) without the permission of the registered owner, or of an authorised user, of the trade mark; and

(b) without being required or authorised to do so by this Act, a direction of the Registrar or an order of a court.

Patents for software and internet processes

A patent provides protection for inventions. The invention may be a device, a substance, a composition, a living organism, a method or a process.[37] A patent will be granted to protect any novel and non-obvious technological development. Patents evolved from mere legal instruments protecting intellectual property into valuable corporate assets and competitive commercial weapons. Businesses have relied upon the application of copyright law to protect computer programs and other software-related intellectual property. However, businesses are now more likely to choose patent protection as their most powerful offensive and defensive weapon in protecting intellectual capital on the internet. Recent amendments to the *Patents Act 1990* (Cth) have included patent protection for business processes and other technology.[38]

Developments in the United States

For many years, methods of doing business were not patentable. In the United States, there has been a deluge of applications for cyberpatents as a result of the 1998 decision in *State Street Bank and Trust Co. v Signature Financial Group Inc.*[39] The court held that 'methods of doing business' are patentable and legitimate, provided that the other statutory criteria are met. The decision applies to electronic commerce, internet technology, banking, insurance and finance.

The US Patent Act requires that patentable inventions must be new, non-obvious and useful. Potentially patentable internet inventions since the State Street decision include communications protocols, data compression schemes,

37 See, generally, www.ipaustralia.gov.au/patents/.
38 For the New Zealand equivalent see the *Patents Act 1953* (NZ).
39 149 F. 3d 1368; 47 USPQ 2d (Fed Cir 1998).

encryption and security procedures, hardware, computer software (including servers, browsers and search engines), user interfaces and methods of conducting business online.

In response to the dramatic increase of applications for cyberpatents, the US Business Method Patent Improvement Act[40] was passed by Congress in 2001. The Act was intended to regulate business method patents and the conditions of issue and to make sure that the Patent Office issued these patents only where the method was new and innovative. Business method patent applications are to be published for 18 months, allowing the public the opportunity to present information about prior inventions or request a hearing to determine whether an invention was known, used by others, or was in public use. The Act aims to provide a speedy and less costly alternative to litigation. Where the business method invention is merely a computer implementation of an existing practice, a presumption of obviousness applies and the invention is not patentable. The Act also reduces the burden of proof for challenges to the validity of a patent.

Developments in Europe

Article 52(1) of the European Patent Convention (EPC) provides that European patents may only be granted for inventions susceptible of industrial application which are new and which involve an inventive step.[41] The term 'invention' is not defined in the EPC, but Article 52(2) expressly excludes methods of doing business, mathematical methods, presentation of information and programs for computers. However, decisions of the Board of Appeal of the European Patent Office have allowed patents for computer programs. This was contingent upon the program being able to produce a 'technical effect' (that is, something beyond normal physical interactions between software and hardware) when the hardware executes the program's instructions.

These decisions bring the European Union's position on the patentability of internet and other software-related inventions closer to that of the United States, Japan and Australia. The UK Patents Act has identical provisions to the EPC. However, in *Merrill Lynch's Application*,[42] the UK Court of Appeal held that a computerised trading system for stocks and shares was capable of being patented. This confirmed that inventions whose novelty and non-obviousness reside in unpatentable subject matter such as computer programs are still capable of being patented.

Developments in Australia

In Australia, section 18 of the *Patents Act 1990* (Cth) provides that a patentable invention must be a manner of manufacture, novel, involve an inventive step,

40 HR 5364.
41 The EPC originally came into force in 1973. A revised version has been in operation from December 2007.
42 [1989] RPC 561.

be useful and not secretly used before the priority date of a patent claim. The Australian Patent Office (APO) followed the United States in rejecting patent applications for software-related inventions because they were not a 'manner of manufacture' within the meaning of the Act.[43] Software-associated inventions were unpatentable because they were perceived as schemes for operating a known machine, or abstract ideas, or intellectual processes or algorithms.[44]

IBM Corporation v Commissioner of Patents[45] concerned a 'method and apparatus for generating curves on computer graphics displays'. One objection by the Commissioner of Patents was that it was a mathematical algorithm, not a 'manner of manufacture'. The court considered an algorithm to be 'a procedure for solving a given type of mathematical problem'.[46] In applying the High Court decision of *National Research Development Corp. v Commissioner of Patents* (NRDC),[47] the Federal Court determined the 'manner of manufacture' test to be whether the invention 'belongs to a useful art as distinct from a fine art . . . that its value to the country is in the field of economic endeavour'.[48] It was held that while the mathematics of the invention was not new, its application to computers was commercially useful in the field of computer graphics: IBM was successful.

CCOM Pty Ltd v Jiejing Pty Ltd[49] followed the IBM decision. At issue was the validity of a 'petty patent'[50] relating to the input of Chinese characters into a word processing system. The Full Federal Court distinguished 'manner of manufacture' from novelty and inventive step in section 18 of the Patents Act, and confirmed the NRDC 'manner of manufacture' test. Further, the court held that the invention was useful in the economic endeavour of using word processing to assemble text in Chinese language characters.[51]

In recognition that patents for business processes and other technology should be protected, the Patents Act was amended in 2000 to significantly lower the inventive threshold. Patentable inventions must demonstrate an 'innovative step' when compared to the prior art base, rather than the current 'inventive step' and be 'obvious to a skilled person in the relevant field'. An innovative step involves a variance from the prior state of knowledge in a way which is meaningful in terms

43 See *British Petroleum Co. Ltd's Application* (1968) 38 AOJP 1020.
44 See Alan Davidson, 'Patents for software and computers processes', (2001) 21 *Proctor* 1, 33.
45 (1991) 22 IPR 417.
46 (1991) 22 IPR 417 at 419.
47 (1959) 102 CLR 252.
48 (1991) 22 IPR 417 at 423.
49 (1994) 27 IPR 481.
50 'The innovation patent was introduced in 2001 following an Advisory Council on Intellectual Property (ACIP) review of the petty patent system. The petty patent system was designed to provide a form of protection that was quick and easy to obtain, was relatively inexpensive and provided short term protection especially for inventions that had a short commercial life. Although the majority of users of the petty patent system were small to medium sized enterprises (SMEs), the system had limited success in meeting its intended objectives': Australian Government/IP Australia, 'Review of the innovation patent issues paper', September 2005.
51 (1994) 27 IPR 481 at 514.

of how the invention actually works. The amendment is intended to include cheese and wine making, brewing and certain industrial processes as patentable inventions.[52]

In *Grant v Commissioner of Patents*,[53] Justice Branson of the Federal Court considered the validity of a business method-type patent. The claim related to a method for protecting an asset, comprising the following steps:

> (a) establishing a trust [and] having a trustee,
> (b) the owner making a gift of a sum of money to the trust,
> (c) the trustee making a loan of said sum of money from the trust to the owner, and
> (d) the trustee securing the loan by taking a charge for said sum of money over the asset.[54]

Justice Branson stated that 'an invention should only enjoy the protection of a patent if the social cost of the resulting restrictions upon the use of the invention is counterbalanced by resulting social benefits'. Her Honour found that the trust structure proposal lacked the appropriate:

> value to the country . . . in the field of economic endeavour . . . The performance of the invention will not add to the economic wealth of Australia or otherwise benefit Australian society as a whole. For this reason, in my view, the invention the subject of the Patent is not a proper subject of letters patent according to the principles which have been developed for the application of s6 of the Statute of Monopolies.[55]

In this case, the patent applicant was required to demonstrate that the performance of the invention 'adds to the economic wealth of Australia or otherwise benefits Australian society as a whole' and 'advances the public interest'. This may be of particular concern for foreign applicants. This approach is in line with the European approach, which requires a 'technically useful' effect.[56]

Australian developments in software-related patents have paralleled those of the United States. Both nations consider software inventions as patentable, and have formulated similar tests for determining patentable subject matter. Australia regards software-related internet inventions in much the same way as other software-related inventions. Many Australian internet patents mirror US patents through the medium of the Patent Co-operation Treaty.

Patents and hardware

Electronic hardware is most suitable for patent application. The criteria are typically novelty and inventiveness. Provided the invention works in an

52 See the Patents Amendment (Innovation Patents) Bill 2000, Explanatory Memorandum.
53 [2005] FCA 1100.
54 [2005] FCA 1100, para 2.
55 [2005] FCA 1100, paras 20–21. The original law was the *Imperial Statute of Monopolies 1623* (UK); its section 6 is the equivalent of *Patents Act 1990* (Cth) s18(1A).
56 See also *State Street v Signature* 149 F. 3d 1368 (1998), which raised a test of 'a useful, concrete and tangible result'.

improved manner over previously known devices, the patent application will be granted. Examples include computer systems, computer components and accessories.

Circuit layout rights

A circuit layout is a two-dimensional representation of a three-dimensional integrated circuit (these are also referred to as computer chip designs or semiconductor chips).[57] Circuit layouts are highly complex and the intellectual input required to create them may be of significant value. Integrated circuits are an integral part of modern electronics, such as computers and medical devices like pacemakers.

Integrated circuits and circuit layouts are both protected under the *Circuit Layouts Act 1989* (Cth).[58] The purpose of the Act is to protect the intellectual property component of circuit layouts and integrated circuits, to give the owner rights, especially the right to make an integrated circuit from the plans. The Act assists in preventing billions of dollars being lost from unauthorised replication and distribution. Like copyright, there is no requirement for registration for the granting of rights to owners.

The Act became operative in 1990. It is part of Australia's obligation under GATT (the General Agreement on Tariffs and Trade). Member nations of the World Trade Organization Agreement are required by the Agreement on Trade-Related Aspects of Intellectual Property Rights (TRIPS) to protect circuit layout designs in accordance with the Treaty on Intellectual Property in Respect of Integrated Circuits adopted at Washington DC in 1989.[59]

The Act protects original circuit layouts made by an Australian citizen or an Australian corporation, or first commercially exploited in Australia.[60] A layout originating in a country formally declared in the Circuit Layout Regulations will also be given protection by the Act. The Act excludes layouts where the making involved no creative contribution by the maker, or was commonplace at the time it was made.[61]

The Act grants the owner of an original circuit layout rights to copy the layout, directly or indirectly, in a material form, to make an integrated circuit in accordance with the layout (that is, a three-dimensional copy of the layout), and to exploit the layout commercially in Australia.[62] The rights include

57 *Circuit Layouts Act 1989* (Cth) s5.
58 For the New Zealand equivalent see the *Layout Designs Act 1994* (NZ).
59 Available at www.wipo.int/treaties/en/ip/washington/trtdocs_wo011.html.
60 See definition of 'eligible layout', *Circuit Layouts Act 1989* (Cth) s5.
61 *Circuit Layouts Act 1989* (Cth) s11.
62 *Circuit Layouts Act 1989* (Cth) s17.

commercial exploitation, such as import and export, sale, hire and distribution. Where another person commercially exploits a layout, if the person knew or ought reasonably to have known that they did not have licence to do so, the rights of the owner will have been infringed.[63] The holder's rights are not infringed where a person copies the layout for his or her private use, where copying is for research or teaching purposes, where copying in the process of evaluation or analysis of a layout or where the uses of layouts are for Commonwealth defence or security.[64]

If the layout is not commercially exploited, the protection period runs for 10 years after the calendar year in which the layout was made. If the layout is first commercially exploited within 10 calendar years after it was made the protection period runs until the end of the 10th year after the calendar year in which the layout was first commercially exploited.[65] This can result in a maximum period of 20 years.

The holder of the rights may take civil action for infringement: possible remedies are an injunction, damages, an account of profits or 'additional damages'.[66] Where there is a flagrant infringement or the infringer has gained some benefit, a court has discretion to award additional damages.[67] Action must be commenced within six years of the alleged infringement.

Circuit layout designs made before 1 October 1990 are not protected by the *Circuit Layouts Act 1989* (Cth). The owners of such must resort to the law of copyright and designs protection.

Member nations of the WTO Agreement must protect circuit layouts and give the same level of protection to foreigners who make or exploit circuit layouts in their country as they do to their own citizens.[68] Layouts made or originating outside Australia in listed countries are given the same level and kind of protection in Australia as layouts made in Australia.[69]

63 *Circuit Layouts Act 1989* (Cth) s19(3).
64 *Circuit Layouts Act 1989* (Cth) ss21(1), 22, 23 and 25 respectively.
65 *Circuit Layouts Act 1989* (Cth) s5.
66 *Circuit Layouts Act 1989* (Cth) s27(1) and (2).
67 *Circuit Layouts Act 1989* (Cth) s27(4).
68 Required by the TRIPS Agreement, Art. 3.
69 The listed countries are: Antigua and Barbuda, Argentina, Austria, Bahrain, Bangladesh, Barbados, Belgium, Belize, Bolivia, Botswana, Brazil, Brunei Darussalam, Burkino Faso, Burundi, Canada, Central African Republic, Chile, Colombia, Costa Rica, Cote d'Ivoire, Cuba, Cyprus, Czech Republic, Denmark, Djibouti, Dominica, Dominican Republic, Egypt, El Salvador, Finland, France, Gabon, Germany, Ghana, Greece, Guatemala, Guinea, Guinea Bissau, Guyana, Honduras, Hong Kong, Hungary, Iceland, India, Indonesia, Ireland, Israel, Italy, Jamaica, Japan, Kenya, Korea, Republic of, Kuwait, Lesotho, Liechtenstein, Luxembourg, Macau, Malawi, Malaysia, Maldives, Mali, Malta, Mauritania, Mauritius, Mexico, Morocco, Mozambique, Myanmar, Namibia, Netherlands (for the Kingdom in Europe and the Netherlands Antilles), New Zealand, Nicaragua, Nigeria, Norway, Pakistan, Paraguay, Peru, Philippines, Poland, Portugal, Romania, Saint Lucia, Saint Vincent and the Grenadines, Senegal, Sierra Leone, Singapore, Slovak Republic, Slovenia, South Africa, Spain, Sri Lanka, Suriname, Swaziland, Sweden, Switzerland, Tanzania, Thailand, Togo, Trinidad and Tobago, Tunisia, Turkey, Uganda, United Kingdom, United States of America, Uruguay, Venezuela, Zambia and Zimbabwe.

Further reading

Katia Bodard, Bruno de Vuyst and Gunther Meyer, 'Deep linking, framing, inlining and extension of copyrights: Recent cases in common law jurisdictions', (2004) *MurUEJL* 2.
IP Australia (Patents): www.ipaustralia.gov.au/patents/.
IP Australia (Trade marks): www.ipaustralia.gov.au/trademarks/.
WIPO Treaties: www.wipo.int/treaties/en/.

8

Domain names

Mapping cyberspace

Cyberspace is chaos. More than a billion users[1] trawl unreal space for the gold, titbits and junk released by a multitude of undisciplined hosts. Cyberspace contains a massive amount of valuable and useless information. There are several billion web pages at users' disposal. Some are freely available and easy to locate. Others are hidden and security protected. The domain name system forms a map in cyberspace, indexing and listing material. Domain names have unwittingly become significant business identifiers and valuable assets. This chapter examines the nature of domain names, property rights, structure and regulations. Chapters 9 and 10 deal with domain name disputes and their resolution through the courts and through the use of domain name dispute resolution policies.

Information in cyberspace is so vast it cannot be properly catalogued. It is metamorphic in nature, growing, twisting and changing continuously. A semblance of order emerges in part by using meaningful identifiers for human interaction. Domain names permit millions of possibilities for cataloguing, indexing and searching. Their value in terms of both knowledge and commercial interests is unprecedented. The internet recognises no boundaries, yet existing laws are territorially based. The use and misuse of domain names have raised conflicts previously unknown in law. Existing legal remedies have proven inadequate, expensive and inconvenient, and often they have been unenforceable. Compulsory arbitration clauses have been placed into domain name licence agreements

1 Internet World Stats reports that internet users passed the one billion mark (approximately 16 per cent of the world's population) in 2005, a 30-fold increase in 10 years – www.internetworldstats.com.

in an attempt to address these deficiencies. But new problems continue to arise, resulting in inconsistent determinations.

The use of similar or identical standard business identifiers in different geographical locations may never give rise to disputes or be of concern. However, domain names are systemically international. The problem of duplication is partly resolved by the use of country codes and various generic top-level domain name suffixes.

The laws and regulations regarding domain names and their use are as new as the internet. Their foundation and understanding test jurisprudential norms in the true sense. Domain name conflicts and rights cannot be satisfactorily pigeonholed with other legal precepts. Some domain names are bought and sold like property. Some people regard domain names as a new form of intellectual property; others see them as having the characteristics of mere licence rights.

Tim Berners-Lee is regarded by many as the inventor of the internet.[2] In 1989, as a scientist at CERN (Centre Européan pour la Recherche Nucléaire; European Laboratory for Particle Physics), he invented the World Wide Web. The underlying principle was to merge the technologies of personal computers, computer networking and hypertext into an information system. This would be modest at first, but it had the potential to be worldwide, subject to network connections. The initial purpose was to communicate among colleagues and access large quantities of data. During 1990 and 1991 Berners-Lee developed the underpinnings of the system.

This system became operative within CERN, then expanded into physics research and hypertext programming communities, and ultimately into the internet. In a short time internet administration was transferred to the Internet Corporation for Assigned Names and Numbers (ICANN).

Business identifiers

Electronic commerce technologies have advanced with such momentum that designers have taken little time to consider their legal implications. While the majority of electronic commercial situations can be resolved using existing legal principles, there are occasionally new situations which the general law has been unable to predict and, initially at least, is unable to cope with. Courts and legislatures have had to be flexible and innovative in their approach to providing remedies in relation to domain name rights. The ICANN and domain administrators' 'first come first served' approach has proven unfair and inappropriate in many circumstances.

It is common to see identical or similar business names and identifiers in different geographic locations. The majority of businesses operate within a suburb,

2 See Ann Gaines, *Tim Berners-Lee and the development of the World Wide Web (Unlocking the secrets of science)*, Mitchell Lane Publishers, Hockessin DE, 2001.

city, state or country. 'Smith Dry Cleaning' could be registered in every major city without any concern regarding misleading or deceptive conduct, passing off or misrepresentation. But there can only be one smith.com. Further, Smith Dry Cleaning may conflict with all international 'Smith' businesses: Smith Motors, Smith Music and so forth.

Commercial entities place great importance on their trade name, spending considerable time and money on research and marketing to make a selection which will reflect the nature of the business and yield the greatest profits. Individuals have the right to use their own name. Most jurisdictions have enacted rules in relation to registering business names and company names.

Trade mark legislation usually includes provisions from international conventions and agreements to assist in standardising the global position. A trade mark is a distinctive sign which identifies certain goods or services as those produced or provided by a specific person or enterprise.[3]

In the absence of trade mark registration, the laws of passing off and misrepresentation usually protect the use of a name within a specific geographic area.[4]

The nature of domain names

Each internet page has a unique address, referred to as a Uniform Resource Locator (URL). Protocols exist for various parts of each URL. Each URL can have a number of parts: for example, the protocol, login, host, port, path, query, file, anchor/fragment. Here is a typical URL with an html file: http://www.uq.edu.au/davidson/index.html.

Protocol:	***http***
Domain name:	*www.uq.edu.au*
Directory(ies) (path):	*davidson*
File:	*index.html*

There are a range of possible protocols, but internet users are most familiar with Hypertext Transfer Protocol (http). Domain names are centrally organised and registered and must be unique. There may be a number of directories or none at all (in the latter case, files are on the root directory). Files are typically in hypertext markup language (html) but may be any type of file. Where there is no file specified, a default file will apply.

Underlying domain names is a numeric Internet Protocol (IP) address. The current version is IPv4, the fourth iteration. The IPv4 means that every individual domain name is in fact a 32 bit number made up of four sets of numbers joined by dots: for example, 3.151.89.14. This permits 2^{32} – or 4,294,967,296 – unique IP addresses. IPv6 is planned to replace IPv4. IPv6 has a larger address space

3 For more, see the WIPO home page and www.wipo.int/about-ip/en/trademarks.html. See also Chapter 7.

4 For more on misrepresentation and passing off in a domain name context see supra.

to permit greater flexibility in allocating addresses and in routing traffic. For the human user a name is assigned to each specific group of numbers, and this name is referred to as the domain name. The number of possible domain names depends in part on the protocols used and the number of unique meaningful letter and numeric combinations. Each domain name must be unique so that when the user types in a particular address the system will only point to one source.

The domain name has prescribed protocols which expand and evolve according to demand and usage. That part of the domain name selected by the individual may be referred to as the identifier. In the example above it is 'uq'. In addition the protocols may involve the top level domain name and second, third and fourth level domain name.

Top Level Domain names (TLDs)

There are two types of top-level domains – generic (gTLD) and country code (ccTLD) – plus a special top-level domain (.arpa) for internet infrastructure. Generic domains were created to be used by the internet public; country code domains were created to be used by individual countries. There are some 240 ccTLDs, such as .au for Australia, .nz for New Zealand, .in for India and .uk for the United Kingdom. Initially the United States did not have an assigned ccTLD and users simply omitted the code. However, this disadvantaged US users as many domain names without a ccTLD were registered internationally. The code .us has now been added to the list.[5]

Generic Top Level Domain names (gTLDs)

Registrations may be made through dozens of competing registrars. These are the current open gTLDs:

.asia (for the Pan-Asia and Asia Pacific community)
.biz (business)
.cat (for Catalan language/culture)
.com (commercial)
.info (unrestricted general use)
.jobs (human resource managers)
.mobi (mobile products)
.name (individuals)
.net (internet organisations)
.org (miscellaneous organisations)

5 The list of country codes is available at www.iana.org/cctld/cctld-whois.htm.

.pro (professionals)
.tel (for services involving connections between the telephone network and the internet)
.travel (travel industry)

The .aero, .coop, and .museum TLDs are sponsored TLDs and are designed for use within a specified community. Registration restrictions for these TLDs have been developed by the sponsor with input from the community.

The .gov domain is reserved exclusively for the US government.

The .edu domain is reserved for postsecondary institutions accredited by an agency on the US Department of Education's list of Nationally Recognised Accrediting Agencies.

The .mil domain is reserved exclusively for the US military.

The .int domain is used only for registering organisations established by international treaties between governments.

In 2008 the ICANN Board approved a recommendation to introduce new gTLDs into the internet's addressing system, to commence in 2009. This is intended to foster choice and competition in domain registration services, which is part of the ICANN's role. ICANN Chairman Dr Twomey referred to the change as 'a massive increase in the "real estate" of the internet'. The proposal allows applicants to self-select their domain name.[6]

Applicants may apply for 'community strings' (like the existing .travel for the travel industry and .cat for the Catalan community). Consortiums have already expressed an interest in city-based gTLDs such as .nyc (New York City), .berlin and .paris. The permitted names can be anything – occupations, products, services and places, for instance.

There will be a limited application period where any established entity from anywhere in the world can submit an application. All applications will go through an evaluation process. Trade marks will not be automatically reserved. Offensive names will be subject to an objection-based process.

From time to time ICANN has considered applications for an expansion of the gTLDs. ICANN had previously considered: .nyc for New York City; .kids for children; .post for postal services; and .xxx for sexually explicit sites. On three occasions since 2000 ICANN has rejected the proposal for the .xxx gTLD. The Governmental Advisory Committee of ICANN has given this serious consideration in an attempt to control the proliferation of sexually explicit websites. The committee's determination is not binding on the ICANN Board, but does carry significant weight. The implementation of a .xxx domain raises issues of access, free speech, government censorship and industry legitimisation of sexually explicit material. More than 100 000 domain names have been pre-registered in anticipation of the possible release of the .xxx gTLD. ICANN regards the issue as controversial and 'polarising'.

6 See ICANN website for the latest information: www.icann.org.

The expansion of gTLDs is long overdue. The reasons for limiting domain names to those with the appendage .com, .org and the like are purely historical. There have been no technical or logistical reasons, other than size, not to expand domain names to almost any combination or language. The restrictive approach continues in countries such as Australia and the United Kingdom, where there are a limited number of choices, such as .com.au, .org.au, .co.uk, .org.uk and so forth. The United States, on the other hand, has no such limitation, requiring only .us. India permits the gTLD .in as well as other categories (such as .com.in and .org.in).[7]

Country Code Top Level Domain names (ccTLDs)

Each country code has its own structure, independent from gTLDs.[8] Many jurisdictions have a limited second level domain name (2TLD) structure; others have no restrictions. Some jurisdictions have preferred .co to .com (for example .co.uk and .co.in), and others have preferred .ac to .edu (for example .ac.uk and .ac.in). The United States, through its considerable influence in ICANN, can still affect which country codes are permitted and who will run each. Some nations, such as Pakistan and Brazil, have been outspoken critics of the United States' influence online.

.au

In Australia there are 12 .au 2TLDs. The non-profit company .au Domain Administration Ltd (auDA) has operated the .au domain since gaining government endorsement. In September 2001, ICANN recognised auDA as the .au operator and transferred control of the domain.[9] auDA have put in place a system of registrars who compete with each other to provide a variety of packages to registrants.

.com.au	commercial
.net.au	network services
.edu.au	education
.gov.au	government
.id.au	limited individual use
.info.au	major information resources
.oz.au	ACSNet members
.telememo.au	X.400 Gateway

7 See www.icann.org/en/topics/new-gtld-program.htm.
8 It is most impractical to deal with the vicissitudes of more than 240 domain administrators. This chapter will use Australia, New Zealand, India, the United Kingdom and the United States as examples. See www.wipo.int/amc/en/domains/cctld_db/index.html.
9 See www.icann.org and www.auda.org.au.

.csiro.au	CSIRO
.conf.au	conferences and exhibitions
.org.au	miscellaneous (for registered organisations)
.asn.au	associations and non-profit organisations

.nz

The New Zealand Domain Name Commissioner (DNC) is responsible for the day-to-day oversight of the .nz domain name registration and management system, and for the authorisation of registrars, and the transfer of management of specific domain names. Under the .nz Shared Registry System, authorised registrars can register and manage .nz domain names directly with the registry. Registrars are authorised by the DNC. The DNC is an operational office of InternetNZ (the Internet Society of New Zealand Inc.).[10]

NZ domain names cannot be owned by any party and are regarded by the DNC as licensed. A registrant must be an identifiable individual over 18 or a properly constituted organisation.[11]

.ac.nz	tertiary educational institutions and related organisations
.co.nz	organisations pursuing commercial aims and purposes
.cri.nz	Crown research institutes
.gen.nz	individuals and other organisations not covered elsewhere
.geek.nz	for people who are concentrative, technically skilled and imaginative who are generally adept with computers
.govt.nz	national, regional and local government organisations operating with statutory powers
.iwi.nz	a traditional Maori tribe, Hapu that belongs to a traditional Maori Iwi or taura-here Iwi group operating with the permission of the main Iwi
.maori.nz	Maori people, groups, and organisations
.mil.nz	military organisations of the NZ Government
.net.nz	organisations and service providers directly related to the NZ internet
.org.nz	not-for-profit organisations
.school.nz	primary, secondary and pre-schools and related organisations

.us

In the United States there is one top-level domain: .us. There are no secondary categories such as .com.us. Other than reserved words, there is no restriction on the use of a 2TLD. The .us domain is administered by NeuStar and was launched in April 2002.[12]

10 See www.internetnz.net.nz.
11 Rule DNC policy on Registering, Managing, and Cancelling Domain Names: dnc.org.nz/content/registering_managing_cancelling.pdf.
12 See www.neustar.biz.

The .us structure is a locality based hierarchy modelled on the geography of the United States. Branches in the locality space are overseen by delegated managers known as delegees or locality delegees. This hierarchical design provides structure, name uniqueness and a geographic reference point for all registrants.

.uk

The .uk TLD was first used during the 1980s. At that time a voluntary group called the Naming Committee managed the registrations. This was replaced in 1996 by Nominet UK, a private non-profit company limited by guarantee. Nominet has more than 130 staff, and a turnover in excess of £12 million. The Policy Advisory Board develops proposals for policies and rules for consideration by the Council of Management. Day-to-day operations are carried out by three departments: operations, technical and legal.[13] Nominet manages more than four million domain names, making it the fourth largest internet registry in the world.

.co.uk	commercial enterprises
.me.uk	personal domains
.org.uk	non-commercial organisations
.ltd.uk	registered company names
.plc.uk	registered company names
.net.uk	Internet Service Providers
.sch.uk	schools
.ac.uk	academic establishments
.gov.uk	government bodies
.nhs.uk	NHS (National Health Scheme) organisations
.police.uk	UK police forces
.mod.uk	Ministry of Defence establishments

The .co.uk domain is a fully open domain. Nominet does 'not impose restrictions' on the 'status as applicant for the registration' of the domains .co.uk, .me.uk or .org.uk.[14]

.in

The National Internet Exchange of India[15] (NIXI) is the official .in registry. The INRegistry was created by NIXI.[16] It functions as an autonomous body maintaining the .in ccTLD and ensuring its operational stability, reliability and security. It implements the policies set out by the government of India through its Ministry of Communications and Information Technology and Department of Information Technology. NIXI and INRegistry assumed responsibility in January 2005.

13 See www.nominet.org.uk.
14 Rule 4.4 of the Rules of Registration and Use of Domain Names, Nominet UK.
15 See www.nixi.in.
16 See www.registry.in.

At the same time the National Informatics Centre (NIC) became the registrar for .gov.in domains, ERNET the registrar for .res.in and .ac.in domains, and the Ministry of Defence the registrar for .mil.in domains. INRegistry does not carry out registrations itself: it accredits registrars.

.in	general
.co.in	commercial
.net.in	Internet Service Providers
.org.in	other organisations
.firm.in	firms
.gen.in	general
.ind.in	individuals
.ac.in	academic
.res.in	Indian research institutes
.edu.in	education
.gov.in	Indian government
.mil.in	Indian military

TLD rationale

The rationale for gTLD protocols is unconvincing. The requirement to include dots and suffixes such as 'com' and 'org' is now a historical anomaly. There are no technical reasons why an open domain name cannot be the full company, personal or business name, including spaces. The 'www' similarly remains an anachronism. The current protocols are reminiscent of the early naming protocols of computer files, when operating systems could only accept a maximum of eight characters with a dot and three more characters and spaces were not permitted. Naming files with cryptic names was often problematic. Subsequent protocols cater for long meaningful names, and spaces.

Many national domain name administrators have following the protocol requiring .com, .org, .edu for example in addition to the country code. However, some jurisdictions, such as the .us regime, remain completely open and free from such protocols. The identifier in front of .us does not have to include any particular secondary suffix, which allows US registrants greater freedom in the selection of their domain names than registrants in other countries.

Permitting domain names of any composition and length would reduce the numbers of conflicts of names, and thus also potentially costly and time-consuming court action and dispute resolution. Regulatory authorities are vested in the current processes, though, as resources and authority are based on them. From ICANN through to the many registrars, processes are highly developed and many vested interests may be compromised by redevelopment. The attempt to expand gTLDs and implement .biz, for example, was intended partly to alleviate the congestion in the .com protocol. It has been a failure. The public is familiar with .com. First guesses are typically <guess>.com. Attempts in the other protocols are typically not contemplated. The removal of the obligatory suffixes would change users' attempts to inputting the identifier alone.

Conversely, the country code TLDs allow local authorities to regulate usage, to control or minimise abuses such as cybersquatting, and to impose conditions appropriate to their own jurisdiction. The ccTLDs also allow for local registrations (and thus increased diversity) and for multinational corporations to have specialist local websites while retaining the same identifier. There is no necessity to overhaul the ccTLDs in line with any gTLD review.

ICANN

ICANN is the international private/public non-profit corporation responsible for managing and coordinating the domain name system (DNS), the Internet Protocol (IP) address space allocation, protocol identifier assignment, generic top level domain names (gTLDs), country code top level domain name (ccTLD) systems and root server system management. Its role is to ensure that each IP address and domain name is unique and that internet users are able to locate all addresses. ICANN maintains each domain name map.[17]

ICANN's stated objectives are to preserve the operational stability of the internet; to promote competition; to achieve broad representation of global internet communities; and to develop policy appropriate to its mission through bottom-up, consensus-based processes. Its mission is technical coordination.

ICANN accredits the domain name registrars by setting minimum standards for the performance of registration functions, vetting applicants in relation to those standards and managing accreditation agreements. While ICANN supervises registrars and the allocation of gTLDs, the actual registration of domain names is performed by specialist registrars.[18]

ICANN is not responsible for transactions and dealings on the internet such as financial transactions, consumer protection, internet content standards, spam, internet gambling, data protection or privacy, but is responsible for coordinating the management of the technical elements of the DNS to ensure 'universal resolvability': making sure that all users of the internet can find all valid addresses. This is the critical feature of the internet. Without it, competing systems with conflicting maps of the internet could evolve, giving rise to inconsistent results and mass confusion.

17 These services were originally performed under US government contract by the Internet Assigned Numbers Authority (IANA) and certain other bodies. See www.icann.org for more detail on this entire section of text.

18 The .biz domain is operated by NeuLevel. The .com and .net domains are operated by VeriSign Global Registry Services. The .info domain is operated by Afilias Limited. The .name domain is operated by Global Name Registry. The .org domain is operated by Public Interest Registry. The .pro domain is operated by RegistryPro. The .aero domain is sponsored by Société Internationale de Télécommunication Aéronautiques. The .coop domain is sponsored by DotCooperation LLC. The .museum domain is sponsored by the Museum Domain Management Association.

ICANN's board of directors includes representatives of more than a dozen nations. Its role is to oversee the policy development process and to respond to rapidly changing technologies and economies. ICANN's President directs an international staff, working from three continents. More than 80 governments closely advise the board of directors via the ICANN Governmental Advisory Committee.

ICANN has established market competition for gTLD registrations, partly by lowering domain name costs by 80 per cent, saving registrants more than US$1 billion annually in fees. ICANN has adopted guidelines for the Internationalised Domain Names (IDN), which is a map permitting the registration of domains in languages other than English. Participation in ICANN is open to all with an interest in global internet policy. ICANN has set up online forums through its website and holds public meetings throughout the world. Its Supporting Organisations and Advisory Committees have active mailing lists for participants.

ICANN implemented the Uniform Domain Name Dispute Resolution Policy (UDRP) in 1999. This is an efficient and cost-effective method of resolving disputes over domain names.

ICANN maintains relationships with governments, international treaty organisations, businesses, associations and individuals, and is the leading example of an alliance by the many constituents of the internet society.

InterNIC

The initial body coordinating the network was the NIC (Network Information Centre) at SRI (Stanford Research Institute in Menlo Park, California). This role was transferred to the Internet Assigned Numbers Authority (IANA) in 1972. In 1993 the US National Science Foundation created InterNIC to take over and manage the allocation of addresses and associated databases. InterNIC (Internet Network Information Center) remained the internet governing body primarily responsible for domain name and IP address allocations until September 1998, when this role was assumed by ICANN. The InterNIC website is now operated by ICANN to provide the public with information regarding domain name registration services.

InterNIC provides four significant services. First is the creation of the Whois registry.[19] This registry provides contact details for all domain names registered using TLDs. Second, it provides registrar contact details in an Accredited Registrar Directory. Third, users can file a complaint using an online Registrar Problem Report Form. Fourth, it allows the Whois data to be monitored: the public can report inaccuracies using a Registrar Problem Report Form.

19 See, respectively, www.internic.net/whois.html; www.internic.net/regist.html; reports.internic.net/cgi/registrars/problem-report.cgi and reports.internic.net/cgi/rpt_whois/rpt.cgi.

Whois

Whois is an internet protocol for users to find details of the owner of a domain name, an IP address or an autonomous system number (ASN) on the internet. It is the policy of ICANN and other domain name administrators to provide and maintain registers of domain name licensees and all registrars. This facility better informs the public, facilitates legitimate dispute resolution concerns, allows rapid resolution of technical problems and permits enforcement of consumer protection, trade mark and other laws. Where a potential conflict in use of domain name arises, initial contact can be made using the Whois database. Failure to maintain accurate details is regarded as a breach of the licence agreement, which may result in forfeiture of the domain name.

For TLDs, InterNIC maintains a Whois site at www.internic.net/whois.html. The country code Whois database is at www.iana.org/cctld/cctld-whois.htm.

Whois databases are maintained by auDA in Australia,[20] NeuStar in the United States,[21] Nominet in the United Kingdom,[22] the Office of the Domain name Commissioner in New Zealand[23] and INRegistry in India.[24]

ICANN Ombudsman

The first ICANN Ombudsman was appointed in November 2004.[25] The primary function of the Ombudsman is to provide an independent internal evaluation of complaints by members of the ICANN community about ICANN's staff or board or an ICANN constituent body. The Ombudsman serves as an objective advocate and can resolve complaints about unfair or inappropriate treatment, using ADR (alternative dispute resolution) techniques such as negotiation, facilitation, and shuttle diplomacy.

The Ombudsman's secondary functions include promoting an understanding of pertinent issues in the ICANN community; raising awareness of administrative fairness; and allowing the community to see the results of previous cases, by posting complaints and resolutions to a dedicated portion of the ICANN website. The postings will be generic, to protect the confidentiality and privilege of communicating. The Ombudsman also conducts outreach programs to increase consumer awareness of and raise the level of understanding of the Ombudsman process, and to encourage its use.[26]

The Office of Ombudsman reports only to the ICANN Board. The Ombudsman can only be removed from office by a 75 per cent majority vote of the board. The

20 See www.mywebname.com.au and www.auda.org.au/whois.
21 See www.whois.us.
22 See www.nic.uk.
23 See dnc.org.nz/whois.php.
24 See www.inregistry.in/whois_search.
25 Frank Fowlie; see www.icann.org/ombudsman. In relation to the Ombudsmen generally, see www.usombudsman.org.
26 See www.icann.org/ombudsman.

Ombudsman has access to all ICANN documentation and files and can interview staff and board members. The Ombudsman provides an annual report to the board, which is posted on the ICANN website.[27]

The Ombudsman charter provides that the Ombudsman will adhere to the standards of practice adopted by The Ombudsman Association.[28] The Ombudsman does not have jurisdiction over complaints concerning internal administrative matters, personnel issues, or issues relating to membership on the board, or over issues relating to vendor/supplier relationships.

The Ombudsman may decline jurisdiction over a complaint based on an action or inaction more than 60 days old in various circumstances, including where the complainant does not have sufficient personal interest in it, the complaint is repetitive, trivial, vexatious, frivolous, non-substantive, otherwise abusive or not made in good faith, or where the complainant engages in an outside legal process.[29]

Nexus requirements

Before registering a domain name some national domain administrators require a nexus with the jurisdiction or affiliation with the type of domain used.

gTLD nexus requirements

The single 'nexus' requirement for the open domain such as .com, .org and .net is that the applicant must complete a declaration that there is no infringement of another's rights in respect of the domain name. However, there is no process to check the legitimacy of such declarations. The .biz domain is restricted to businesses. The .com domain is a gTLD originally intended for commercial businesses around the world. The .info domain is an unrestricted domain for websites containing information about individuals, organisations and products. The .name domain is reserved for individuals. The .net domain is a gTLD for many types of organisations and individuals globally. Historically it was intended for and is still commonly used by Internet Service Providers (ISPs). The .org domain is unrestricted, but was intended to serve the non-commercial community. The .pro domain is restricted to certified professionals and related entities. The .aero domain is exclusively reserved for the aviation community. The .coop domain is restricted to use by *bona fide* cooperatives and cooperative service organisations that subscribe to specific cooperative principles (such as member ownership and control). The .museum domain was developed exclusively for the museum community.

27 See www.icann.org/ombudsman.
28 See www.usombudsman.org.
29 See www.icann.org/ombudsman/framework.html.

Persons who claim that they have been aggrieved have only two courses of action: they may use ICANN's UDRP[30] or take the issue to a court of law and rely on established legal principles.[31]

There have been multiple illegitimate registrations of domain names in the .com TLD. The main reason is the ability to register to anyone with any combination of letters and numbers without there being any effective nexus requirement. This has led to considerable speculation in registering domain names. Bad faith registrations are referred to as cybersquatting.[32]

.au nexus requirements

Of the 12 .au 2TLDs, the domains .com.au, .net.au, .org.au, .id.au and .asn.au are open to the general public. These domain name licences are allocated on a first come, first served basis and are fixed at two years. Renewal is automatic provided there is compliance with the nexus (eligibility) and allocation rules.[33]

It is not possible to reserve a domain name. Domain names must be at least two characters long, contain only letters (a–z), numbers (0–9) and hyphens (-) (or a combination of these), start and end with a number or a letter, and not contain hyphens in the third and fourth position.

The auDA directors have made a serious attempt to minimise and reduce the cybersquatting and typosquatting that are so prevalent in the .com domain by requiring some form of pre-existing interest in the name before registration is permitted.

To be eligible for a .com.au or .net.au domain name a registrant must be:

- an Australian registered company;
- trading under a registered business name in Australia;
- an Australian partnership or sole trader;
- a foreign company licensed to trade in Australia;
- an owner or applicant of an Australian Registered Trade Mark;
- an incorporated association in Australia; or
- an Australian commercial statutory body.

Additionally, the domain name must exactly match, be an acronym or abbreviation of the name or be closely and substantially connected to the registrant by referring to a product that the registrant manufactures or sells, a service that the registrant provides, an event that the registrant organises or sponsors, an activity that the registrant facilitates, teaches or trains, a venue that the registrant operates or a profession that the registrant's employees practise.

To be eligible for a .org.au domain name a registrant must be a charity or non-profit organisation operating in Australia. To be eligible to use .asn.au a

30 See www.icann.org/udrp.
31 See infra. There are serious concerns with the directions taken by the UDRP arbitrators. The remedies available by the legal system were not designed to meet the particular exigencies of domain names.
32 See infra.
33 See www.auda.org.au for more detail on this section of text.

registrant must be an incorporated association, a registered political party, an organisation registered under Workplace Relations legislation or a sporting or special interest club operating in Australia. For both .org.au and .asn.au the domain name must exactly match, be an acronym or abbreviation of the name or be closely and substantially connected to the registrant by referring to a service that the registrant provides, a program that the registrant administers, an event that the registrant organises or sponsors, an activity that the registrant facilitates, teaches or trains, a venue that the registrant operates or a profession that the registrant's members practise.

To be eligible for a .id.au domain name a registrant must be an Australian citizen or resident. The domain name must exactly match, be an acronym or abbreviation of the personal name or be closely and substantially connected to the registrant because the domain name is derived from one or more words of the registrant's personal name or includes one or more words of the registrant's personal name.

These nexus requirements have substantially reduced the number of domain name disputes in Australia. The rules cannot eliminate such disputes, as there may remain legitimate yet conflicting claims to a particular domain name. Bad faith registrations also cannot be eliminated. A potential registrant needs only to register a business name which technically complies with the auDA rules to nevertheless register a bad faith domain name.

A .au domain name may be revoked where the registrant is found to be in breach of the relevant policy rules: if the registrant makes a false warranty to the registrar, for instance.[34]

No test is made to determine if the same or a similar domain name already exists. Once other criteria are met, the domain name is issued. There are many registrars who publicly state that the nexus requirements and restriction ought to be lifted in favour of ICANN's more liberal approach.

The auDA Dispute Resolution Process (auDRP) for the open .au domain names was adopted in August 2002. It substantially mirrors the ICANN UDRP.

.nz nexus requirements

A registrant needs to be an identifiable individual over 18 years of age or properly constituted organisations. Registrants do not need to be based in New Zealand; nor does their domain name need to be hosted in New Zealand.[35]

.us nexus requirements

A registrant in the .us TLD must be:

- a natural person who is a US citizen, permanent resident or has a US primary place of domicile;

34 auDA policies are available at www.auda.org.au/policies.
35 DNC policy on Registering, Managing, and Cancelling Domain Names, dnc.org.nz/content/registering_managing_cancelling.pdf.

- a US corporation; or
- a foreign entity or organisation that has a *bona fide* presence in the United States.

It is a continuing requirement that all .us TLD registrants remain in compliance with the above requirements. NeuStar, the .us TLD administrator, requires initial certification, conducts a scan of selected registration request information and conducts spot checks on registrant information. The information required from registrants for the Whois database is:

- the domain name registered;
- the IP address and corresponding names of the primary and secondary name servers for the registered name;
- the registrar name and URL or, where appropriate, the identity of the delegated manager under whom the name is registered;
- the original creation date and term of the registration;
- the name and postal address of the domain name registrant;
- the name, postal address, email address, voice telephone number and where available the fax number of the name holder, technical contact and administrative contact for the name registered.

Failure to comply with these requirements results in the domain name being placed on a 30-day hold. If compliance is not corrected within the 30-day period, the registration will be cancelled and the name will be returned to 'available' status.[36]

.uk nexus requirements

The .uk rules rely entirely on the expressed intentions of the registrant. The .uk SLD Charter for the open SLDs sets out certain intentions regarding the class of applicant or use of registrations of the domain name. The Charter also states:

> However, we do not forbid applications, and will take no action in respect of registrations that do not comply with the SLD Charters. We may request certain information from you regarding your legal identity when you make an application for or seek to amend the registration of a Domain Name in the Open SLDs.[37]

.in nexus requirements

A registrant does not need to be based in India; nor does their domain name need to be hosted in India.[38]

36 See www.neustar.us/policies/docs/ustld_nexus_requirements.pdf.
37 See www.nominet.org.uk/registrants/legal/rules.
38 See www.registry.in/register/.

9

Domain name disputes

Domain names disputes arise for a number of reasons. The uniqueness of each domain name leads to the potential for conflicts with businesses and individuals with similar names. In addition to these relatively accidental conflicts, some parties deliberately register names to hijack businesses, extort money from or disrupt the operations of established organisations. Such an action is known as cyberpiracy, cybersquatting or typosquatting. This chapter analyses the nature of domain name disputes and remedies using national courts. The next chapter deals with the application of the Uniform Dispute Resolution Policy (UDRP) of ICANN and selected national domain name administrators, and with selected dispute resolution policies.

Cybersquatting

Speculating on the resale value of internet domains names has become a profitable pastime for internet devotees. For example, loans.com sold for US$3 million, cinema.com for US$800,000 and HappyBirthday.com for US$55,000.[1] These transactions are part of commerce and offend no legal precepts. ICANN and most National Domain Administrators accept the principle of first come first served. Also, many country code domain administrators prohibit the selling of domain names; those who breach such a rule risk forfeiting the domain name.[2] However, a person who registers a domain name identical to a well-known

1 For example, see www.greatdomains.com, domains.biz.com, www.dotcomagency.com and www.domainbarn.com.

2 For an example, see the policy by auDA at www.auda.org.au/policies.

or famous name for the purpose of subsequently demanding an exorbitant fee for 'transfer' (that is, sale) is referred to as a cybersquatter. The value to the cybersquatter lies in the fact that every domain name is unique. The term cybersquatting originates from 'squatting', typically used to refer to physically taking over 'tenements' and refusing to move. In many jurisdictions squatters attain real proprietary interests through continuous possession and the passage of time.[3]

Some registrations offend particular sensitivities: when the names of eminent political, scientific and religious persons, and the names of countries, cities or indigenous peoples are registered by people with no actual association with those names, for instance. The possibility of registering these identifiers as domain names is a consequence of the highly automated and efficient first-come, first-served system used for domain name registration, a system that does not involve any screening of domain name applications, but that has allowed tremendous growth in the use of the internet, and helped preserve universal connectivity.

In 1994 Joshua Quittner, a journalist, contacted the McDonalds Corporation to ask why they had not registered mcdonalds.com. Quittner then registered the domain name, in part to generate a story. He created the email address ronald@mcdonalds.com and asked readers for comments. Some suggested he use the mcdonalds.com site to promote vegetarianism, others suggested requesting an exorbitant price for its sale. Quittner published an article offering the name back to McDonalds in exchange for computer equipment for a local school. Under pressure from the corporate giant, InterNIC first agreed to revoke the registration, then changed its mind, leaving the registration with Quittner. McDonalds ultimately agreed to donate $3500 to purchase the equipment.[4] This is the first known instance of a person obtaining an economic benefit from a willing registrant.[5]

Dispute resolution

Paralleling the growth of the internet, the demand for domain names rose and disputes became inevitable. Many were innocent conflicts, such as Pitman's case (see below). However, cybersquatting became rampant. Many organisations were slow to recognise the importance and impact of a domain name as a business identifier. While some cybersquatters were a nuisance asking for sums too small to consider a formal legal challenge, others insisted on large sums amounting to extortion.

Prior to 1999, claimants challenging a particular registration only had recourse to the legal system. From 1999, starting with ICANN, domain name

3 Squatting can lead to the establishment of possessory title or historical title and is well established in common law jurisdictions.

4 Joshua Quittner, *You deserve a break today*, Newsday, October 1994, A05.

5 See James W. Marcovitz, 'ronald@mcdonalds.com – Owning a bitchin' corporate trademark as an internet address – infringement?', (1995) 17 *Cardozo L Rev* 85.

administrators initiated a compulsory Uniform Dispute Resolution Policy (URDP) for registrants. These two options remain in most jurisdictions today.

Remedies using the court process

Electronic commerce has been embraced by a broad range of bodies and individuals, for commercial and personal reasons. In the majority of situations standard legal principles can resolve the issues. However, new technologies may permit new relationships and interactions at a scale not previously anticipated. Domain name rights and issues are such areas, and inevitably disputes arise concerning interests in given domain names. The law did not have an obvious remedy for this, and courts have considered trade mark rights, misrepresentation issues and the tort of passing off. The latter particularly requires consideration, as the courts have moulded and shaped pre-existing elements to suit new factual situations.

Domain name rights were effectively unheard of prior to the 1990s. Courts were asked to apply existing legal principles which were not designed for and were often ill suited to this new world. Those who hold established business identifiers, plus goodwill, plus trade marks demanded the right to domain names which they regarded as belonging to them.

Cause of action

Whether innocently registered, as in *Pitman Training Limited v Nominet*,[6] or deliberately registered with the aim of extortion, as in *Panavision International v Toeppen*,[7] claimants maintained some form of proprietary right.[8]

In Pitman's case the court ruled that in the absence of any established cause of action and all other factors being equal, the right to a domain is first come first served. This concept is well established in most jurisdictions.

The first step in registering business names, company names and trade marks has traditionally involved determining whether or not the name has been previously registered. Typically, within the jurisdiction of a state or country, to operate a business under any name other than a personal name, the business or corporate name must be registered. Names that are identical to an existing business or corporation name anywhere within the jurisdiction will not be registered. A determination may also be made on whether the name is too similar, perhaps deceptively so, to an existing business or corporation name. Additional steps include whether the proposed name is misleading or offensive – there tend

6 [1997] EWHC Ch 367.

7 District Court California Case No. CV-96–03284-DDP (1996) and on appeal (1998) 141 F. 3d 1316 (9th Cir).

8 For early examples see *The Princeton Review Management Corp. v Stanley H Kaplan Educational Centre Ltd* 84 Civ 1604 (MGC) (SD NY 1994); *The Comp Examiner Agency Inc. 25th Century Internet Publishers v Juris Inc.* No. 96–0213-WMB US Dist LEXIS 20259 (CD Cal 1996).

to be words and phrases that cannot be used without consent.[9] Some jurisdictions will search and consider the trade marks register; others leave this to the applicant.

The major drawback to this system is that the allocation of names is jurisdictionally specific. An application for registration in the New South Wales will result in searches being made in every state and territory of Australia, but not elsewhere. However, today there can be identical names used in other countries. Expanding enterprises must deal with their encroachment on other jurisdictions, and find out whether or not their business identifier is already in use. For example, Burger King is an international chain of more than 11,000 of fast food restaurants in more than 60 countries. However, when it attempted to expand into Australia, it found that its business name was already trademarked by a person running a small takeaway food shop. In Australia Burger King is branded as Hungry Jacks.[10]

Domain name registration authorities have taken a simplistic first-come first-served approach to the allocation of the domain names (though they must also ascertain nexus requirements[11]). However, the internet encompasses one jurisdiction, and each domain name is a unique international address. There can be no identical domain names in different countries. This problem is exacerbated by the typically short domain names most registrants prefer. For example, there are businesses all over the world with names such as Smith Bros, Smith Dry Cleaning, Smith Cars, Smith Electronics and so on. Each may wish to shorten the domain name to the logical choice 'smith', but there can only be one smith.com, though there can be a number of smith.com.cc,[12] and full or extended use of the actual business name. The problem of limited names is partly resolved by registration protocols, such as country codes, nexus requirements and other options.

In 1997 the case of *Pitman Training Limited v Nominet*[13] was the first decision on internet domain names in the United Kingdom and described domain name competition:

> The Internet is a network of computer networks. A computer which is attached to an appropriate network can use appropriate software to communicate and exchange information quickly with any other computer on the network. In order to receive or to make available information on the Internet a domain name is needed. A domain name can be likened to an address. It identifies a particular Internet site. A particular domain name will only be allocated to one company or individual. It represents that company's computer site and is the means by which that company's customers can find it on the Internet.[14]

9 For example: Olympic, Red Cross, United Nations, UNICEF, Queen, Prince, Royal, Charity, Police and University.

10 There is a significant role for patent attorneys, and intellectual property lawyers in the planning of a proposed international product, service or business.

11 See supra.

12 Where cc can be any country code. Some nations do not use .com. For example, the United Kingdom and New Zealand use .co. The availability of .biz was intended to reduce this concern.

13 [1997] EWHC Ch 367.

14 [1997] EWHC Ch 367.

Two UK organisations held the right to use the trading name 'Pitman' and wanted to use the domain name pitman.co.uk. One had been established for 150 years and the other for 12 years. Due to an administrative error, the domain name was wrongly allocated to the Pitman Training Ltd. After receiving a complaint from Nominet, the UK naming committee transferred the domain name back. The plaintiff then commenced proceedings on the grounds of passing off, tortious interference with contract and (later) abuse of process. The court also considered a number of additional causes of action which might justify granting relief. The court concluded that no wrong had been committed, and no relief could be justified. The court held that, given that both parties had a prima facie right to use 'Pitman' in their business name, the first to register had the better interest in the domain name. Other jurisdictions have taken the same approach.[15] The law did not recognise the plaintiff's 150-year-old use as giving it any greater right than the registration by the defendant.

On its own, *Pitman v Nominet* was an open invitation to domain name squatters. An avalanche of cybersquatting occurred. Knowledgeable internet users made a grab for well-known word-based corporate marks, symbols and logos, relying on the initial ignorance of the future power and significance of domain names.

The majority of domain names are ordinary words, personal names, acronyms and abbreviations, a substantial number include a registered trade mark. In Australia, section 120(3) and (4) of the *Trade Marks Act 1995* (Cth) provides that a person infringes a registered trade mark if 'the trade mark is well known in Australia', 'is substantially identical with, or deceptively similar to' goods or services and is 'likely to be taken as indicating a connection between' the goods or services. Also, 'one must take account of the extent to which the trade mark is known within the relevant sector of the public, whether as a result of the promotion of the trade mark or for any other reason'.[16]

US experience

In *Panavision International v Toeppen*[17] in 1996, Toeppen had registered the domain name panavision.com before the multinational corporation of the same name had done so. Judge Trott stated: 'Panavision accuses Dennis Toeppen of being a "cyber pirate" who steals valuable trademarks and establishes domain names on the internet using these trademarks to sell the domain names to the rightful trademark owners.' Toeppen displayed 'visions of Pana', a town in Illinois, on the website. The court considered this action to be a devious ploy

15 See *Prince plc v Prince Sportswear Group* [1998] FSR 21 and *Fry's Electronics v Octave Systems Inc.* No. 95-CV-02525 (N.D. Cal. 1997).

16 See, generally, *Koninklijke Philips Electronics NV v Remington Products Australia Pty Ltd* (2000) 100 FCR 90.

17 District Court California Case No. CV-96–03284-DDP (1996) and on appeal (1998) 141 F. 3d 1316 (9th Cir).

designed to mask Toeppen's real intention: to extort a substantial payment from Panavision. Toeppen's argument – that his actions amounted to legitimate commerce – was rejected. Toeppen offered to settle the dispute for a payment of $13,000 and promised not to acquire any other internet address which may be claimed by Panavision. When Panavision refused this offer Toeppen registered panaflex.com, displaying the word 'Hello' on the website. Panaflex was another trade mark belonging to Panavision. The court upheld Panavision's 'right' to the domain name. Toeppen had also registered many other names, such as: aircanada.com; anaheimstadium.com; arriflex.com; australiaopen.com; deltaairlines.com; eddiebauer.com; flydelta.com; frenchopen.com; lufthansa.com; northwestairlines.com; and yankeestadium.com. Toeppen had also previously attempted to 'sell' domain names for other trade marks, such as intermatic.com to Intermatic Inc. for $10,000 and americanstandard.com to American Standard Inc. for $15,000 clearly increased the court's cynicism of Toeppen's defence.[18]

Panavision's cause of action was based on the US Federal *Trademark Dilution Act 1995*,[19] so its principles are not universally applicable. The legislation made a specific distinction between trade mark infringement and trade mark dilution. Trade mark infringement typically arises where a third party uses the trade mark in a manner likely to cause confusion, most usually where the same class of goods is involved. The trade mark could be used for a different class of goods without causing confusion. However, where particularly well-known trade marks are involved, consumers may perceive an association between the two uses. The use by the third party may be said to impact the reputation of the trade mark. Secondly, the consumer will begin to dissociate the trade mark from its original use, thus lessening the strength of the established association between the trade mark and the original goods. Trade mark infringement is universally incorporated into trade mark law; there are varying degrees of trade mark dilution law in other jurisdictions.[20] The United States adopted specific legislation.

The identifier components of most domain names are not registered trade marks. Complainants must rely upon some other cause of action to persuade a court to provide relief. US cybersquatting cases since 1999 have had the benefit of the *Anti-Cybersquatting Consumer Protection Act* (US).[21] This Act is a US federal statute that took effect on 29 November 1999. It is intended to give the owners of trade marks, service marks, personal names and business identifiers legal remedies against persons who register or use domain names, 'in bad faith', that are identical or confusingly similar to a trade mark or service mark. A trade mark may be a mark, word, name, symbol, colour, sound, smell etc which identifies particular products or services. A service mark is a form of trade mark typically

18 The opinion stated that there were more than 100 such registrations by Toeppen.

19 15 USC § 1125(c) and also the California anti-dilution statute, California Business and Professions Code § 14330. See also *Playboy Enterprises Inc. v Calvin Designer Label* 985 F. Supp. 2d 1220 (1997).

20 For example *Trade Marks Act 1995* (Cth) s120; *Trade Marks Act 2002* (New Zealand) s89(1)(d); *Trade Marks Act 1999* (India) s29; *Trade Marks Act 1994* (UK) s10.

21 15 USC s1125(d)(2)(C).

used in the sale or advertising of services. If a service mark is famous, the same remedies are available if the domain name is identical to, confusingly similar to or dilutive of the mark. In the US, trade mark and service mark rights can be acquired by registration or through usage at common law. Generally, the first to either use a mark in commerce or file an intent to use application with the Patent and Trademark Office has the ultimate right to use and registration. The Act amends the Lanham Act[22] by adding cyberpiracy prevention:

> A person shall be liable in a civil action by the owner of a mark, including a personal name which is protected as a mark under this section, if, without regard to the goods or services of the parties, that person– (i) has a bad faith intent to profit from that mark, including a personal name which is protected as a mark under this section; and (ii) registers, traffics in, or uses a domain name . . . [23]

The circumstances required are typically where the domain name is identical or confusingly similar to a mark. The provision has an inclusive definition of 'bad faith' aimed at cybersquatting practices. It covers lack of a sufficient *bona fide* purpose, an intent to extort money from a person with legitimate rights and interests, and any intent to divert consumers, harming goodwill, or to tarnish or disparage the mark, and the likelihood of confusion as to the source, sponsorship, affiliation, or endorsement of the website. Under this legislative remedy a court may order the forfeiture or cancellation of the domain name or the transfer of the domain name to the owner of the mark.

The Anti-Cybersquatting Consumer Protection Act also included a provision relating to the protection of individuals from cyberpiracy:

> Any person who registers a domain name that consists of the name of another living person, or a name substantially and confusingly similar thereto, without that person's consent, with the specific intent to profit from such name by selling the domain name for financial gain to that person or any third party, shall be liable in a civil action by such person.[24]

The remedies that a court may award are injunctive relief, including the forfeiture or cancellation of the domain name or the transfer of the domain name to the plaintiff, and costs.

International approaches

In *Sydney Markets Limited v Sydney Flower Market Pty Limited*,[25] the Federal Court of Australia stated:

22 *Lanham (Trademark) Act* (15 USC 22) is US legislation that contains the federal statutes of trade mark law. The Act prohibits trade mark infringement, trade mark dilution, and false advertising. The Lanham Act defines the statutory and common law boundaries to trade marks and service marks.
23 *Trademark Act of 1946* (15 USC 1125) s43.
24 15 USC 1129.
25 [2002] FCA 124.

> where two domain names are sufficiently similar so as to make it difficult for a member of the public to know in advance exactly which site they will be taken to (as, for example, where the only difference between them is the presence or absence of '.au'), there is considerable scope for the public to be misled.[26]

In *Campomar Sociedad, Limitada v Nike International Limited*[27] the High Court of Australia (in *obiter*) stated that subsections 120(3) and (4) have:

> extended the infringement action to restrain activities which are likely adversely to affect the interests of the owner of a 'famous' or 'well-known' trade mark by the 'dilution' of its distinctive qualities or of its value to the owner. The 'dilution' theory of liability 'does not require proof of a likelihood of confusion'; rather, what is protected is 'the commercial value or 'selling power' of a mark by prohibiting uses that dilute the distinctiveness of the mark or tarnish the associations evoked by the mark.[28]

The *Trade Marks Act 2002* (NZ) s89(1)(d) provides:

> (1) A person infringes a registered trade mark if the person does not have the right to use the registered trade mark and uses in the course of trade a sign . . .
>
> (d) identical with or similar to the registered mark in relation to any goods or services that are not similar to the goods or services in respect of which the trade mark is registered where the mark is well known in New Zealand and the use of the sign takes unfair advantage of, or is detrimental to, the distinctive character or the repute of the mark.

This may be contrasted with the United Kingdom, Singapore and Australian approaches.[29] New Zealand and Singapore use the expression 'well-known' trade marks; the UK Act refers to a trade mark which has a 'reputation' in the United Kingdom. The Australian provision deals with marks that are well known 'within the relevant sector of the public'. It may be reasoned that a mark may only be well known in New Zealand where a significant part of the general public know of the trade mark. The words 'take unfair advantage of, or is detrimental to the distinctive character or the repute of the mark' are from the UK Act. As a result, the UK case law may be most relevant. For example, the UK courts have held that detriment may be caused by erosion of distinctiveness or tarnishing of distinctiveness.[30] In Singapore the plaintiff must prove that the interests 'of the proprietor are likely to be damaged by such use'.[31] In Australia the plaintiff must prove that the interests 'of the registered owner are likely to be adversely affected'.[32] In New Zealand and the United Kingdom the emphasis is on the harm to the character or reputation of the mark. In Australia and Singapore it is the interests of the owner of the well-known mark that are primary.

26 [2002] FCA 124, para 149.
27 [2000] HCA 12.
28 [2000] HCA 12, paras 42–43.
29 *Trade Marks Act 1994* (UK), *Trade Marks Act 1998* (Singapore) and *Trade Marks Act 1995* (Cth).
30 *Premier Brands v Typhoon* [2000] RPC 477.
31 *Trade Marks Act 1998* (Singapore) s27(3)(d).
32 *Trade Marks Act 1995* (Cth) s120(3)(d).

In India, the Supreme Court has held that Indian trade mark laws provides direct protection for domain names. In *Satyam Infoway Ltd v Siffynet Solutions Pvt Ltd*,[33] the plaintiff was one of India's largest internet service providers (ISPs) and claimed to have formulated the word 'SIFY' from its corporate name, 'Satyam Infoway', and to have established a global reputation and goodwill in the name. Its domain name was www.sify.com. The defendant used the domain names www.siffynet.net and www.siffynet.com. The plaintiff obtained a temporary injunction in the District Court on the grounds of prior user of the trade name. The Bangalore High Court reversed this, stating that the court must consider where the balance of convenience lies and that as there was no similarity between the two businesses there was no question of confusion. Satyam Infoway appealed to the Supreme Court.

The Supreme Court considered that domain names could qualify as services and obtain protection under paragraph 2(zb) of the Indian *Trade Marks Act 1999*, which, like the UK Trade Marks Act, defines a trade mark as 'a mark capable of being represented graphically', 'distinguishing goods or services of one person from those of others'. A mark is defined to include a name and an abbreviation of a name. The Indian Act widens the definition of 'services' so that it includes business, commercial and industrial activities. The Supreme Court reinstated the injunctive relief. It regarded the defendant's actions as an 'albeit *prima facie* . . . dishonest adoption' of the plaintiff's trade name.

The courts' decisions in trade mark domain name cases have been criticised for construing the principles of infringement, passing off and dilution too generously. It is argued that general common law principles have been either modified or negated to suit the requirement of trade mark owners.[34] For example, there is a general rule that a person has the right to use his or her own name for the purpose of honest trading.[35] This natural right has been consistently negated in favour of the reputation of the commercial enterprises.[36] Decided cases imply that the use of a domain name will override an individual's right if a prominent, well-known business possesses a similar name.[37]

In *Re Krupp*,[38] the defendant had registered the domain name krupp.com for his internet services in 1995, using his first name, 'krupp'. The word 'krupp' is also used as a trade mark by leading German steel company who wished to obtain the domain name 'krupp.com'. The defendant's use of his first name was not allowed: the commercial reputation of the steel manufacturer was recognised as overruling his right to use his own name. The court found that the outstanding commercial reputation of the plaintiff meant it had a right not to tolerate other

33 (2004) (28) PTC 566 (SC).
34 Adrian Wolff, 'Pursuing domain name into uncharted waters: Internet domain names that conflict with corporate trade marks', (1997) 34 *San Diego Law Review* 1463, 1468.
35 See *Marengo v Darly Sketch & Sunday Graphics* Ltd (1948) RPC 242.
36 Anri Engel, 'International domain name disputes: Rules and practice of the UDRP', (2003) 25 *EIPR* 351, 354.
37 See Andrea F. Rush, 'Internet domain protection: A Canadian perspective', (1997) 11 *IPJ* 1, 11.
38 [1999] *EIPR* N24.

firms bearing the same name weakening the value of its mark. This decision, like many other cases, also highlighted the fact that in most of the cases, courts failed to consider even the basic principles of passing off – such as the effects or likelihood of confusion – and upheld the plaintiff's claim even though parties were in completely different industries (in this case a steel producer and an ISP).

Domain name passing off

The expression 'domain name passing off' was coined to refer to a new hybrid of the traditional tort of passing off. The tort of passing off is concerned with the protection of business reputation and with the protection of consumers from deceptive and misleading trading activities. The underlying elements are active conduct by a defendant which causes actual damage to the plaintiff's goodwill or business. However, in an attempt to provide a remedy for domain name disputes, the courts have exercised great flexibility to reshape the passing off concept so as to yield a remedy.

Passing off involves appropriating a reputation that belongs to another and thus misrepresentation of goods, services or goodwill, typically in a business situation. In 1896 Lord Halsbury, in *Reddaway v Banham*, stated, 'nobody has any right to represent his goods as the goods of somebody else'.[39] In *Reckitt & Colman Products Ltd v Borden Inc. (No. 3)*[40] Lord Oliver of Aylmerton said, 'The law of passing off can be summarised in one short general proposition – no man may pass off his goods as those of another.'[41]

The leading case on passing off has been the House of Lords decision of *Warnink v Townsend & Sons*,[42] where Lord Diplock defined the necessary elements:

(1) A misrepresentation,
(2) made by a trader in the course of trade,
(3) to prospective customers of his or customers or ultimate consumers of goods or services supplied by him,
(4) which is calculated to injure the business or goodwill of another trader (in the sense that this is a reasonably foreseeable consequence) and
(5) which causes actual damage to a business or goodwill of the trader by whom the action is brought or (in a *quia timet* action) will probably do so.[43]

In the High Court of Australia Isaacs J said, in *Orange Crush (Australia) Ltd v Gartrell*:

39 [1896] AC 199, 204.
40 [1990] 1 WLR 491.
41 [1990] 1 WLR 499.
42 [1979] AC 731; known as the Advocaat case. See also *Reckitt & Colman Products Ltd v Borden Inc. (No. 3)* [1990] 1 WLR 491 and *Hodgkinson & Corby Ltd v Wards Mobility Services Ltd* [1994] 1 WLR 1564.
43 [1979] AC 731 at 742.

> the right protected by the tort of passing off is not a property in the mark, name or get-up, but in the business or goodwill likely to be injured by the misrepresentation conveyed by the defendant's use of the mark, name or get-up with which goods or services of another are associated in the minds of a particular class of the public. The plaintiff's property in the goodwill is in its nature transitory and exists only so long as the name is distinctive of the plaintiff's services in the eyes of the class of the public.[44]

This judgment was written in 1928, long before the internet and domain name rights were even contemplated.

Oggi Advertising Ltd v McKenzie

In 1998 the High Court of New Zealand in *Oggi Advertising Ltd v McKenzie*,[45] used passing off as the underlying cause of action. McKenzie registered oggi.co.nz. Oggi Advertising Ltd was an established advertising company preparing to set up a website. The company sought an injunction claiming a breach of the tort of passing off. McKenzie argued that he had registered the domain name on behalf of a person he allegedly met on the net named Elliott Oggi. However, no such person could be found and no further evidence of this person's existence was presented. The website displayed phases connected with the plaintiff's business (advertising): 'Open your eyes – 80 million people can drive past – every day – the changing face of advertising'. This site was removed immediately after the complaint was made. The initial registrant's name was fictitious, but was changed to 'Elliott Oggi' four days after service of the legal proceedings.[46]

Baragwanath J stated that as a matter of law the plaintiff, in seeking the injunction to restrain the use of the domain name by the defendant, needed to establish the five elements of passing off listed by Lord Diplock. His Honour held that first, McKenzie had 'floated into Cyberspace' a misrepresentation that McKenzie was associated with Oggi Advertising Ltd. Second, the defendant projected a 'clear business implication'. Third, New Zealand users of the web were prospective customers. Fourth, the defendant's actions diverted business from Oggi Advertising Ltd, damaging its goodwill. Fifth, the defendant's conduct 'would probably cause actual damage' to Oggi Advertising Ltd.

McKenzie was ordered to assign the domain name to Oggi and remove any link whereby users could access his site using the name Oggi.

This may have provided a remedy on the facts of this case, but upon any analysis it could prove to severely limit the possible remedies for other domain name disputes. His Honour made no attempt to extend passing off concepts any further. A significant proportion of cybersquatters may not intend to trade off the goodwill of the name of the target. They may not intend to trade, or to promote

44 (1928) 41 CLR 282, 292.

45 (1999) 44 IPR 661; 6 NZBLA 102, 567, High Court of New Zealand, per Baragwanath J.

46 It has been suggested that the name 'Elliott Oggi' was deliberately chosen as a play on both the plaintiff's name, Oggi, and the plaintiff's counsel's name, Clive Elliott. See C Elliott and B Gravatt, 'The Oggi case: New Zealand domain name grabber dealt to', (1998) *Trademark World*, August, 19.

their activities on a website, or to set up a web page after registration. Should Baragwanath J's approach in Oggi's case be strictly applied, few plaintiffs would have a remedy.

Should a cybersquatter simply register a 'well-known' name and take no other steps, none of the five passing off elements would be met. First, there would be no 'floating into Cyberspace' of a misrepresentation of association with the plaintiff. Second, there would be no projection of a 'clear business implication'. Third, there would be no prospective customers. Fourth, there would be no diversion of business from the plaintiff, damaging goodwill. Fifth, there would be no probability of actual damage to the plaintiff.

Such a factual situation arose in the UK case *Harrods Ltd v UK Network Services Ltd*.[47] Harrods challenged the registration of harrods.com by a third party. Despite the absence of any trading on the goodwill of Harrods Ltd, the court ignored the elements of passing off and granted summary judgment, ordering the transfer of the domain name. However, this was an uncontested case, and thus devoid of argument.

Marks & Spencer v One in a Million

In *Marks & Spencer v One in a Million*[48] the English High Court was faced with a similar situation to the Oggi case, and pioneered the extension of the tort of passing off in a manner not previously contemplated. This is an example of the courts finding and moulding a remedy to suit new circumstances. As in *Harrods Ltd v UK Network Services Ltd*, the defendant had made no attempt to use the domain name. It had not placed any material online to be accessed via the domain name. The English High Court refused to allow the absence of a precedent to foil the plaintiff's 'right' to a remedy. The defendant registered several domain names in the hope of on-selling them to interested parties. One such domain name was marksandspencer.com, Marks & Spencer being a well-known chain of department stores. A cynical observer may remark that English judges would not permit such an established name as Marks & Spencer to be usurped. The defendant did not use the domain name in any form of trade; nor did it set up a website. It had merely registered the domain name. A strict application of the first-come first-served rule effectively blocked Marks & Spencer from acquiring registration of the name. Some of the other names registered by the defendant included: britishtelecom.co.uk, macdonalds.co.uk, burgerking.co.uk, marksandspencer.co.uk, motorola.co.uk, sundaytimes.co.uk and thetimes.co.uk.

47 (1997) EIPR D-106.

48 *Marks & Spencer plc, Ladbrokes plc, J Sainsbury plc, Virgin Enterprises Ltd, British Telecommunications plc, Telecom Securicor Radio Ltd v One in a Million* [1997] EWHC Patents 357; and on appeal [1999] 1 WLR 903; [1998] EWCA Civ 1272. This case is also known as *British Telecom v One in a Million*, as the defendant, in a manner reminiscent of Toeppen, had speculatively registered a number of domain names. See also *Hoath v Connect Internet Services* [2006] NSWSC 158, *Australian Stock Exchange Ltd v ASX Investor Services Pty Ltd* (QSC Burchett J (1999)) and *connect.com.au Pty Ltd v GoConnect Australia Pty Ltd* [2000] FCA 1148.

Sumpton DJ,[49] at first instance, considered and discussed Lord Diplock's five passing off elements. To provide a remedy, his Honour expanded the meaning of passing off and its application to a level not previously contemplated. In particular, his Honour applied the principles of a *quia timet* injunction to restrain the '*threat* of passing off'. None of the five passing off elements was met. The judgment contained the broadest proposition for challenging a domain name under the tort of passing off.

His Honour described the 'essence of the tort of passing off' as a misrepresentation to the public, whether or not intentional, liable to lead them to believe that the goods and services offered are those of the plaintiff. He added that the tort is also committed 'by those who put or authorise someone to put an "instrument of deception" into the hands of others'.[50] However, His Honour added that the mere creation of this 'instrument of deception', without some further act, is not passing off: 'There is no such tort as going equipped for passing off.'[51] The mere registration of 'a deceptive Internet domain name is not passing off',[52] but a *quia timet* injunction may be ordered to restrain the threat of passing off rather than the actual tort: '[E]ven a final injunction does not require proof that damage will certainly occur. It is enough that what is going on is calculated to infringe the Plaintiff's rights in future.'[53]

Quia timet injunctions are distinguished from others in that they may be granted against apprehended wrongs. In *Fletcher v Bealey*,[54] Pearson J stated there are 'at least two necessary ingredients' for equity courts to grant a *quia timet* injunction:

> There must, if no actual damage is proved, be proof of imminent danger, and there must also be proof that the apprehended damage will, if it comes, be very substantial. I should almost say it must be proved that it will be irreparable, because, if the danger is not proved to be so imminent that no one can doubt that, if the remedy is delayed, the damage will be suffered, I think it must be shown that, if the damage does occur at any time, it will come in such a way and under such circumstances that it will be impossible for the plaintiff to protect himself against it if relief is denied to him in a *quia timet* action.[55]

With respect, it is questionable whether the application of these principles should lead to the granting of a *quia timet* injunction. Actual damage was not imminent. The registrant may never have used the domain name registration at all and the evidence indicated that there was no intention to activate any website. His Honour reasoned that there was 'only one possible reason' to register the marksandspencer domain address: 'to pass himself off as part of that group or his products

49 The court comprised Jonathan Sumpton QC sitting as a Deputy Judge of the High Court.
50 At para 16, his Honour, citing *Singer v Loog* (1880) 18 Ch D 395, 412, cited with approval by Lord Macnaughten in *Camel Hair Belting* [1896] AC 199, 215–16.
51 *Marks & Spencer plc v One in a Million* [1997] EWHC Patents 357, para 19.
52 *Marks & Spencer plc v One in a Million* [1997] EWHC Patents 357, para 19.
53 *Marks & Spencer plc v One in a Million* [1997] EWHC Patents 357, para 19.
54 (1885) 28 Ch D 688.
55 (1885) 28 Ch D 688, 698.

off as theirs'.[56] His Honour then contradicted this statement by acknowledging counsel's submission of two additional uses not involving passing off: the resale of the domain name to Marks & Spencer, and blocking the use of the name by Marks & Spencer in order to induce them to pay. Neither of these activities constitutes passing off. His Honour regarded the domain names as worthless to the defendant except as a negotiating tactic to extort payment from the plaintiff. He expressed concern that there is a danger of deception simply through registration, but determined that no relief is available for passing off which has not yet occurred:

> Someone seeking or coming upon a website called *marksandspencer.co.uk* would naturally assume that it was that of the Plaintiffs . . . Any person who deliberately registers a domain name on account of its similarity to the name, brand name or trade mark of an unconnected commercial organisation must expect to find himself on the receiving end of an injunction to restrain the threat of passing off, and the injunction will be in terms which will make the name commercially useless to the dealer.[57]

His Honour ordered a *quia timet* injunction against the defendant. Note that the arguments of Sumpton DJ are directed at famous and well-known identifiers.

Not surprisingly, One in Million appealed this new interpretation of the threat of passing off. The Court of Appeal dismissed the appeal, stating that merely registering the domain name has the potential to amount to passing off:

> [Counsel] submitted that mere registration did not amount to passing-off. Further, Marks & Spencer Plc had not established any damage or likelihood of damage. I cannot accept those submissions. The placing on a register of a distinctive name such as marksandspencer makes a representation to persons who consult the register that the registrant is connected or associated with the name registered and thus the owner of the goodwill in the name.[58]

The Court of Appeal regarded the mere registration of the domain name as an erosion of the exclusive goodwill in the name which damaged or was likely to damage Marks & Spencer.

In an attempt to fashion a remedy the court drew a parallel with the cases to grant injunctive relief where a defendant is equipped with or is intending to equip another with an instrument of fraud. As with Sumpton DJ, the Court of Appeal limited its reasoning to famous and well-known domain names (which would 'inherently lead to passing-off'). The factors to consider included the similarity of the identifier to the plaintiff's name, the intention of the defendant, and the type of trade. If the intention of the defendant was to appropriate the goodwill, a clear case of passing off would be made out 'even if there is a possibility that such an appropriation would not take place':[59]

56 *Marks & Spencer plc v One in a Million* [1997] EWHC Patents 357, para 20.
57 *Marks & Spencer plc v One in a Million* [1997] EWHC Patents 357, paras 20 and 21.
58 Lord Justice Stuart-Smith, Lord Justice Swinton Thomas and Lord Justice Aldous.
59 [1998] EWCA Civ 1272.

> If, taking all the circumstances into account the court should conclude that the name was produced to enable passing-off, is adapted to be used for passing-off and, if used, is likely to be fraudulently used, an injunction will be appropriate. It follows that a court will intervene by way of injunction in passing-off cases in three types of case. First, where there is passing-off established or it is threatened. Second, where the defendant is a joint tortfeasor with another in passing-off either actual or threatened. Third, where the defendant has equipped himself with or intends to equip another with an instrument of fraud. This third type is probably mere *quia timet* action.[60]

The court proceeded on the basis that Marks & Spencer denotes Marks & Spencer plc and nobody else. A search on the 'whois' database reveals the defendant as the registrant. This, the court considered, would lead a substantial number of persons to conclude that the defendant must be connected or associated with Marks & Spencer plc, and that 'amounts to a false representation which constitutes passing-off'.[61] The court reasoned that the defendant's purpose was to threaten use and disposal, sometimes explicitly and on other occasions implicitly.

Developments

Architects (Australia) Pty Ltd (trading as Architects Australia) v Witty Consultants Pty Ltd[62] applied the Marks & Spencer case. The defendant registered architectsaustralia.com.au but unlike the Marks & Spencer case, the defendants argued that the words 'Architects Australia' are general dictionary terms combining a profession and location without any necessary reference to the plaintiff. Chesterman J of the Supreme Court of Queensland considered the issue of domain name passing off. His Honour stated:

> The plaintiff must establish:
> 1. That it holds goodwill or reputation in a specific trade or business;
> 2. That the defendant has misrepresented, intentionally or unintentionally, that a connection exists between the defendant or the defendant's goods, services or business, and the plaintiff or the plaintiff's business; and
> 3. That the plaintiff has suffered, or is under threat of, damage either by diversion of custom, diminished reputation or some other like form of damage.[63]

His Honour regarded the essence of an action for passing off to be the protection of goodwill 'as embodied in the warmth of public sentiment towards its product, service, name or other feature'.[64] The words 'Architects Australia' were held to be 'sufficiently fancy' to be distinctive of the plaintiff's business to amount to an instrument of fraud, and the court ordered deregistration. The placement

60 [1998] EWCA Civ 1272.
61 [1998] EWCA Civ 1272.
62 [2002] QSC 139.
63 [2002] QSC 139 para 13.
64 [2002] QSC 139 para 14.

of a disclaimer on the defendant's website was ineffective because in the view of the court it referred only to the plaintiff's company name, not to its trading name.

In *Yahoo Inc. v Akash Arora*[65] the defendants registered yahooindia.com. The plaintiff was the owner of the well-established internet site Yahoo, as well as the trade mark and the domain name yahoo.com. The Yahoo India site had also copied Yahoo's format, contents, layout, colour scheme and source code. Based on this usage the Delhi High Court found that the defendant had 'appropriated' the identifier 'Yahoo', and granted an injunction based on the tort of passing off and trade mark infringement restraining the defendant from dealing in the identifier.[66]

A single judge of the Bombay High Court took the domain name passing off developments a step further. In *Rediff Communication Ltd v Cyberbooth*[67] the Court stated that a 'domain name is more than an Internet Address and is entitled to equal protection as trade mark'. This was an example of typosquatting. The plaintiff was the owner of the well-known portal rediff.com. The defendant had registered radiff.com. The Court held that there was a clear 'intention to deceive' and that the only purpose of the registration was to trade on the goodwill and reputation of the plaintiff.

In *Qantas Airways Limited v The Domain Name Company Limited*,[68] the defendant registered the domain name qantas.co.nz. It then attempted to sell the name to Qantas. The New Zealand High Court was quick to condemn this and ordered the defendant to de-register the name. The Court found that such actions were a deliberate blocking of the lawful exploitation of the name and a fraudulent use of Qantas's goodwill.

In *Investment India Ltd v ICIC*[69] the plaintiff had registered investmartindia.com; the defendant had registered investsmartindia.com. Notwithstanding the similarity, the court refused to enjoin the defendant's use. The court noted that the identifier was a combination of three descriptive terms, 'invest', 'smart' and 'India' and could be distinguished from such cases as the Yahoo India case where the unique identifier 'Yahoo' was at the centre of the issue.

Trade Practices Act relief

A statutory form of relief, similar to the tort of passing off, is contained in the *Trade Practices Act 1974* (Cth), sections 52 and 53:

65 (1999) PTC (19) 210 (Delhi).
66 See *Titan Industries Ltd v Prashanth Koorapati & Ors* (Application 787 of 1998 in action 179), Delhi High Court, finding a *prima facie* case of passing off where the trade mark application was pending.
67 AIR (2000) Bom 27.
68 (2000) 1 NZECC 70005.
69 Mumbai High Court, Action 1040 of 2000.

52(1) A corporation shall not, in trade or commerce, engage in conduct that is misleading or deceptive or is likely to mislead or deceive.

53 A corporation shall not, in trade or commerce, in connexion with the supply or possible supply of goods or services or in connexion with the promotion by any means of the supply or use of goods or services: . . .

(c) represent that goods or services have sponsorship, approval, performance characteristics, accessories, uses or benefits they do not have;

(d) represent that the corporation has a sponsorship, approval or affiliation it does not have . . .

Section 52 applies without intent and gives rise to civil action. Section 53 requires a form of intent. Civil action may be based on the section. In addition, a breach of section 53(c) and (d) conduct is an offence.[70]

State Fair Trading legislation contains similar provisions in relation to misleading or deceptive conduct by non corporations such as individuals.

Where a person claims loss or damage as a result of misleading or deceptive conduct, that person may commence an action under section 82, apply for orders to rescind or modify a contract under section 87 or apply for an injunction under section 80.

In *Kailash Center for Personal Development Inc. v Yoga Magik Pty Limited*[71] the Federal Court of Australia followed the Marks & Spencer case. Jonn Mumford, also known as Swami Anandakapila Saraswati, had developed a reputation with regard to yoga and Eastern spirituality. He was approached to lend his name and reputation to the promotion of several websites. After a falling out between the parties Mumford demanded the removal of meta tags for websites which included his names and the cancellation of the registration the domain names omkarakriya.com and jonnmumford.com. Despite the prior authorisation and agreement, the Federal Court held that that there had been misleading and deceptive conduct within the meaning of section 52 of the *Trade Practices Act 1974* (Cth) and ordered deregistration of the domain names. The court regarded the use of Mumford's name and pseudonyms as a misappropriation.

In *CSR Limited v Resource Capital Australia Pty Limited*,[72] Hill J, in the Federal Court of Australia, stated that anyone seeing the domain name would assume that CSR was the real owner of the domain name. CSR is a well-known Australia icon. It produces 4 per cent of the raw sugar traded on the world market and 40 per cent of Australia's total raw sugar. 'CSR' and 'CSR Sugar' are registered trade marks. The defendant registered the domain names csrsugar.com and csrsugar.com.au and wrote to CSR offering to sell the domain names. The defendant had previously written to another sugar company offering to sell domain

70 See *Trade Practices Act 1974* (Cth) s75AZC.
71 [2003] FCA 536.
72 [2003] FCA 279.

names for $5000 to $10 000 for 'introductory purposes'.[73] Referring to the One in a Million case, the court stated:

> a cyber squatter who registers a name intending to sell that name to the owner of a trade mark or threaten a sale to a competitor if the owner does not come up with the money may have registered the name as an instrument of fraud, and thereby be guilty of the tort of passing off as was held by the Court of Appeal in England.[74]

His Honour held that 'the act of obtaining registration' of both domain names amounted to misleading and deceptive conduct pursuant to section 52 of the *Trade Practices Act 1974* (Cth), and constituted a representation that CSR and the defendant were affiliated. Further, his Honour held that the sole director of RCA was in breach of the equivalent provisions in *Fair Trading Act 1987* (NSW). Hill J ordered that the domain names be transferred to CSR and restrained both the defendant and the director from registering any other domain name where 'CSR' appears or in connection with sugar. His Honour ordered that a copy of the judgment be forwarded to the Australian domain name registration authority.

Australian Competition and Consumer Commission v Chen[75] demonstrates international enforceability with domain name disputes. The defendant was a resident of the United States and had registered, among others, sydneyopera.org. Chen falsely represented that the site was affiliated with the Sydney Opera House. Persons used their credit cards, believing they were purchasing tickets for events at the Sydney Opera House. The plaintiff did not raise passing off but relied on the consumer protection provisions of the *Trade Practices Act 1974* (Cth). Chen's conduct was held to be misleading or deceptive or likely to mislead or deceive (in contravention of section 52) and to include misleading representations (in contravention of subsections 53(c) and (d)). His Honour expressed concern that 'cross-border fraud and misleading conduct, particularly through the Internet, is a growing problem for the international community'. Consumer protection and law enforcement agencies have established mechanisms for international cooperation to protect consumers. The Australian Competition and Consumer Commission (ACCC) informed the court that it would bring the orders to the attention of the US Federal Trade Commission and request its assistance. The extra-territorial enforcement of federal orders is ordinarily a matter for the domestic law of the country in which the orders are sought to be enforced. At common law, four conditions must be satisfied if a foreign judgment is to be recognised by an Australian court: first, the foreign court must have exercised a jurisdiction which Australian courts will recognise; second, the foreign judgment must be final and conclusive; third, there must be an identity of parties;

73 A separate dispute against the defendant (RCA) dealing with www.tullochwines.com had been resolved before the Administrative Panel of WIPO Arbitration and Mediation Centre at the complaint of JY Tulloch Pty Ltd. The panel found that RCA acted in bad faith and that there was 'no clearer case of cybersquatting'.
74 [2003] FCA 279, para 42.
75 [2003] FCA 897.

and fourth, if based on a judgment *in personam*, the judgment must be for a fixed debt.

In *Hoath v Connect Internet Services*,[76] White J approved the contention that the right to sue for passing off was not dependent upon any title to a domain name, but on whether the plaintiff had developed goodwill or a reputation such that the defendant's use of the domain name, without the plaintiff's consent, misrepresented to the public that the defendant's business was the plaintiff's business, or that there was a trade connection between them.[77]

In *Satyam Infoway Ltd v Siffynet Solutions Pvt Ltd*,[78] the Supreme Court of India held that domain names are business identifiers and should be protected by the law of passing off as well as by trade mark legislation.

Fraud

Several domain name cases have raised fraud as grounds for relief. In *Marks & Spencer plc v One in a Million*,[79] the English Court of Appeal held that the domain names registered by the defendant were instruments of fraud. The value of the domain names, the court said, 'lay in the threat that they would be used in a fraudulent way':

> The registrations were made with the purpose of appropriating the [plaintiffs'] property, their goodwill, and with an intention of threatening dishonest use by them or another. The registrations were instruments of fraud and injunctive relief was appropriate just as much as it was in those cases where persons registered company names for a similar purpose.

The classic statement of fraud, from the common law world, comes from *Derry v Peek*:[80] fraud exists 'when it is shown that a false representation has been made (1) knowingly, or (2) without belief in its truth, or (3) recklessly, careless whether it be true or not'.[81] Fraud includes equitable fraud. In equity, the term 'fraud' also embraces conduct which falls below the standard demanded in equity. There is no exhaustive definition of equitable fraud, but undue influence and unconscionability are applicable considerations.

In *Powell v Birmingham Vinegar Brewing Co. Ltd*,[82] Lord Halsbury LC said:

> A person who puts forward this 'Yorkshire Relish,' made as it is by the present defendants, is representing it as being a particular manufacture. It may be true that the

76 [2006] NSWSC 158.
77 See *Fletcher Challenge Ltd v Fletcher Challenge Pty Ltd* [1981] 1 NSWLR 196 at 204, and *Nicholas v Borg* (1986) 7 IPR 1.
78 (2004) (28) PTC 566 (SC).
79 [1999] 1 WLR 903; [1998] EWCA Civ 1272.
80 (1889) 14 App Cas 337, 374 per Lord Herschell.
81 In *Peek v Gurney* (1873) LR 6 HL 377, 403 Lord Cairns considered that fraud existed where there was a partial statement of fact in such a manner that the withholding of what is not stated 'makes that which is stated absolutely false'.
82 [1897] AC 710.

> customer does not know or care who the manufacturer is, but it is a particular manufacture that he desires. He wants Yorkshire Relish to which he has been accustomed, and which it is not denied has been made exclusively by the plaintiff for a great number of years. This thing which is put into the hands of the intended customer is not Yorkshire Relish in that sense. It is not the original manufacture. It is not made by the person who invented it. Under these circumstances it is a *fraud* upon the person who purchases to give him the one thing in place of the other.[83]

The primary relief sought by plaintiffs is the discontinuation of the use of the offending domain name; in cases of cybersquatting it is the transfer of the domain name. However, courts have made awards of damages – compensatory, aggravated and punitive – and have considered making an order of account of profits.[84] Punitive damages require a separate actionable wrong.

Typosquatting is a variation of cybersquatting but retains the bad faith and insidious features. Typosquatters typically register domain names which are spelling variations of existing sites that have substantial traffic. The offending site often directs browsers to sites offering adult entertainment, gambling or recursive advertising. Some count the accidental hits and then sell advertising in proportion to the traffic generated. Reliance is placed on the fact that a percentage of accidental hits respond positively to the diversion.

In *Yahoo! Inc. and GeoCities v Data Art*[85] the claimant, who owned geocities.com, had 1.7 billion hits in one month. The defendant had registered 36 domain name variations, including eocities.com and gocities.com. Even a small percentage of hits resulting from mistyping resulted in considerable traffic.[86] The arbitration panel ordered the transfer of these bad faith registrations. Similarly in *Eddie Bauer Inc. v Paul White*[87] the misspelt domain name was eddibower.com and the panel restrained its use.

Typosquatting includes taking advantage of the accidental omission of the dot after the initial www, such as wwwdowjones.com.[88]

In 2001 Telstra Australia succeeded in forcing the transfer of the domain name telsra.com from a Melbourne-based typosquatter.[89] Upon accessing the site, the user was automatically redirected to an online gambling website. Telstra submitted that the practice caused embarrassment and tarnished Telstra's reputation and marks. Telstra previously won other domain name disputes relating to telstrashop.com[90] and telstra.org.[91]

83 [1897] AC 710, 713–14, emphasis added; see also Lord Herschell at 715.
84 *Hoath v Connect Internet Services* [2006] NSWSC 158, para 209.
85 WIPO Case No. D2000–0587.
86 Should one in a thousand visitors make the appropriate typo, the defendant would receive 1.7 million visits per month, an amount of traffic which would attract considerable advertising interest.
87 (2000) *eResolution* AF-204.
88 *Dow Jones v Powerclick Inc.*, WIPO Case No. D2000–1259.
89 *Telstra Corporation Ltd v David Whittle*, WIPO Case No. D2001-0434.
90 *Telstra Corporation Ltd v Barry Cheng Kwok Chu*, WIPO Case No. D2001-0423.
91 *Telstra Corporation Ltd v Nuclear Marshmallows*, WIPO Case No. D2000-0003.

Conclusion

Most problems and conflicts involving electronic commerce, the internet or cyberspace can be satisfactorily addressed and resolved using traditional established legal principles. However, there are several unusual and unique circumstances requiring law makers and law enforcers to rethink and reconceptualise their field. Domain name disputes is one such area. The domain name system was invented in 1983. The common law and statutes did not anticipate this new form of business identifier, and indeed debate continues about whether or not a domain name should be considered a new form of intellectual property.

In Australia the *Trade Practices Act 1974* (Cth) provides a generalist remedy adopted in some domain name cases. In other cases an allegation of fraud can be made. However, disputants have seized upon the tort of passing off as the most favourable judicial cause of action to resolve their conflict. But even then, the traditional and strict approach required by the judiciary fails to provide a remedy in all situations. The English Court of Appeal moulded and fashioned the tort in a manner not previously anticipated or expected.[92] The result may be classified as a new tort, or at least a subset known as 'domain name passing off'. This kind of modification is not new; it is in the spirit of evolving law. To paraphrase Learned Hands, common law 'stands as a monument slowly raised, like a coral reef, from the minute accretions of past individuals, of whom each built upon the relics which his predecessors left, and in his turn left a foundation upon which his successors might work'.[93] New and novel creations such as domain names call to the judiciary, as custodians of the common law, to do just this in order to serve law and society.

Further reading

Anri Engel, 'International domain name disputes: Rules and practice of the UDRP', (2003) 25 *EIPR* 351.
Megan Richardson, 'Trade marks and language',(2004) *Syd L Rev* 9.
Andrea F. Rush, 'Internet domain protection: A Canadian perspective', (1997) 11 *IPJ* 1.
Adrian Wolff, 'Pursuing domain name into uncharted waters: Internet domain names that conflict with corporate trade marks', (1997) 34 *San Diego Law Review* 1463.

92 As in the Marks & Spencer case above.
93 Billings Learned Hand's review of Cardozo's 'The Nature of the Judicial Process', (1932) 35 *Harv LR* 479.

10

Uniform domain name dispute resolution policies

By the 1990s, ICANN was in need of a solution to the rising dispute resolution problem. The legal proceedings in court were slow, expensive and involved a minefield of jurisdictional issues. There were moves to put in place an international treaty, but this process was too slow and would not be all-encompassing. The US legislative solution in 1999 was only a partial solution and few jurisdictions expressed the need to follow their lead; the result could well be a plethora of divergent responses. This chapter deals with the dispute resolution policies of ICANN and selected national domain name administrators.

In 1999 ICANN issued its Uniform Dispute Resolution Policy (UDRP)[1] as an alternative to legal proceedings before courts. The UDRP has become the international standard for resolving domain name disputes. It is intended to also discourage abusive registrations. The complainant is required to demonstrate that the disputed domain name is identical or confusingly similar to theirs, that the registrant does not have a right or legitimate interest in the domain name, and that the registrant has registered and used the domain name in bad faith.[2]

ICANN provides that domain name disputes must be resolved by agreement, court action, or arbitration before a registrar will cancel, suspend, or transfer a domain name. The UDRP has been adopted by all ICANN-accredited registrars in the gTLDs .aero, .biz, .cat, .com, .coop, .info, .jobs, .mobi, .museum, .name, .net, .org, .pro, .tel and .travel. It does not extend to any domain name ending with a country code: these are separately administered. However, a significant number of domain name administrators have issued their own Dispute Resolution Policies, which largely reproduce the ICANN policy.

1 See www.icann.org/udrp.
2 In addition, ICANN has issued a number of specialist policies for particular TLDs. See below.

The UDRP operates as a contractual term between the registrar and its customer and is included in all registration agreements. Disputes alleged to arise from abusive registrations of domain names may be addressed by proceedings initiated by a complaint and brought before an approved dispute-resolution service provider. These disputes may involve cybersquatting, but can also be between two legitimate traders.

WIPO internet domain name reports

At the instigation of concerned member nations, the World Intellectual Property Organisation (WIPO) conducted extensive consultations with the internet community and published a report containing recommendations in relation to domain name issues. WIPO promotes the protection, dissemination and use of intellectual property throughout the world for economic, cultural and social development. It is a forum for the development and implementation of intellectual property policies internationally, through treaties and other policy instruments. It was founded by treaty, with 183 member states.[3] The WIPO domain name process involved 17 consultation meetings in 15 cities and the receipt of written submissions from 334 governments, intergovernmental organisations, professional associations, corporation and individuals.

The investigation included a detailed review of the Administrative Procedure Concerning Abusive Domain Name Registrations. The Final Report recommended that ICANN adopt a dispute resolution policy with a uniform procedure for domain name disputes in all gTLDs. The Interim Report had recommended that applicants should be required to submit any intellectual property dispute arising out of their application, but the Final Report recommended that the administrative procedure be limited to cases of bad faith. The procedure is intended to be 'quick, efficient, cost-effective and conducted to a large extent on-line', and its remedies are limited to orders for the cancellation or transfer of domain name registrations and the allocation of the costs of the procedure (excluding legal fees) against the losing party.

WIPO recommended that there should be 'Exclusions for Famous and Well-known Marks': marks that exist over a widespread geographical area and across different classes of goods or services. The mechanism would be akin to the special protection established for famous and well-known marks in the Paris Convention for the Protection of Industrial Property and the TRIPS Agreement, and would cover only the exact name of the mark. However, cybersquatters also register close variations of the mark; this is called typosquatting. An evidentiary presumption should arise which would place the burden of proving justification for the use of the domain name at issue on the domain name holder where the

3 For the latest particulars see www.wipo.org.

domain name is identical or misleadingly similar to the famous or well-known mark.

The 2nd WIPO Internet Domain Name Process began in August 2000. The issues to be addressed included bad faith and abusive domain name registrations that violate: (1) intellectual property rights, including Geographical Indications (such as wine-producing regions) and personality rights (such as the names of celebrities); (2) the names and acronyms of international intergovernmental organisations (such as the United Nations); (3) International Nonproprietary Names set down by the World Health Organisation for pharmaceutical substances, so as to protect patient safety worldwide; and (4) unfair competition law and the rights established under international treaties.

The WIPO report resulting from this process found considerable evidence of the registration and use of identifiers as domain names by persons who might be considered not to be properly entitled to use the identifiers in question. Some registrations offend many sensitivities – the registration of the names of eminent political, scientific or religious persons, or the names of countries, cities or indigenous peoples, by parties with no association at all, for instance.

The ease of registering these identifiers as domain names is a consequence of the first-come, first-served system of registration (see Chapter 9).

UDRP rules

There were three approved dispute resolution providers at first.[4] WIPO became the first, effective on 1 December 1999. The National Arbitration Forum (NAF), effective on 23 December 1999, and the Asian Domain Name Dispute Resolution Centre (ADNDRC),[5] effective on 28 February 2002, were the other two.

The UDRP has been adopted by ICANN-accredited registrars in all gTLDs: it is included as a term in registration agreements between the applicant and the registrar.[6] Dispute proceedings arising from alleged abusive registrations of domain names may be initiated by parties claiming trade mark or service mark rights.

The purpose of the URDP is to create global uniformity, reduce the costs for resolving domain name disputes and enable the laws of nation states to continue to operate.

In applying for an ICANN open domain name the registrant must represent and warrant that the statements made in the application are complete and accurate

4 There have been other approved dispute resolution providers which have ceased to provide services. The Disputes.org/eResolution consortium (DeC) was approved on 1 January 2000. On 16 October 2000 the approval of DeC was transferred to eResolution (eRes). On 30 November 2001 eResolution ceased accepting proceedings. The CPR Institute for Dispute Resolution accepted domain name disputes from June 2000 until January 2007.

5 Offices in Beijing, Hong Kong and Seoul.

6 In the domains .aero, .biz, .cat, .com, .coop, .info, .jobs, .mobi, .museum, .name, .net, .org, .pro, .tel and .travel.

and that to the registrant's knowledge the registration does not infringe upon or violate the rights of any third party.[7] The numerous examples of cybersquatting are a testament to the fact that a considerable number of registrants ignore this contractual undertaking. The registrant further declares that the domain name will not be used for an unlawful purpose and that the domain name will not knowingly be in violation of any applicable laws or regulations. ICANN makes no attempt to require its registrars to make any determination about whether the applicant may be in breach. Clearly there are examples of well-known and famous trade marks being registered without there having been any attempt to question or otherwise limit the applications. Registrars often rely, for their income, on turnover of domain names, so restricting registration would be an additional cost and thus against their own interests.

Registrants are required to submit to a mandatory administrative proceeding in the event that a third party makes a complaint:

> ... to the applicable Provider, in compliance with the Rules of Procedure, that
> (i) your domain name is identical or confusingly similar to a trademark or service mark in which the complainant has rights; and
> (ii) you have no rights or legitimate interests in respect of the domain name; and
> (iii) your domain name has been registered and is being used in bad faith.[8]

The complainant must prove that all three elements are present. The UDRP's administrative panel's decisions cannot be enforced in court.

Identical or confusingly similar

Many cases simply involve the exact trade or service mark of the complainant, and there is little need for analysis. In *Blue Sky Software Corp. v Digital Sierra Inc.*[9] the registrant had registered robohelp.com, which was identical to the complainant's trade mark 'ROBOHELP'. First, the WIPO panel made the somewhat obvious statement, 'The addition of .com is not a distinguishing difference.' Similarly, there is little scrutiny or examination required when, for example, 'i' is added to 'Toyota' to create itoyota.com[10] or 'girls' is added to 'nokia' to form nokiagirls.com.[11] In the latter case the WIPO panel stated, 'The ability of this word "girls" to distinguish the Domain Name from the trademark of the Complainant is limited. As a general noun, "girls" is indeed a rather neutral addition to this trademark.'[12]

The words 'identical or confusingly similar' are meant to be separate concepts. On occasion parties have attempted to argue that the use of an identical identifier

7 It should be reiterated that the UDRP applies to the open gTLD and not directly to any of the approximately 240 country code domain names, such as .us, .au, .nz, .uk and .in. The majority of the national domain name administrators have incorporated parallel if not identical rules into there own domain name agreements.
8 UDRP, paragraph 4(a).
9 WIPO Case No. D2000–0165.
10 *Toyota Jidosha Kabushiki Kaisha v S & S Enterprises Ltd*, WIPO Case No. D2000–0802.
11 *Nokia Corporation v Nokiagirls.com,* WIPO Case No. D2000–0102.
12 See also *The Stanley Works and Stanley Logistics Inc. v Camp Creek Co. Inc.*, WIPO Case No. D2000–0113.

should be identical *and* confusingly similar. The case of *BWT Brands Inc. and British American Tobacco (Brands) Inc. v NABR*[13] involved a well-known brand of cigarettes – Kool – and the registration of kool.com. The WIPO panel rejected the registrant's argument that an identical domain name must simultaneously be confusingly similar for the provision to become operative:

> The burden on the Complainant under this head is merely to demonstrate that the disputed domain name is similar or identical to the Complainants' trademark, without reference to the way in which the domain name is being used. Plainly enough, the disputed domain name kool.com is identical to the Complainant's registered trademark.[14]

The test of confusing similarity is limited to a comparison between the domain name identifier and the mark alone. The panel does not need to consider other trade mark or passing off considerations. In a majority of cases the panels have concluded that the domain names were confusingly similar: for example, walmartcanadasucks.com,[15] salvationarmysucks.com,[16] guinness-really-sucks.com,[17] adtsucks.com,[18] accorsucks.com,[19] autotradersucks.com,[20] standardcharteredsucks.com[21] and wal-martsucks.com.[22] However, the opposite conclusion was reached with wallmartcanadasucks.com,[23] mclanenortheast-sucks.com[24] and lockheedmartinsucks.com.[25]

The panels have considered a 'mark for name comparison'. This involves the potential and the actual use of the domain name for confusion.[26] While actual deceptive use is relevant to the second and third requirements, it is not for the first requirement. In *Koninklijke Philips Electronics NV v In Seo Kim*[27] the panel considered that phillipssucks.com was confusingly similar to the trademark 'Philips'. The panel stated:

> Not all Internet users are English speaking or familiar with the use of 'sucks' to indicate a site used for denigration. Furthermore, it is not unknown for companies to establish complaint or comment sites or areas of sites to obtain feedback on their products; accordingly, some people might suppose that a website of this nature at the Domain Name was operated by the Complainant.[28]

13 WIPO Case No. D2001–1480.
14 WIPO Case No. D2001–1480, para 6.
15 *Wal-Mart Stores Inc. v Walsucks and Walmarket Puerto Rico*, WIPO Case No. D2000–0477.
16 *The Salvation Army v Info-Bahn Inc.*, WIPO Case No. D2001–0463.
17 *Diageo plc v John Zuccarini*, WIPO Case No. D2000–0996.
18 *ADT Services AG v ADT Sucks.com*, WIPO Case No. D2001–0213.
19 *Société Accor contre M Philippe Hartmann*, WIPO Case No. D2001–0007.
20 *TPI Holdings Inc. v AFX Communications*, WIPO Case No. D2000–1472.
21 *Standard Chartered PLC v Purge IT*, WIPO Case No. D2000–0681.
22 *Wal-Mart Stores Inc. v Richard MacLeod*, WIPO Case No. D2000–0662. See also *Direct Line Group Ltd v Purge IT*, WIPO Case No. D2000–0583 (*directlinesucks.com*); *Dixons Group PLC v Purge IT*, WIPO Case No. D2000–0584 (*dixonssucks.com*); *Freeserve PLC v Purge IT*, WIPO Case No. D2000–0585 (*freeservesucks.com*); *National Westminster Bank PLC v Purge IT*, WIPO Case No. D2000–0636 (*natwestsucks.com*); and *Vivendi Universal v Jay David Sallen*, WIPO Case No. D2001–1121 (*vivendiuniversalsucks.com*).
23 *Wal-Mart Stores Inc. v wallmartcanadasucks.com and Kenneth J Harvey*, WIPO Case No. D2000–1104.
24 *McLane Company Inc. v Fred Craig*, WIPO Case No. D2000–1455.
25 *Lockheed Martin Corporation v Dan Parisi*, WIPO Case No. D2000–1015.
26 See *Koninklijke Philips Electronics NV v Cun Siang Wang*, WIPO Case No. D2000–1778.
27 WIPO Case No. D2001–1195.
28 WIPO Case No. D2001–1195, para 6.

However, the panel rightly stated that the addition of sucks.com will not necessarily result in a successful challenge. Where a domain name is genuinely registered and used for the purposes of criticism, the second and third requirements of the rule will not necessarily be met.

Critics have argued that the domain name rules are against free speech; that sucks.com sites are legitimate forms of free expression. Panels have noted that the UDRP is directed against abusive registration, not free speech.[29]

Registrants often argue that the identifier is made up of generic or descriptive words which are not exclusively associated with complainant's mark or business. In *Energy Source Inc. v Your Energy Source*[30] the domain name was yourenergysource.com. The three-member panel focused on the domain name itself, stating that it is 'undoubtedly similar, but not identical' to the complainant's mark, 'Energy Source'. The panel said that 'confusing similarity should be determined by comparing the mark and domain name alone, independent of the other marketing and use factors usually considered in a traditional infringement action'. The panel regarded the words 'energy source' as the salient feature and determined that the addition of 'your' would be likely to be regarded by internet users as something related to 'energy source'.[31] In *Eddie Bauer Inc. v Paul White* (edibower.com)[32] the registrant used the misspelling to direct users to another website, where clothing similar to the complainant's was offered for sale. The panel held that the domain name was confusingly similar to the trade name 'Eddie Bauer'.

The application of these principles continues to prove problematic. Three panel decisions have included dissenting opinions on this issue.[33] However, the decisions of panels do not form precedents in the formal sense, so panels are not bound by them. Nevertheless, parties before panels do refer to earlier decisions, and panels refer to such decisions in their reasons for judgment.

Trade mark or service mark

The ICANN rules fit the US concept of common law trade mark and service marks much more neatly than they do the laws of other jurisdictions. Many jurisdictions define trade marks as the exclusive purvey of trade mark legislation. The extended meaning of 'service mark' has been used to deal with unregistered marks which might otherwise fit within the US approach.

29 WIPO Case No. D2001–1195, para 6.
30 NAF No. FA0096364.
31 See *Sony Kabushiki Kaishal v Sin Eonmok*, WIPO Case No. D2000–1007 and *MP3.com Inc. v Sander*, WIPO Case No. D2000–0579, dealing with the addition of 'my'.
32 (2000) eResolution AF-204.
33 For example, *Vivendi Universal v Jay David Sallen and GO247.com Inc.*, WIPO Case No. D2001–1121. See also *Bosley Medical Institute Inc. v Kremer* 403 F. 3d 672 (9th Cir 2005).

Common law trade mark rights are usually confined to the jurisdiction where the mark is used. In *ESAT Digifone Ltd v Colin Hayes*[34] the complainant had failed to register its own name, though there was a trade mark application pending in respect of other marks. The panel resolved the dispute in favour of the complainant. This Irish based case referred to local trade mark law, which hitherto had not recognised common law trade marks.[35] In *Fiber-Shield Industries Inc. v Fiber Shield Ltd* (fibershield.net), the respondent had operated a legitimate, non-competing business under the name for more than 10 years but had nevertheless lost the domain registration.[36] In *State of the Netherlands v Goldnames Inc.*,[37] the complainant alleged that there was a common law mark Staten-Generaal (States General), representing the Dutch state, even though in the Netherlands no common law trade mark exists. The panel held that the complainant's mark, Staten-Generaal, was reasonably well known throughout the world, as well as within the Netherlands, and that the complainant might thus be viewed as having legally protectable rights.

The panels have ruled that registration as a registered trade mark or service mark was not necessary, and that even individual names have sufficient secondary association for complainants to maintain that 'common law trademark rights' exist. For example, well-known novelist Jeanette Winterson reclaimed her name, which had been registered as Jeanettewinterson.com (as well as .org and .net).[38] The registrant, Hogarth, was a research fellow at Cambridge University and claimed that he intended to develop websites with reviews, biographies and forthcoming works of authors. However, Hogarth also has stated that he wanted to make money, and had asked for a 3 per cent share of profits of books sold via the website. The panel stated that the 'Complainant does not rely upon any registered trade marks but on her common law rights in her real name'. She has achieved international recognition and critical acclaim for her works and 'use of that Mark has come to be recognised by the general public as indicating an association with words written and produced exclusively by the Complainant'. The panel was satisfied that Jeanette Winterson had established trade mark rights in the mark for the purpose of the UDRP and ordered that all three domain names be transferred.

The well-known singer and performer Madonna challenged Dan Parisi, who purchased madonna.com for US$20,000. Parisi argued that this complainant was named after the Virgin Mary, as was her mother and hundreds of thousands of other people throughout the world over the past 2000 years, and had no greater right than any other person so named. However, for the purposes of the panel decision, the identifier was identical, Parisi had no legitimate interest in the name and he had acquired it and was using it in bad faith. This complainant

34 WIPO Case No. D2000–0600.

35 The Panel referred to *Coca-Cola v F Cade & Sons Limited* [1957] IR 196. See also *PA Consulting Services Pty Ltd v Joseph Barrington-Lew*, Case No. DAU2003–0002.

36 NAF (2000) No. FA0001000092054.

37 WIPO Case No. D2001–0520.

38 See *Jeanette Winterson v Mark Hogarth*, WIPO No. D2000–0235.

had registered 'Madonna' as a trade mark, establishing her rights in the name.[39] In a not dissimilar case the registrant had registered the domain name mickjagger.com. The panel first noted that the domain name was identical except for the .com, the capitalisation and spacing. The singer had had 'continuous commercial use of that mark for more than thirty-five (35) years'.[40] The panel stated that the UDRP 'does not require that the Complainant have rights in a registered trademark or service mark' and decided that the complainant held a common law trademark in his famous name, Mick Jagger.[41]

Other celebrities who have succeeded in having their name transferred include Julia Roberts,[42] Isabelle Adjani,[43] the Wiggles,[44] Jethro Tull[45] and Jimi Hendrix.[46] In addition, Time Warner has been successful in seeking orders to have several domain names which include the fictitious name 'Harry Potter'.

On the other hand, another well-known singer, Sting, failed in his attempt.[47] In the Sting case the panel agreed that unregistered or common law marks are sufficient, but the complainant did not provide any documentary evidence in support of his assertion that he was the owner of the unregistered trademark or service mark. The complainant did assert – and the panel took the equivalent of judicial notice – that the complainant is a world-famous entertainer known by the name Sting. However, after considering the fact that this identifier has common dictionary meanings, the panel stated:

> Although it is accepted that the Complainant is world famous under the name STING, it does not follow that he has rights in STING *as a trademark or service mark* . . . the personal name in this case is also a common word in the English language, with a number of different meanings.[48]

In *Re Eilberg*,[49] a trade mark application was filed for www.eilberg.com in the class of intellectual property. The applicant used the mark on letterheads and business cards. The claim was made that by not including the prefix http://, the use operated as a trade mark rather than merely as an indicator of a way to contact the law firm. These contentions were rejected by the USPTO Trademark Trial and Appeal Board. It held that displaying the domain name on the letterhead did not function as a service mark.

39 *Madonna Ciccone v Dan Parisi*, WIPO Case No. D2000–0847, (*madonna.com*).
40 *Mick Jagger v Denny Hammerton*, NAF FA0095261.
41 *Mick Jagger v Denny Hammerton*, NAF FA0095261.
42 *Julia Fiona Roberts v Russell Boyd*, WIPO Case No. D2000–0210.
43 *Isabelle Adjani v Second Orbit Communications Inc.*, WIPO Case No. D2000–0867.
44 *The Wiggles Touring Pty Ltd v Thompson Media Pty Ltd*, WIPO Case No. D2000–0124.
45 *The Ian Anderson Group of Companies Ltd v Denny Hammerton*, WIPO Case No. D2000–0475; referring to the musical group.
46 By the estate of the late Jimi Hendrix, *Experience Hendrix LLC v Denny Hammerton and The Jimi Hendrix Fan Club*, WIPO Case No. D2000–0364.
47 *Gordon Sumner, Sting v Michael Urvan*, WIPO Case No. D2000–0596.
48 *Gordon Sumner, Sting v Michael Urvan*, WIPO Case No. D2000–0596, emphasis added.
49 49 USPQ 2d 1955 (1998).

Registrant has no rights or legitimate interests

The second limb of the UDRP[50] requires the complainant to show that the registrant has no rights or legitimate interests in the domain name. The mere assertion that the respondent has no such rights does not constitute proof. The panel is free to make reasonable inferences. Paragraph 4(c) of the UDRP sets out specific circumstances to assist the registrant in demonstrating legitimate rights or legitimate interests in the domain name. The circumstances are inclusive and useful in considering this issue. The title to the paragraph is 'How to Demonstrate Your Rights to and Legitimate Interests in the Domain Name in Responding to a Complaint':[51]

> 4(c) Any of the following circumstances . . . shall demonstrate your rights or legitimate interests to the domain name for purposes of Paragraph 4(a)(ii):
> (i) before any notice to you of the dispute, your use of, or demonstrable preparations to use, the domain name or a name corresponding to the domain name in connection with a bona fide offering of goods or services; or
> (ii) you (as an individual, business, or other organization) have been commonly known by the domain name, even if you have acquired no trademark or service mark rights; or
> (iii) you are making a legitimate noncommercial or fair use of the domain name, without intent for commercial gain, to misleadingly divert consumers or to tarnish the trademark or service mark at issue.[52]

In relation to paragraph 4(c)(i), in *Libro Ag v NA Global Link*[53] (libro.com) the registrant argued that the complainant's trade mark 'libro' is common Spanish and Italian word for 'book'. However, although the registrant claimed it had registered the domain name to establish an online virtual book store, in fact users were redirected to restaurants.com. The panel held that the mere assertion of making preparations to use the domain name for the *bona fide* offering of goods is insufficient to demonstrate rights or legitimate interests.

In *Bruce Springsteen v Jeff Burgar and Bruce Springsteen Club*,[54] both the complainant and the registrant had rights in the name. The panel described Bruce Springsteen as 'the famous, almost legendary, recording artist and composer . . . his name is instantly recognisable in almost every part of the globe' and accepted the complainant's common law rights in the name. The registrant was the Bruce Springsteen Club, with Jeff Burger as the point of contact. When accessing the website users were immediately transferred to celebrity1000.com. In relation to 4(c)(i), the panel found there was no *bona fide* commercial use.

50 UDRP paragraph 4(a)(ii) provides the second of three assertions by the complainant: 'the registrant has no rights or legitimate interests in respect of the domain name'.
51 The terms 'your' and 'you' refer to the registrant.
52 UDRP paragraph 4(c).
53 WIPO Case No. D2000–0186.
54 WIPO Case No. D2000–1532.

In relation to 4(c)(ii), the panel considered the meaning of 'commonly' and 'known by'. It found that the use of the name 'Bruce Springsteen Club' would not give rise to an impression in the minds of internet users that the proprietor was effectively 'known as' Bruce Springsteen, let alone 'commonly' recognised in that fashion. Accordingly the panel found that this requirement too was not met.

In applying 4(c)(iii), the panel considered the meaning of 'non-commercial' and 'fair use'. It observed that 'Bruce Springsteen', when inputted into a search engine, would yield thousands of hits, from which it would be apparent to users that some sites were official or authorised sites and others were not. Due to the number of sites, the panel noted that users would be 'unsurprised' to arrive at celebrity1000.com via the name 'Bruce Springsteen'. Accordingly, they stated that 'it is hard to infer from the conduct' of the registrant that the intent was for commercial gain or 'to misleadingly divert consumers'. In determining that such conduct does not 'tarnish' the common law rights of Bruce Springsteen by association with the celebrity1000.com website, the panel found there was 'fair use'. They considered that fair use would not exist if the domain name in question connected, for example, to 'sites containing pornographic or other regrettable material'.[55]

The panel found that the complainant, Bruce Springsteen, had not satisfied the second limb of of the three-part test in paragraph 4(a) of the UDRP.

Registration and use in bad faith

The UDRP rules provide an explanation of 'Registration and Use in Bad Faith'. The circumstances include dealing with cybersquatting and the attempted extortion from trade mark and service mark owners:

> 4(b) ... Any of the following circumstances, in particular but without limitation, if found by the Panel to be present, shall be evidence of the registration and use of a domain name in bad faith:
>
> (i) circumstances indicating that you have registered or you have acquired the domain name primarily for the purpose of selling, renting, or otherwise transferring the domain name registration to the complainant who is the owner of the trademark or service mark or to a competitor of that complainant, for valuable consideration in excess of your documented out-of-pocket costs directly related to the domain name; or
>
> (ii) you have registered the domain name in order to prevent the owner of the trademark or service mark from reflecting the mark in a corresponding domain name, provided that you have engaged in a pattern of such conduct; or
>
> (iii) you have registered the domain name primarily for the purpose of disrupting the business of a competitor; or

55 WIPO Case No. D2000–1532, para 6.

(iv) by using the domain name, you have intentionally attempted to attract, for commercial gain, Internet users to your web site or other on-line location, by creating a likelihood of confusion with the complainant's mark as to the source, sponsorship, affiliation, or endorsement of your web site or location or of a product or service on your web site or location.[56]

These rules encompass and go beyond typical cybersquatting and typosquatting situations.[57]

The remedies available to the complainant pursuant to any proceeding before the panel are the cancellation of the domain name or the transfer of the domain name to the complainant. All panel decisions under the UDRP are published in full on the internet; the panel can determine to withhold portions in exceptional circumstances. The UDRP's mandatory administrative proceeding requirements do not prevent legal proceedings in any court of competent jurisdiction from being commenced before or after the UDRP proceedings.[58]

The UDRP allows a period of grace of 10 business days before taking action on any decision to cancel or transfer the domain name. This is to allow the registrant the opportunity to commence legal proceedings. All that is required to do this is notice of the legal proceedings by, for example, a plaint file-stamped by the clerk of the court.

The UDRP decision will not be implemented until ICANN receives: satisfactory evidence of a resolution between the parties; or satisfactory evidence that the legal proceedings have been dismissed or withdrawn; or a copy of an order from the court dismissing the legal proceedings or ordering that the registrant has no right to continue to use the domain name.[59]

In *GlobalCenter Pty Ltd v Global Domain Hosting Pty Ltd*[60] the panel members considered their right to make independent investigations. One took the view that the UDRP rules[61] preclude panel members from conducting their own factual investigations and from visiting relevant websites unless invited by a party to the dispute. Visit to the relevant websites might reveal the parties' compliance with the rules, their state of mind in relation to such issues as bad faith, and the validity of their respective submissions.[62] Another took the view that the panel is entitled to have regard to any freely available online public material 'germane to an issue in dispute'.[63]

56 UDRP paragraph 4(b).

57 See *Miele Inc. v Absolute Air Cleaners and Purifiers*, WIPO Case No. D2000–0005 and *Telaxis Communications Corp. v William E. Minkle*, WIPO Case No. D2000–0756.

58 UDRP paragraph 4(k).

59 See UDRP clause 4(d)–(k).

60 Case No. DAU2002–0001.

61 Rules for UDRP rules 10(a) and 15(a).

62 See also *Jazid Inc. Michelle McKinnon v Rennemo Steinar eResolution*, Case No. AF-0807, where the panel stated that it is not 'the burden of the Panel to seek further evidence (other than judicial knowledge) to sustain the parties' allegations, as this may be disruptive of the arbitration process. Therefore, the burden rests on the parties to either support or sustain their allegations with the appropriate documentation whenever possible.'

63 Case No. DAU2002–0001, para 6.

The proceedings before panels are governed by the Rules for the UDRP.[64] These rules deal with the procedure and aspects such as fairness and evidentiary standards. The general power of the panels are:

10(a) The Panel shall conduct the administrative proceeding in such manner as it considers appropriate in accordance with the Policy and these Rules.
(b) In all cases, the Panel shall ensure that the Parties are treated with equality and that each Party is given a fair opportunity to present its case.
(c) The Panel shall ensure that the administrative proceeding takes place with due expedition. It may, at the request of a Party or on its own motion, extend, in exceptional cases, a period of time fixed by these Rules or by the Panel.
(d) The Panel shall determine the admissibility, relevance, materiality and weight of the evidence.
(e) A Panel shall decide a request by a Party to consolidate multiple domain name disputes in accordance with the Policy and these Rules.

On occasion proceedings may be by teleconference, videoconference or web conference.[65] There are also rules concerning the timing, method and form of the panel's decision, and about how it deals with dissenting opinions and with bad faith complaints:

15(a) A Panel shall decide a complaint on the basis of the statements and documents submitted and in accordance with the Policy, these Rules and any rules and principles of law that it deems applicable.
(b) In the absence of exceptional circumstances, the Panel shall forward its decision on the complaint to the Provider within fourteen (14) days of its appointment pursuant to Paragraph 6.
(c) In the case of a three-member Panel, the Panel's decision shall be made by a majority.
(d) The Panel's decision shall be in writing, provide the reasons on which it is based, indicate the date on which it was rendered and identify the name(s) of the Panelist(s).
(e) Panel decisions and dissenting opinions shall normally comply with the guidelines as to length set forth in the Provider's Supplemental Rules. Any dissenting opinion shall accompany the majority decision. If the Panel concludes that the dispute is not within the scope of Paragraph 4(a) of the Policy, it shall so state. If after considering the submissions the Panel finds that the complaint was brought in bad faith, for example in an attempt at Reverse Domain Name Hijacking or was brought primarily to harass the domain-name holder, the Panel shall declare in its decision that the complaint was brought in bad faith and constitutes an abuse of the administrative proceeding.[66]

In *Mary-Lynn Mondich and American Vintage Wine Biscuit Inc. v Big Daddy's Antiques,*[67] the panel held that merely offering to sell the domain name to owner of the trade mark can be sufficient evidence for registration and use in bad faith where the registrant claimed more than any underlying costs. However, where

64 Available at: www.icann.org/en/udrp/udrp-rules-24oct99.htm.
65 Rules for UDRP rule 13.
66 Rules for UDRP rule 15.
67 WIPO Case No. D2000–0004.

the complainant initiated transfer discussions and the registrant had no prior plans to sell the domain name, the panel can find no bad faith.[68] Where the registrant concealed their identity by operating under a false name the panel can find bad faith.[69]

In a few cases the sale of domain names has been considered a legitimate interest,[70] but in others, such as *Microsoft Corporation v Amit Mehrotrata*[71] (Microsoft.org), it has been recognised as a factor in proving of bad faith.

UDRP process

The advantage of avoiding the court system – cutting down on the number of times the case may be run – is sometimes regarded as a drawback. There is only one bite at the cherry. There are no appeals. Decisions are typically made in a timely manner, with the average time to resolve disputes 37–45 days. NAF now permits additional written submissions within five calendar days of the respondent's response.[72] The large number of disputes constitutes a tempting market for dispute resolution providers. NAF markets its services prominently, distributing press releases about its decisions, which are often pro complainant.

The Rules for UDRP stipulate that a respondent has 20 days to file a response. Because this is not sufficient time to consider a course of action, obtain expert opinion, legal or otherwise, research precedents and procedures, then draw up and file a response, approximately half the total number of applications go unanswered. This means that half of the thousands of cases are decided on the basis of the complainants' assertions only. This may be a disturbing consequence of the swift process. Alternatively, it may be evidence of the extent of cybersquatting and bad faith use. It is most likely to be a combination of the two. There has been a considerable amount of abuse by registrants, but the breakneck pace of the dispute resolution process will leave many legitimate registrants intimidated and wary.

Critics also argue systemic bias in the results. However, the high percentage of decisions in favour of the complainant can be explained in part by three underlying factors. The first is the effect on these figures of the number of default cases. Second, complainants have no time restriction when they prepare an application. Third, while the reasons for commencing legal proceedings are in general many and varied, the reasons for engaging in the expeditious UDRP

68 See *City Utilities v Ed Davidson*, WIPO Case No. D2000–0004 (*cityutilities.com*). See also *Blue Cross and Blue Shield Association and Trigon Insurance Company v Interactive Communications Inc.*, WIPO Case No. D2000–0788.

69 See *Telstra Corporation Ltd v Nuclear Marshmallows*, WIPO Case No. D2000–0003 (*Telstra.org*).

70 See *Car Toys Inc. v Informa Unlimited, Inc.* (2000) NAF 93682 (*cartoys.net*); *Philippe Tenenhaus v Telepathy Inc.* (2000) NAF 94355 (*daf.com*); and *Allocation Network Gmbh v Steve Gregory*, WIPO Case No. D2000–0016 (*allocation.com*).

71 WIPO Case No. D2000–0053.

72 With an additional payment of US$250.

process are more likely to be a sincere and greater belief of success. The process is thus self-selecting. However, claiming bias on the basis of statistics only is injudicious.

The applicant can choose a panel of one or three members. Statistics show that three-member panel decisions are more often favourable to the applicant. Some may argue inherent bias in this too, but again the process may be self-selecting – perhaps those with a greater confidence choose the larger panel.

Interestingly, the former provider eResolution not only had a lower success rate for the complainants but also a higher percentage of contested disputes.[73]

The application of the three elements to make a finding should be a simple and uncomplicated exercise. All three elements must be present. Yet there have been cases where elements two and three are absent or questionable, but the panel has nevertheless found reason to order a cancellation or transfer. These decisions seem to have involved the consideration of elements absent from the rules: they have recognised established rights in the relevant identifier, sometimes by trade mark, registered or otherwise, and sometimes by usage.

The two most popular providers have been WIPO and NAF. It is not likely to be a coincidence that both these providers have more positive outcomes for applicants than for respondents. The success rate for applicants is 82 per cent and 83 per cent respectively. The results have been analysed and interpreted from a number of points of view.

This interpretive twisting has encouraged forum-shopping by complainants and promoted a bias toward large trade mark holders. It seems that dispute resolution providers that recognise these additional non-specified rights receive a greater share of the lucrative market. WIPO and NAF share the majority of the applications, with 58 per cent and 34 per cent respectively.

ccTLD dispute resolution policies

A significant number of domain name administrators have implemented domain name dispute resolution policies based on the UDRP. Most require the same or similar three elements to be established. A few use related principles.

auDRP

In Australia the Australian Domain Name Administrator, auDA adopted the .au Dispute Resolution Policy (auDRP) in 2002. This policy is modelled on the UDRP and applies to all open 2TLDs, namely asn.au, com.au, id.au, net.au and org.au. The relevant provisions state:

> You are required to submit to a mandatory administrative proceeding in the event that a third party (a 'complainant') asserts to the applicable Provider, in compliance with the Rules of Procedure that:

73 See International Trade Mark Association (INTA) Internet Committee reports: www.inta.org.

(i) your domain name is identical or confusingly similar to a name, trademark or service mark in which the complainant has rights; and
(ii) you have no rights or legitimate interests in respect of the domain name; and
(iii) your domain name has been registered or subsequently used in bad faith.[74]

There are three differences between the provisions of the UDRP and those of the auDRP. First, in paragraph (i), the auDRP added 'name'. The auDRP applies to domain names that are identical or confusingly similar, not only to a trademark or service mark, but also to any 'name' in which the complainant has rights. This includes the complainant's company, business or other legal or trading name, as registered with the relevant Australian government authority, and the complainant's personal name.

Second, in paragraph (iii) the word 'and' has been replaced with 'or'. Under the auDRP, it is sufficient to prove that either registration 'or' subsequent use of the domain name by the registrant occurred in bad faith, whereas the UDRP requires the complainant to prove both elements.

Third, under the second circumstance of bad faith of the UDRP a pattern of conduct is required. Under the auDRP no such pattern is required.[75]

.nz DRSP

The registry responsible for the .nz (New Zealand) domain name extension instituted a new Dispute Resolution Service Policy (DRSP) in 2006 (applicable to .nz domains). It is similar to the .uk DRSP. However, it uses the expression 'Unfair Registration' in place of 'Abusive Registration':

> This Policy and Procedure applies to Respondents when a Complainant asserts to the DNC according to the Procedure, that:
> 4.1. The Complainant has Rights in respect of a name or mark which is identical or similar to the Domain Name; and
> 4.1.2 The Domain Name, in the hands of the Respondent, is an Unfair Registration.[76]

An 'Unfair Registration' is set out as non-exclusive circumstances in paragraph 5.1:

> 5.1. A non-exhaustive list of factors which may be evidence that the Domain Name is an Unfair Registration is set out in paragraphs 5.1.1–5.1.5:
> 5.1.1. Circumstances indicating that the Respondent has registered or otherwise acquired the Domain Name primarily:
> (a) for the purposes of selling, renting or otherwise transferring the Domain Name to the Complainant or to a competitor of the Complainant, for valuable consideration in excess of the Respondent's

74 auDRP clause 4(a).
75 See *GE Capital Finance Australasia Pty v Dental Financial Services Pty Ltd*, Case No. DAU2004–0007; *University of Melbourne v union melb*, Case No. DAU2004–0004; and *The National Office for the Information Economy v Verisign Australia Limited* LEADR, Case No. 02/2003.
76 New Zealand DRSP clause 4.1.

documented out-of-pocket costs directly associated with acquiring or using the Domain Name;

(b) as a blocking registration against a name or mark in which the Complainant has Rights; or

(c) for the purpose of unfairly disrupting the business of the Complainant; or

5.1.2. Circumstances demonstrating that the Respondent is using the Domain Name in a way which is likely to confuse, mislead or deceive people or businesses into believing that the Domain Name is registered to, operated or authorised by, or otherwise connected with the Complainant;

5.1.3. The Complainant can demonstrate that the Respondent is engaged in a pattern of registrations where the Respondent is the registrant of domain names (under .nz or otherwise) which correspond to well known names or trade marks in which the Respondent has no apparent rights, and the Domain Name is part of that pattern;

5.1.4. The Complainant can demonstrate that the Respondent has knowingly given false contact details to a Registrar and/or to the DNC; or

5.1.5. The Domain Name was registered arising out of a relationship between the Complainant and the Respondent, and the circumstances indicate that it was intended by both the Complainant and the Respondent that the Complainant would be entered in the Register as the Registrant of the Domain Name.

usTLD Dispute Resolution Policy

The usTLD Dispute Resolution Policy is also based on the UDRP. The operative provisions state:

> You are required to submit to a mandatory administrative proceeding in the event that a third party (a 'Complainant') asserts to the applicable Provider, in compliance with the Rules, that:
>
> i. Your domain name is identical or confusingly similar to a trademark or service mark in which the Complainant has rights;
> ii. You have no rights or legitimate interests in respect of the domain name; and
> iii. Your domain name has been registered in bad faith or is being used in bad faith.[77]

.uk DRSP

The .uk Dispute Resolution Service Policy (DRSP) is part of the contract of registration for .uk domains. It sets out the policy ideas and a general overview of the structure and purpose of the DRS. The .uk DRSP replaces the UDRP's concept of bad faith with the expression 'abusive registration' and provides a definition. Circumstances indicating abusive registration include that the registrant has registered or acquired the domain name primarily:

> for the purposes of selling, renting or transferring the Domain Name to the Complainant or competitor for an amount valuable in excess of out-of-pocket costs;

77 usTLD DRP clause 4(a).

> as a blocking registration against a name or mark in which the Complainant has rights; or
>
> for the purpose of unfairly disrupting the business of the Complainant.
>
> There shall be a presumption of Abusive Registration if the Complainant proves that [the] Registrant has been found to have made an Abusive Registration in three or more Dispute Resolution Service cases in the two years before the Complaint was filed.[78]

The operative provisions state:

> A Respondent must submit to proceedings under the Dispute Resolution Service if a Complainant asserts to us, according to the Procedure, that:
> i. The Complainant has Rights in respect of a name or mark which is identical or similar to the Domain Name; and
> ii. The Domain Name, in the hands of the Respondent, is an Abusive Registration.[79]

INDRP

In India the .in Domain Name Registry has published the .in Dispute Resolution Policy (INDRP). It has been formulated in line with internationally accepted guidelines, the UDRP and the relevant provisions of the *Information Technology Act 2000* (India). The operative provisions state:

> Any Person who considers that a registered domain name conflicts with his legitimate rights or interests may file a Complaint to the .IN Registry on the following premises:
> (i) the Registrant's domain name is identical or confusingly similar to a name, trademark or service mark in which the Complainant has rights;
> (ii) the Registrant has no rights or legitimate interests in respect of the domain name; and
> (iii) the Registrant's domain name has been registered or is being used in bad faith.[80]

Additional policies

The Eligibility Reconsideration Policy (ERP) is incorporated into agreements with registrants concerning domain name registrations in .aero. It sets out terms and conditions in connection with any challenge to a decision by the sponsor concerning eligibility to register in .aero. This policy was developed by the sponsor of .aero and is not an ICANN policy.[81]

The Eligibility Requirements Dispute Resolution Policy (ERDRP) deals with the TLD .name. Registrations in .name must comprise an individual's

78 .uk DRSP clause 3.
79 .uk DRSP clause 2.
80 INDRP clause 4.
81 See www.icann.org/udrp.

personal name or the personal name of a fictional character provided the registrant holds a trade mark or service mark in the character's personal name. Numeric characters may be used in combination with the personal name. Disputes in the .name domain are filed under the ERDRP. Defensive registrations and second level domain email address registrations may be disputed under the ERDRP.

The .cat Eligibility Requirements Dispute Resolution Policy (.cat ERDRP) applies to the TLD .cat. Registrations in .cat are restricted to members of the Catalan linguistic and cultural community. Challenges to a registration in .cat on the grounds that it does not meet the eligibility requirements are filed under the ERDRP.

The Intellectual Property Defensive Registration Challenge Policy (IPDRCP) applies to intellectual property defensive registrations in the .pro TLD. This TLD is restricted to certified practising members of the medical, legal and accounting professions, although extensions may be considered. An intellectual property defensive registration may be registered only by the owner of an eligible trade mark or service mark registration. The IPDRCP provides an avenue for challenges concerning whether such registrant meets the Registration Qualifications.

The Qualification Challenge Policy (QCP) also deals with the TLD .pro. Challenges to a registration on the grounds that the registrant did not meet the registration qualifications may be made under the QCP. A challenge to a registration under the Qualification Challenge Policy may be brought by any interested party.

The Restrictions Dispute Resolution Policy (RDRP) applies to the restricted TLD .biz. The .biz TLD is intended for *bona fide* business or commercial purposes. Challenges to a registration or use of a given domain name on the grounds that it is not being or will not be used primarily for such a purpose are filed under the RDRP. Challenges under the RDRP may be initiated by any party filing a complaint with an approved dispute resolution service provider.[82]

The Transfer Dispute Resolution Policy (TDRP) applies to transactions where the registrant transfers or attempts to transfer the domain name to a new registrar. The TDRP deals with registrar disputes under the Inter-Registrar Transfer Policy, which is followed by the .biz, .com, .info, .name, .net, .org, and .pro TLDs. Proceedings under the TDRP must be filed either with the appropriate registry operator or with an independent dispute resolution provider. Any ICANN-accredited registrar may initiate a TDRP proceeding against another registrar.[83]

82 Note that the Start-Up Trademark Opposition Policy (STOP) was available to intellectual property owners who enrolled in the IP Claim Service during the start-up phase of the .biz registry during 2001. STOP is no longer available. Disputes can be brought under the UDRP, the RDRP or through the courts.

83 For all the policies, see, generally, www.icann.org/udrp.

Practical ramifications

An aggrieved party who wishes to initiate proceedings may choose a dispute resolution process available from the domain name administrators, or legal proceedings before a court. These actions are not mutually exclusive, and a party may fail in one action, and succeed in the other. Their procedures, rules and approaches are for the most part unrelated.

The legal process is typically slow and the cost of lawyers is often prohibitive. The proceedings are in person and rely on a range of laws and malleable common law and general law principles. An aggrieved party may be able to establish the elements of passing off, but the same circumstances may not give rise to a remedy under the dispute resolution processes.

It is up to aggrieved parties to determine which process is best in their situation.

Conclusion

The nature of domain names gives rise to an inevitable conflict between registered users of domain name and holders of trade marks. The reason for conflict is that the domain name must be unique, while trade marks and other business and social identifiers may be shared: there can be only one holder of domain name 'identifier.com', for example, but there can be multiple users of the trade mark 'identifier' in various countries and jurisdictions.

Keeping this uniqueness of domain names in mind, a legal mechanism for its allocation is most necessary. The policy of 'first come first served' was adopted and consistently followed by various domain name administrators, but in recent times it has been frequently negated by the various courts, UDRP panels and various local DRP panels in favour of other rights (not exclusively those of trade mark owners). The rejection of the policy in cases of 'cybersquatting' is quite justified in my view, but rejection of the policy in order to protect senior users of the trade mark is uncalled for.[84] Domain name registration policy rewards trade mark owners who have the foresight to register their trade marks as domain names.[85]

The initial run on established stakeholders who failed to record their interests on domain names registers has expired. The ignorance of the 1990s has passed. Nevertheless, new interests emerge. As evidence of this, just note the run of various combinations of the footballer Beckham, his number and club when he changed clubs. Cyberpiracy continues to take on various forms, not all of which are readily apparent. ICANN and its various constituents must maintain a constant vigil. There has been a reluctance to make changes to the UDRP, but after a decade of use and experience a review of its application would be wise.

84 See Spyros M. Maniatis, 'Trade mark law and domain name: Back to basics' (2002) 24 *EIPR* 397.
85 *Data Concepts v Digital Consulting Inc. and Network Solutions Inc.* 150 F. 3d 620, 1998 US App. LEXIS 17758 (6th Cir 1998).

Further reading

Colm Brannigan, 'The UDRP: How do you spell success?', (2004) *Digital Technology Law Journal*.

JR Dupre, 'A solution to the problem? Trade-mark infringement and dilution by domain names: Bring the cyber world in line with the "real" world', (1997) 87 *Trade Mark Reporter* 613.

Michael Geist, 'Fair.com?: An examination of the allegations of systemic unfairness in the ICANN UDRP', (2002) 27 *Brooklyn Law School Journal of International Law* 903.

Julia Hörnle, 'Disputes solved in cyberspace and the rule of law', 16th BILETA Annual Conference, Edinburgh, 9–10 April 2001; www.bileta.ac.uk/01papers/horle.html.

Annette Kur, 'UDRP: A study by the Max-Planck Institute for Foreign and International Patent, Copyright and Competition Law, Munich': see Summary and Proposals at p. 72. Available at www.intellecprop.mpg.de/Online-Publikationen/2002/UDRP-study-final-02.pdf.

David S. Wall, *Cyberspace crime*, Dartmouth, London, 2003.

11

Jurisdiction in cyberspace

> *When a radically new situation is presented to the law it is sometimes necessary to think outside the square . . . this involves a reflection upon the features of the Internet that are said to require a new and distinctive legal approach.*[1]

Cyberspace is an illusion. There is no such place. Many terrestrial norms do not and cannot apply to such a fictitious construct. Nevertheless cyberspace users perceive metaphorical chat rooms, folders, files, shops, libraries and so forth. They live digital lives with digital personas in 'places' such as Second Life and Facebook.[2] The reality is that each step of the digital experience is rooted terrestrially. Traditional legal principles are applicable to the majority of electronic commerce disputes. Nevertheless, the operation of electronic commerce in cyberspace results in new circumstances to which legal jurists cannot readily apply established legal rules.

The borderless nature of the internet often hides or disguises the origin of particular websites and corresponding information. Questions sometimes arise as to the country or state whose courts have jurisdiction to adjudicate on a matter, and as to which law is to be applied. Courts also have to determine issues such as where conduct occurs – at the computer, the server, the place of business or residence or somewhere else? – and thus which time zone applies. This area of law is referred to as conflict of laws or private international law,[3] and its principles are well established.

Rules of private international law

It is possible that an Australian court will determine that the law which applies to a particular situation is that of another country. In this case the Australian court will

1 Kirby J, *Dow Jones v Gutnick* [2002] HCA 56, para 112.
2 See secondlife.com and www.facebook.com.
3 See Peter Nygh and Martin Davies, *Conflict of laws in Australia*, 7th edn, LexisNexis, Sydney, 2002.

apply that other country's law. Typically, the parties to an international contract will specifically identify the law which applies to the contract. The parties may also identify which country's courts have jurisdiction to hear disputes arising under the contract.

Contractually, parties may include an express choice of law clause under the principle of party autonomy.[4] However, there are exceptions.[5] Where there is no express choice of law the courts will attempt to find an implication of choice from the language of the documents, the prior conduct of the parties, the place of performance and other surrounding circumstances.[6] Where there is no express choice and one cannot be implied, the contract will be governed by the law of the jurisdiction 'most closely connected' to the contract. This may depend upon the subject matter, nationality or domicile of the parties, the place of performance and other factors.[7] Once the proper law is determined, the rules of that system of law are applied to the contract.

Forum non conveniens

The issue of choice of forum in which to hear an action is known as *forum non conveniens*. Each nation state determines its own laws and procedures. These laws are termed *lex fori* or the law of the forum. Each nation's court system must determine, through its own civil procedure, when and how it should accept jurisdiction. It will take into account numerous factors, such as residence and nationality of the parties, the place or places of business, and the subject matter (when the law suit begins). Often the issue is routine and uncontroversial, but where one or more of the parties reside outside the nation state or the subject matter of harm occurs elsewhere the question of jurisdiction must be settled.

The English approach to determining the forum, recognised by most common law jurisdictions, is the test of 'the clearly more appropriate forum'.[8] The Australian courts have developed a broader test, that of 'the clearly inappropriate forum'. This has resulted in the taking of jurisdiction more often.[9]

The High Court reaffirmed its position in *Regie National des Usines Renault SA v Zhang*:[10] '[a]n Australian court cannot be a clearly inappropriate forum merely by virtue of the circumstance that the choice of law rules which apply in the forum

4 See *Vita Food Products Inc. v Unus Shipping Co.* [1939] AC 277, per Lord Wright.

5 For example, see *Golden Acres Ltd v Queensland Estates Pty Ltd* [1969] Qd R 378, and *Minnesota v Granite Gate Resorts Inc.* 568 NW 2d 715.

6 See *Atlantic Underwriting Agencies Ltd v Compagnia di Assicuranzione di Milano SpA* [1979] 2 Lloyds Rep 240.

7 See *Re United Railways of the Havana and Regla Warehouses Ltd* [1960] Ch 52 and *Mendelson-Zeller Co. Inc. v T & C Providores Pty Ltd* [1981] 1 NSWLR 366.

8 See *Spiliada Maritime Corporation Ltd v Cansulex Ltd* [1987] AC 460.

9 See *Oceanic Sun Line Special Shipping Co. v Fay* (1988) 62 ALJR 389; *Voth v Manildra Flour Mills Pty Ltd* (1990) 65 ALJR 83; *Henry v Henry* (1996) 185 CLR 571.

10 [2002] HCA 10.

require its courts to apply foreign law as the *lex causae*'. Although superficially the English and Australian tests look similar, an analysis demonstrates that the Australian test will result in the courts taking jurisdiction more often.

Dow Jones v Gutnick

A number of foreign cases have raised novel problems associated with jurisdiction in cyberspace.

The application of the internet vis à vis jurisdiction and cyberspace was explained in *Dow Jones v Gutnick*.[11] Gleeson CJ, McHugh, Gummow and Hayne JJ explained:

> The World Wide Web is but one particular service available over the Internet. It enables a document to be stored in such a way on one computer connected to the Internet that a person using another computer connected to the Internet can request and receive a copy of the document . . . the terms conventionally used to refer to the materials that are transmitted in this way are a 'document' or a 'web page' and a collection of web pages is usually referred to as a 'web site'. A computer that makes documents available runs software that is referred to as a 'web server'; a computer that requests and receives documents runs software that is referred to as a 'web browser'.
>
> The originator of a document wishing to make it available on the World Wide Web arranges for it to be placed in a storage area managed by a web server. This process is conventionally referred to as 'uploading'. A person wishing to have access to that document must issue a request to the relevant server nominating the location of the web page identified by its 'uniform resource locator (URL)'. When the server delivers the document in response to the request the process is conventionally referred to as 'downloading'.[12]

Dow Jones v Gutnick[13] is the most significant Australian authority on internet jurisdiction to date. Dow Jones published the *Barrons Magazine* (*Barrons*), which contained an article entitled 'Unholy Gains' and sub-headed 'When stock promoters cross paths with religious charities, investors had better be on guard'. The article alleged, among other things, that Victorian businessman Joseph Gutnick was 'masquerading as a reputable citizen when he was a tax evader who had laundered large amounts of money'.[14] The magazine sold 305,563 hard copies (14 of those in Victoria) and was available on the website wsj.com to some 550,000 subscribers (some 300 of whom were in Victoria). Dow Jones operated in New York and its internet servers were located in New Jersey.

Dow Jones v Gutnick did not involve personal jurisdiction, as the defendant was not within jurisdiction. Personal jurisdiction is typically satisfied where the defendant is physically present in the jurisdiction or has significant contacts with

11 [2002] HCA 56; (2002) 210 CLR 575.
12 [2002] HCA 56, [15]–[16]; (2002) 210 CLR 597–98.
13 [2002] HCA 56; on appeal from *Gutnick v Dow Jones* [2001] VSC 305 and *Dow Jones Company Inc. v Gutnick* [2001] VSCA 249. See, generally, 'A "category-specific" legislative approach to the internet personal jurisdiction problem in US law', (2004) 117 *Harv LR* 1617.
14 [2001] VSC 305, para 3.

the jurisdiction, such as a branch office or registered interest. The nature of internet interaction will lead to many instances where personal jurisdiction will not arise.

The case attracted significant international attention. Media outlets were concerned about the implications for them if nations like Australia accepted jurisdiction for publications which were ostensibly intended for the domestic US market, but which were sold in part or peripherally internationally. The High Court of Australia permits pleadings in appeals by interveners in matters of general public importance if the issue relates to maintaining some particular right, power or immunity which affects the interveners.[15] An intervener has the same rights and obligations as the other parties to the action, including the ability to appeal, tender evidence and participate fully in all aspects of the argument. The interveners included Amazon.com Inc.; Associated Press; Cable News Network LP LLLP (CNN); Guardian Newspapers Ltd; The New York Times Company; News Limited; Time, Inc.; Tribune Company; The Washington Post Company; Yahoo! Inc.; and John Fairfax Holdings Ltd.

The matter came before Hedigan J at first instance. His Honour particularly examined jurisdiction in relation to the internet connection, as this could affect the award of damages.

Geoffrey Robertson QC for Dow Jones made several unsuccessful submissions. Robertson submitted that imposing liability where downloading occurred 'would have a serious "chilling effect" on free speech'. He argued that 'a narrow rule was appropriate for the age of globalisation' and that 'the Internet offer[ed] Australians the greatest hope of overcoming the tyranny of distance'. He submitted that Hedigan J had 'a national duty to decide that there was no jurisdiction in Australia even if [the judge] had a legal view to the contrary, and that it [wa]s [his Honour's] duty publicly to declare that Mr Gutnick's action against Dow Jones take place in New Jersey'. Robertson suggested that downloading is 'self-publishing' and that 'the process is akin to taking a book out of a library in New Jersey and taking it home to Victoria to read'. He 'flirted with the idea' that cyberspace should be a defamation-free zone,[16] and argued that the article was published 'in America for Americans' and that the events constituted 'a tort having an indelibly American complexion'.[17] The plaintiff submitted that publishing occurs when data is made 'intelligible or manifest to a third party' and argued that that has been the law for 400 years in other contexts.

Hedigan J rejected counsel's 'bold assertions', commenting that the unique nature of the internet must not lead to the abandonment of the analysis that the law has traditionally and reasonably followed to reach just conclusions. His Honour expressed concern that Robertson's arguments, 'attractively presented

15 *Australian Railways Union v Victorian Railways Commissioners* (1930) 44 CLR 319 at 33, per Dixon J.
16 See M Kirby, 'Privacy in cyberspace', (1998) *UNSWLJ* 47.
17 [2001] VSC 305, paras 15–20.

as they were, became enmeshed in pop science language' and degenerated into '"sloganeering" which in the end decides nothing'.[18]

Hedigan J referred to the following authorities to formulate his view. First, his Honour politely questioned the value of the decision in *Macquarie Bank v Berg*[19] without making any attempt to directly address issues raised. He then referred to *Digital Equipment Corporation v Alta Vista Technology Inc.*,[20] where to determine jurisdiction the court examined the question of whether the owner and controller of a website can with some degree of certainty 'know' if the content of the site reaches the user. The court felt that:

> Using the Internet . . . is as much knowingly 'sending' . . . as is a *telex, mail or telephonic transmission*; . . . ATI 'knows' that its Website reaches residents . . . who choose to access it, just as surely as it 'knows' any lateral telephone call is likely to reach its destination.

Lee Teck Chee v Merrill Lynch International Bank[21] involved the republication of a Singapore newspaper on the internet. Nathan J, of the High Court of Malaysia, held that publication had not taken place in Malaya because the alleged defamatory words had been published on a Singapore server. However, the judgment is of limited value as there was no evidence presented that any person in Malaya had accessed the website.

In *Kitakufe v Oloya Ltd*[22] the Canadian court assumed jurisdiction for a defamatory statement made in a Ugandan newspaper republished on the internet. The plaintiff was a medical practitioner born in Uganda, but practising in Canada. The defendant argued that Uganda was a more convenient and natural forum, and that the proceedings had numerous real and significant links to Uganda. These included that the plaintiff had assets in Uganda, and that the defendant's concerns regarding malice, ethnic rivalry and the defence's credibility meant that the case would be better dealt with in Uganda. The defence also expressed concern about the significant expense and inconvenience of calling witnesses based in Uganda to a case run in Canada. Despite this multitude of connections with Uganda, Hume J assumed jurisdiction on the basis of access to the website and downloading, concluding that it was 'not satisfied that the plaintiff would not be deprived of a legitimate personal and juridical advantage' if the case were run in Uganda. Hedigan J described this case as one where 'a superior court assumed jurisdiction over a defamation suit on the basis of access to the Website and its reception (that is, downloading) in Ontario, Canada'.

In *Godfrey v Demon Internet Ltd*,[23] Moreland J determined that an internet bulletin board amounted to publishing postings to subscribers. Godfrey was a science lecturer in England and the defendant hosted a news group whereby users could make postings contributing to targeted discussions. One posting

18 [2001] VSC 305, paras 70 and 71.
19 [1999] NSWSC 526.
20 960 F. Supp. 456 (D Mass 1997).
21 [1998] CLJ 188.
22 [1998] OJ No. 2537 QL (Ont Gen Div).
23 [2001] QB 201.

claimed to be Godfrey, but in fact was written by an imposter. The posting was defamatory. Godfrey requested its removal within ten days. The defendant did not and it remained online until it expired in the usual course (two weeks later). Moreland J held that the defendant had published the posting whenever a subscriber accessed the news group:

> In my judgment the defendant, whenever it transmits and whenever there is transmitted from the storage of its news server a defamatory posting, publish[es] that posting to any subscriber to its ISP who accesses the newsgroup containing that posting. Thus every time one of the defendant's customers accesses 'soc.culture.thai' and sees that posting defamatory of the plaintiff there is a publication to that customer.

Hedigan J took particular note of the expression 'sees that posting defamatory of the plaintiff': he regarded this as the equivalent of downloading.

Hedigan J cited *Calder v Jones*[24] with approval, referring to the statement of the US Supreme Court that 'jurisdiction may be exercised over a foreign defendant who directs his or her defamatory message at the forum and the plaintiff suffers harm there'.[25]

From these cases Hedigan J concluded that the place of defamation is 'the jurisdiction where the defamatory material was published and received by the plaintiff, rather than where it was spoken or written', and so the Dow Jones article was published in the state of Victoria when downloaded by Dow Jones subscribers who had met Dow Jones's payment conditions and used their passwords. The defendant's argument that it would be unfair for the publisher to have to litigate in the multitude of jurisdictions in which its statements are downloaded and read, he said, 'must be balanced against the world-wide inconvenience caused to litigants, from Outer Mongolia to the Outer Barcoo'. His Honour stated that 'if you do publish a libel justiciable in another country with its own laws . . . then you may be liable to pay damages for indulging that freedom'.[26]

Hedigan J considered the following additional factors as significant:

- the publication was downloaded in Victoria;
- the plaintiff's residence, business headquarters, family, social and business life are in Victoria;
- the plaintiff seeks to have his Victorian reputation vindicated by the courts of the state in which he lives;
- the plaintiff undertook not to sue in any other place.

His Honour concluded that given these factors 'it would be verging on the extraordinary to suggest that Mr Gutnick's action in respect of that part of the publication on which he sues should be removed for determination to the State of New Jersey'.[27] Dow Jones maintained that the article was 'indelibly American, written by Americans for Americans interested in the stock market and

24 465 US 783 (1984).
25 Counsel also referred to *R v Burdett* (1820) 4 B & Ald 115.
26 [2001] VSC 305, para 75.
27 [2001] VSC 305, para 131.

its affairs'. The weakness in this argument, writes Hedigan J, is that 'the aspect sued on by Mr Gutnick is indelibly Victorian, connected with no other place, and that any documentation or evidence concerning the matter will all be found in Victoria'.[28]

Less than one month after the judgment by Hedigan J, the Victorian Court of Appeal quickly and with little consideration and analysis dismissed the appeal by Dow Jones.[29] The issue then came before the High Court of Australia.

On 10 December 2002 the High Court of Australia handed down its decision in *Dow Jones v Gutnick*. By 7–0, but in four separate judgments, the High Court dismissed the appeal of Dow Jones. The court determined that publication of a defamatory statement for an online subscription website occurs at the place of downloading. Kirby J flagged the danger in finding otherwise, stating:

> To tell a person uploading potentially defamatory material onto a website . . . [that that act] will render that person potentially liable to proceedings in courts of every legal jurisdiction where the subject enjoys a reputation, may have undesirable consequences. Depending on the publisher and the place of its assets, it might freeze publication or censor it or try to restrict access to it in certain countries so as to comply with the most restrictive defamation laws that could apply.[30]

Kirby J dealt with the legal response to the impact of the internet. His Honour described the internet as 'essentially a decentralised, self-maintained telecommunications network' which 'demands a radical reconceptualisation of the applicable common law'.[31] His Honour drew an analogy between the development of such new legal rules and the Law Merchant (*lex mercatoria*), which arose out of the general custom of the merchants of many nations in Europe.

Adventitious and opportunistic

Dow Jones submitted that the applicable law should be that of the place where the web servers were maintained, 'unless that place was merely adventitious or opportunistic'. Dow Jones recognised the argument that publishers could place their servers in jurisdictions most advantageous to them, but argued that it did not 'misuse' this rule, as the location of its servers was not determined for legal or advantageous considerations. Gleeson CJ, McHugh, Gummow and Hayne JJ stated that 'adventitious' and 'opportunistic' are words likely to produce considerable debate, and noted that a publisher may choose the server's location based on a range of factors, including costs of operation, benefits offered for setting up business, security, and continuity of service. They also noted that the publisher may have servers in more than one state or country.

The corollary argument was made that plaintiffs like Gutnick could choose the most favourable jurisdiction to them in which to commence proceedings,

28 [2001] VSC 305, para 129.
29 *Dow Jones v Gutnick* [2001] VSCA 249. This judgment was less than two pages.
30 [2002] HCA 56, para 117.
31 [2002] HCA 56, para 79.

bypassing nations with free speech and freedom of expression. Gutnick's response was that whilst that may be so, on this occasion he sued in the jurisdiction of his business, domicile and residence, and sought no adventitious of opportunistic advantage.

The court's attitude towards either the plaintiff or the defendant choosing the most advantageous jurisdiction or the publisher placing its server in a jurisdiction with no defamation law remains untested.

Gleeson CJ, McHugh, Gummow and Hayne JJ suggested that 'reasonableness of the publisher's conduct' might be considered 'necessary or appropriate' as a common law defence where all the publisher's conduct occurred outside the jurisdiction. Their Honours identified relevant circumstances as 'including where that conduct took place, and what rules about defamation applied in that place or those places'. Their Honours drew an analogy with the developing defence of innocent dissemination.[32]

Kirby J described the dismissal of the appeal as 'contrary to intuition', which 'does not represent a wholly satisfactory outcome.' His Honour seemed to feel that a balance is yet to be determined between the human right of access to information and to free expression, and the human right to protection by law for the reputation and honour of individuals, and that this warrants 'national legislative attention and . . . international discussion in a forum as global as the Internet itself'.[33]

The court determined that the place where damage was inflicted was a most significant factor. The place where the defamation was comprehended and the plaintiff's connection with the locality were important factors.[34]

According to Gleeson CJ, McHugh, Gummow and Hayne JJ, the 'spectre which Dow Jones [seeks] to conjure up' – that a publisher would be forced to consider, for every article it publishes on the internet, 'the defamation laws of every country from Afghanistan to Zimbabwe' – is 'unreal when it is recalled that in all except the most unusual of cases, identifying the person about whom material is to be published will readily identify the defamation law to which that person may resort'.[35]

Effects test

One solution may be the adoption of an effects test. *Calder v Jones*[36] concerned Shirley Jones, an Academy award-winning actress who lived and worked in California. Jones was defamed by an article in the *National Enquirer* which was

32 [2002] HCA 56, para 51.
33 [2002] HCA 56, para 166.
34 Postscript: The High Court of Australia determined that the Supreme Court of Victoria had jurisdiction, but it did not determine liability. The issue of defamation did not reach the court. Dow Jones made an out of court payment of $450,000, but claimed that the payment was not for defamation. Dow Jones claimed it was in the right and that the payment would not cover Gutnick's legal costs. Gutnick claimed vindication.
35 [2002] HCA 56, para 54.
36 465 US 783 (1984).

published in Florida but had a nationwide circulation, including a substantial readership in California. The defendants challenged the jurisdiction of the Californian courts. The US Supreme Court held that California was the 'focal point both of the story and of the harm suffered'. This joint 'target' and 'harm' criteria is often referred to as the 'effects test':

> The allegedly libelous story concerned the California activities of a California resident. It impugned the professionalism of an entertainer whose television career was centered in California. The article was drawn from California sources, and the brunt of the harm, in terms both of respondent's emotional distress and the injury to her professional reputation, was suffered in California. In sum, California is the focal point both of the story and of the harm suffered. Jurisdiction over petitioners is therefore proper in California based on the 'effects' of their Florida conduct in California . . . The mere fact that [the defendant] can 'foresee' that the article will be circulated and have an effect in California is not sufficient for an assertion of jurisdiction . . . [The defendants] are not charged with mere untargeted negligence. Rather, their intentional, and allegedly tortious, actions were expressly aimed at California. Under the circumstances, petitioners must 'reasonably anticipate being hauled into court there' to answer for the truth of the statements made in their article.[37]

This case did not involve the internet or cyberspace, but its approach has been used and quoted in cyberspace jurisdiction cases.[38] The concept that a person ought to be liable for a deliberate act which targets another jurisdiction and causes harm there clearly can be useful for cyberspace quandaries.

The effects test can be compared to the terminatory theory of criminal law espoused by Professor Glanville Williams[39] and approved by Stephen J in *Ward v R.*[40] It is the place where the physical act took effect on its victim, not where the physical act of perpetrator was done, that determines the locus of the crime and, in turn, which courts have jurisdiction in respect of that act.

In *Dow Jones v Gutnick*, Gleeson CJ, McHugh, Gummow and Hayne JJ endorsed the effects test approach:

> Activities that have effects beyond the jurisdiction in which they are done may properly be the concern of the legal systems in each place. In considering where the tort of defamation occurs it is important to recognise the purposes served by the law regarding the conduct as tortious: purposes that are not confined to regulating publishers.[41]

Australian cases

The first Australian case with an element of internet jurisdiction is *Macquarie Bank v Berg*.[42] Berg was a disgruntled ex-employee of Macquarie Bank who

37 465 US 783 (1984), 788–89.
38 For example by Hedigan J in *Gutnick v Dow Jones* [2001] VSC 305 and *Healthgrades.com v Northwest Healthcare Alliance* No. 01-35648 (9th Cir 2002).
39 See G Williams, 'Venue and the Ambit of Criminal Law', Part 3, (1965) 81 *Law Quarterly Review* 518.
40 *Ward v R* [1980] HCA 11, para 4 per Stephen J.
41 [2002] HCA 56, para 24.
42 [1999] NSWSC 526.

placed defamatory material regarding the bank and its senior employees on a website. Berg had moved to California and the material in question was located on a server there, but could be accessed and viewed in New South Wales. Macquarie Bank sought an *ex parte* interlocutory injunction to restrain publication of the material on the internet. Simpson J, bearing in mind the fact that publication on the internet is continuous, 24 hours a day, for as long as the material remains online, described the difficulties of such a request in the following terms:

> . . . once published on the Internet, material is transmitted anywhere in the world that has an Internet connection . . . to make the order as initially sought, would have the effect of restraining publication of all the material presently contained on the website to any place in the world. Recognising the difficulties associated with orders of such breadth, [counsel] sought to narrow the claim by limiting the order sought to publication or dissemination 'within NSW'. The limitation, however, is ineffective.[43]

Simpson J did not accept jurisdiction and refused to grant the injunction. However, her Honour spuriously reasoned:

> The difficulties are obvious. An injunction to restrain defamation in NSW is designed to ensure compliance with the laws of NSW, and to protect the rights of plaintiffs, as those rights are defined by the law of NSW. Such an injunction is not designed to superimpose the law of NSW relating to defamation on every other state, territory and country of the world. Yet that would be the effect of an order restraining publication on the Internet. It is not to be assumed that the law of defamation in other countries is coextensive with that of NSW, and indeed, one knows that it is not. It may very well be that, according to the law of the Bahamas, Tazhakistan, or Mongolia, the defendant has an unfettered right to publish the material. To make an order interfering with such a right would exceed the proper limits of the use of the injunctive power of this court.[44]

This reasoning repeats the erroneous approach of *Inset System Inc. v Instruction Set Inc.* (see below).[45] The mere fact that a website reaches another jurisdiction is not a sufficient ground to grant or assume that jurisdiction. However, Berg clearly had several connections with New South Wales, and the website was dedicated entirely to Macquarie Bank and its senior employees. In terms of an effects test, the target and the harm were entirely in New South Wales, so Berg ought to have expected to be brought before the courts there.

The factors against the grant of jurisdiction included the enforceability of such an order, the undesirability of superimposing the defamation law of New South Wales onto every other state, territory and country of the world, and the interlocutory nature of the application.[46]

In *Airways Corporation of NZ Ltd v PriceWaterhouse Coopers Legal*[47] Simpson J had the opportunity to reconsider her decision in Macquarie Bank. The proceedings arose out of the publication, by email, of certain material capable of

43 [1999] NSWSC 526, para 12.
44 [1999] NSWSC 526, para 14.
45 937 F. Supp. 161 (D Conn 1996).
46 [1999] NSWSC 526, see paras 14–16.
47 [2002] NSWSC 138.

defaming the plaintiffs. The only evidence as to receipt of the material was from two witnesses, each of whom received it in New Zealand. Counsel argued that the evidence disclosed no connection with New South Wales, and that the appropriate forum for the commencement of any defamation proceedings should be New Zealand. Counsel cited the decision of Hedigan J in *Gutnick v Dow* and the Court of Appeal. Simpson J stated:

> In these decisions it was held that *internet publications take place in the location where they are read, and not in the location from which they are transmitted*... For present purposes, I am prepared to accept the conclusions of the Supreme Court of Victoria and the Court of Appeal of Victoria but bear in mind that these conclusions were reached on the basis of evidence placed before Hedigan J. That evidence was not necessarily identical to the evidence on this issue placed before me. Nevertheless, for present purposes I work on the basis that publication by email takes place where the email is received and that, on the evidence in this case, that was in New Zealand.[48]

Thus Simpson J reversed her previous position, though making use of qualifying language.

The NSW Supreme Court had the opportunity to revisit the *Macquarie Bank Limited v Berg* case in mid-2002.[49] Smart AJ dealt with procedural issues, but in light of the Dow Jones decisions in the Victorian Supreme Court,[50] Macquarie Bank's counsel reserved the right to argue whether the placement on the internet in the United States of defamatory material accessible in New South Wales amounted to publication of the material in New South Wales. Berg submitted that the case was 'overwhelmed by the application of Californian law and that the indications were that the plaintiffs could not win in California because of the law as to free speech'. Smart AJ concluded that until 'the facts are found definitively it is too early to say which law applies'.[51]

Smart AJ distinguished the Dow Jones case, noting, without deciding the issue, that 'Gutnick was not concerned with the World Wide Web because Dow Jones only put its material on for subscribers or trial subscribers.'[52]

In *ACCC v Purple Harmony Plates Pty Limited*[53] the Federal Court of Australia considered the difference between websites registered internationally with .com and Australian-administered and registered sites with .au. The defendants were in breach of the Trade Practices Act by advertising on the website in a manner which was misleading and deceptive. The defendants challenged the jurisdiction of the court on several spurious grounds. In relation to the website, the defendant argued that it used a .com and not a .au registration and as such the site should be governed by the laws of Virginia, where the .com sites were administered. Goldberg J rejected this approach: the registrant was Australian, and it was

48 [2002] NSWSC 138, para 12.
49 [2002] NSWSC 254.
50 See below.
51 [2002] NSWSC 254, para 55.
52 [2002] NSWSC 254, para 41.
53 [2001] FCA 1062.

irrelevant that the agreement with the domain name administrator may not be subject to the jurisdiction of Australian courts.

Early US experience

The early cases emanated from the United States, where disputes arose between parties in different states. Although these US cases involved constitutional factors, more than one common law jurisdiction has been guided by them.[54]

In the United States, to determine personal jurisdiction, it is sufficient if the defendant 'purposefully availed' itself of the jurisdiction (such as by repeatedly conducting business there). US courts are required to use due process (under the 14th Amendment to the US Constitution). In *International Shoe Co. v Washington*,[55] the US Supreme Court held that due process requires the defendant to have 'minimum contacts' with the plaintiff's jurisdiction and prescribes that the choice of jurisdiction be consistent with 'traditional notions of fair play and substantial justice'.[56]

The 'purposeful availment' requirement ensures that random and fortuitous contacts do not cause a defendant to be improperly brought into a forum. The 'fair play' requirement permits the court to examine whether the defendant's acts or their consequences have sufficient connection with the state to make jurisdiction reasonable.

Early cyberspace cases mistakenly granted jurisdiction merely on the basis that the defendant's website could be accessed in the plaintiff's jurisdiction. In *Inset System Inc. v Instruction Set Inc.*,[57] the court, on hearing evidence that as many as 10,000 residents of the state could access the website, granted jurisdiction over a defendant who maintained a mere passive website. This number of potential users is now ridiculously conservative.

In *Minnesota v Granite Gate Resorts Inc.*,[58] the Minnesota court examined five factors: the quantity of contacts; the quality of contacts; the connection between the cause of action and the contacts; the state's interest; and the convenience of the parties.

The courts later distinguish between passive and active websites. Passive sites merely display information. Active websites may provide for form filling, and for contracting online, or may simply provide an email link. On this basis, in *Zippo Manufacturing Co. v Zippo Dot Com Inc.*[59] the court formulated a sliding scale approach which has since been generally adopted.

54 For example, see the British Columbia Court of Appeal in *Braintech Inc. v Kostiuk* (1999) 63 BCLR (3d) 156 and the High Court of Delhi in *Yahoo Inc. v Akash Arosa* [1999] FSR 931.
55 326 US 310 (1945).
56 326 US 310 (1945), 316.
57 937 F. Supp. 161 (D Conn 1996). See also *Maritz Inc. v Cybergold Inc.* 947 F. Supp. 1328 (ED Mo 1996).
58 568 NW 2d 715.
59 952 F. Supp. 1119 (WD Pa 1997). See also *Helicopteros Nationales de Columbia SA v Hall* 466 US 408.

At one end of the scale:

> are situations where a defendant has simply posted information on an Internet Web site which is accessible to users in foreign jurisdictions. A passive Web site that does little more than make information available to those who are interested in it is not grounds for the exercise of personal jurisdiction.[60]

For example, in *Bensusan Restaurant Corp. v King*[61] the plaintiff was the owner of the well-known Blue Note jazz night club in New York. The defendants advertised their night club, which had the same name, in Missouri; it catered mainly for local university students. Notwithstanding evidence that New York residents both accessed and indeed made bookings based on the information from the Missouri website, the Appeals Court held that passive websites do not meet the standard of 'purposeful availment'. There was no element of 'targeting' (see Effects test, above) and accordingly no substantial connection was created.[62]

In *Webber v Jolly Hotels*[63] the New Jersey-based plaintiff booked a holiday in Italy online. The defendant ran several hotels in Italy and had no connection with the United States. The website provided photographs of hotel rooms, descriptions of hotel facilities, information about numbers of rooms and telephone numbers. The plaintiff was injured whilst on holiday in the defendant's hotel. She commenced proceedings in New Jersey. The court held that it lacked jurisdiction over the defendant. Access to the website alone was held to be insufficient. The court required continuous and substantial contacts with the forum to establish the defendant's personal jurisdiction: 'exercising jurisdiction over a defendant who merely advertises its services or product on the Internet would violate the Due Process Clause of the Fourteenth Amendment [and] would disrespect the principles established by International Shoe'. The defendant's website provided no more than passive advertising information.[64]

At the other end of the scale:

> are situations where a defendant clearly does business over the Internet. If the defendant enters into contracts with residents of a foreign jurisdiction that involve the knowing and repeated transmission of computer files over the Internet, personal jurisdiction is proper.[65]

The Appeals Court in Zippo's case cited *Compuserve Inc. v Patterson*[66] as an example: there the defendant had purposefully directed his business activities towards residents of another state and the repeated transmissions showed a substantial connection with that state. Acts such as selling goods or services would in all likelihood be sufficient to constitute 'business' on the internet.

60 952 F. Supp. 1119, 1124 (WD Pa 1997).
61 937 F. Supp. 296 (SD NY 1996).
62 See also *Cybersell Inc. v Cybersell Inc.* 130 F. 3d 414 (9th Cir 1997).
63 977 F. Supp. 327 (D NJ 1997).
64 See also *Casio India Co. Ltd v Ashita Tele Systems Pvt Ltd* (2003 (27) PTC 265 (Del)), where the Delhi High Court held that once a website can be accessed from Delhi, it is enough to invoke the territorial jurisdiction of the court.
65 *Zippo Manufacturing Co. v Zippo Dot Com Inc.* 952 F. Supp. 1119 (WD Pa 1997), 1124.
66 89 F. 3d 1257 (6th Cir 1996).

In *Playboy Enterprises Inc. v Chuckleberry Publishing Inc.*[67] the court held that the Italy-based defendant actively solicited US customers to its internet site, and in doing so had distributed its product within the United States in breach of a contempt order. One of defendant's websites was not just a source of passive information, but was a 'pay' site. The defendant knew that US citizens accessed its site and ought to have expected to be held accountable for its breach of the contempt order. Taking jurisdiction was thus appropriate.

> The middle ground is occupied by interactive Web sites where a user can exchange information with the host computer. In these cases, the exercise of jurisdiction is determined by examining the level of interactivity and commercial nature of the exchange of information that occurs on the Web site.[68]

These initial cases are difficult to reconcile as they are complicated by non-internet factors and by the varying degrees of judicial understanding of the technology involved.[69] First, the defendant must have acted deliberately and purposefully. The courts then determine the defendant's level of cognisance. Next, the defendant must have targeted or singled out a particular forum. If the defendant has repeatedly and knowingly engaged in business with the plaintiff, the grant of jurisdiction is more likely.

Universal rights

Australia and the United States are signatories to the UN International Covenant on Civil and Political Rights (ICCPR). In Dow Jones, Kirby J argued that the Australian courts must uphold the right to freedom of speech and freedom of expression as enunciated in article 19 of the ICCPR:

> **Article 19**
>
> 1. Everyone shall have the right to hold opinions without interference.
> 2. Everyone shall have the right to freedom of expression; this right shall include freedom to seek, receive and impart information and ideas of all kinds, regardless of frontiers, either orally, in writing or in print, in the form of art, or through any other media of his choice.
> 3. The exercise of the rights provided for in paragraph 2 of this article carries with it special duties and responsibilities. It may therefore be subject to certain restrictions, but these shall only be such as are provided by law and are necessary:
>   (a) For respect of the rights or reputations of others;
>   (b) For the protection of national security or of public order (ordre public), or of public health or morals.

67 939 F. Supp. 1032 (SD NY 1996).
68 *Zippo Manufacturing Co. v Zippo Dot Com Inc.* 952 F. Supp. 1119 (WD Pa 1997), 1124.
69 Criticism has been levelled at the minimum contacts test in relation to the internet and as to what level of interactivity is required to trigger jurisdiction. The required level of 'interactivity' can be gleaned from *Compuserve v Patterson* 89 F. 3d 1257 (6th Cir 1996) and *Cybersell Inc. v Cybersell Inc.* 130 F. 3d 414 (9th Cir 1997).

Kirby J was the only High Court Justice to address the effect of this article of the covenant. However, his Honour balanced this with the requirements of articles 17.1 and 17.2:

> **Article 17**
> 1. No one shall be subjected to arbitrary or unlawful interference with his privacy, family, home or correspondence, nor to unlawful attacks on his honour and reputation.
> 2. Everyone has the right to the protection of the law against such interference or attacks.

On the one hand Kirby J stated that any development of the common law of Australia should be consistent with the ICCPR principles, to the extent that Australian law does not provide effective legal protection for the honour, reputation and personal privacy of individuals: 'Australia, like other nations so obliged, is rendered accountable to the relevant treaty body for such default'[70] pursuant to the first Optional Protocol to the ICCPR.

On the other hand, he foreshadowed Dow Jones' plea, stating the 'the need for a clear and single rule to govern the conduct in question according to pre-established norms'.

In response to the High Court's decision, Dow Jones issued a press release supporting an action by William Alpert, the author of the article in question. In 2003 Alpert stated that he was filing an action with the Human Rights Commission in Geneva, fearing unjustified restrictions on journalists. Alpert stated:

> I am filing this action with the Human Rights Commission in Geneva because I fear restrictions on the ability of financial journalists such as myself to report truthfully to United States investors on the activities of foreigners who are actively engaged in the U.S. markets. I even fear for our ability to report on U.S. corporations and business people, who might see the High Court's decision as an invitation to attack the U.S. press in a remote forum. Given the differences between the laws of Australia and those of other countries in the Commonwealth and beyond, the impact of Australia's law – as laid out by the High Cour – could harm journalists throughout the world. Powerful and sophisticated plaintiffs could search out overseas jurisdictions willing to help stifle news coverage that was only directed at local readers in those journalists' home markets.
>
> Australia has accepted the jurisdiction of the U.N. Human Rights Committee and is obliged to modify Australia's libel laws, should the Committee find that those laws unduly restrict the right of free speech that's protected under Article 19 of the International Convention on Human Rights.
>
> I hope that the Human Rights Committee will recognize the threat to free speech – and an informed public – posed by Australian laws that allow suit against any journalist, anywhere that an article published on the internet can be downloaded. Perhaps the Government of Australia will recognize the need to modify its laws, even before the Human Rights Committee takes up this case.[71]

70 *Dow Jones v Gutnick* [2002] HCA 56, para 116.
71 Dow Jones press release, April 2003.

This foreshadowed action may have been politic, but it seems that it did not in fact proceed.

Alternative approaches

Three days after the decision was handed down by the High Court in the Dow Jones case, the 4th Circuit Appeals Court in the United States handed down its decision in *Young v New Haven Advocate.*[72] Two Connecticut newspapers published articles on the internet allegedly defaming a Virginia resident. The *New Haven Advocate* was a free newspaper published once a week in New Haven, Connecticut with a small number of subscribers, none from Virginia. The *Hartford Courant* is a daily newspaper distributed in and serving Hartford, Connecticut. The *Courant* had eight mail subscribers residing in Virginia. Neither newspaper solicits subscriptions in Virginia, nor maintains any offices or employees within that state. Both newspapers posted some content on the internet. The articles included unfavourable comments regarding a movement of inmates between prisons and prison conditions. Young was the warden at one of the prisons and commenced action in Virginia against the newspapers and the journalists.

In a statement somewhat reminiscent of the Zippo interactivity test, the court said that it would only accept personal jurisdiction where the defendant:

> (1) directs electronic activity into the State, (2) with the manifested intent of engaging in business or other interactions within the State, and (3) that activity creates, in a person within the State, a potential cause of action cognizable in the State's courts.[73]

On the facts, the Appeals Courts held that as the newspapers were small Connecticut-based publications which served 'almost exclusively Connecticut readers', they would not be subject to jurisdiction in Virginia courts. The courts reasoned that the defendants did not manifest an intent to engage in business or other interactions in Virginia. The effects test in *Calder v Jones* was applied, although more narrowly:

> The facts in this case establish that the newspapers' websites, as well as the articles in question, were aimed at a Connecticut audience. The newspapers did not post materials on the Internet with the manifest intent on targeting Virginia readers. Accordingly, the newspapers could not have 'reasonably anticipate[d] being hauled into court [in Virginia] to answer for the truth of the statements made in their article[s].' Calder, 465 U.S. at 790 (quotation omitted). In sum, the newspapers do not have sufficient Internet contacts with Virginia to permit the district court to exercise specific jurisdiction over them.[74]

72 315 F. 3d 256 (4th Cir 2003)
73 The court was quoting from *Scan Inc. v Digital Service Consultants Inc.* 293 F. 3d 707, 714 (4th Cir 2002).
74 *Young v New Haven Advocate* 315 F. 3d 256, 264 (4th Cir 2003).

In *Healthgrades.com v Northwest Healthcare Alliance*[75] the US Supreme Court declined an opportunity to clarify the jurisdiction test when it denied *certiorari* without opinion. The 9th Circuit Court of Appeals held that Washington State courts may exercise personal jurisdiction over Healthgrades.com, a Delaware corporation with its principal place of business in Colorado, in a defamation action arising out of statements found on defendant's website that caused injury to the plaintiff in Washington. The statements were ratings of home health-care providers, including the Washington-based plaintiff. The Appeals Court approved and applied the effects test of *Calder v Jones*:[76]

> We find that defendant Healthgrades.com has purposefully interjected itself into the Washington state home health care market through its intentional act of offering ratings of Washington medical service providers. This act was expressly aimed at plaintiff's forum state, since defendant was well aware that its rating of Washington home health care providers would be of value primarily to Washington consumers . . . the allegedly defamatory rating received by plaintiff on defendant's website concerned the Washington activities of a Washington resident. Finally, the brunt of the harm allegedly suffered by plaintiff occurred in Washington – where plaintiff is incorporated, where plaintiff has its principal place of business, and where plaintiff's reputation is likely to suffer if in fact it has been injured by defendant's actions. The effects, therefore, of defendant's out-of-state conduct were felt in Washington. Plaintiff's claims arise from that out-of-state conduct, and defendant could reasonably expect to be called to account for its conduct in the forum where it understood the effects of its action would be felt.

While the underlying reasoning in HealthGrades and *Young v New Haven* appears to differ, it can be argued that its application to the facts is identical: a newspaper, even if published on the internet, may be aimed at a local audience, and a healthcare rating service is aimed wherever the rated health care providers are located.

Though the Dow Jones case caused international concern, its impact has been limited, implicitly if not expressly, to subscription websites. In *Ward Group Pty Ltd v Brodie & Stone Plc*[77] the Federal Court of Australia distinguished the Dow Jones case:

> As the allegedly defamatory publication was made available to subscribers in Victoria on the Internet, no issue arose about the publication of that material in Victoria. Therefore that case is of no assistance to the Ward Group.[78]

However, the Federal Court accepted a test implicitly identical to the effects test in *Calder v Jones.* The court held that the use of a trade mark on the internet, uploaded on a website outside of Australia, without more, is not a use by the website proprietor of the mark in each jurisdiction where the mark is downloaded. However:

75 No. 01-35648 (9th Cir 2002).
76 See above.
77 [2005] FCA 471.
78 [2005] FCA 471, para 42.

> if there is evidence that the use was specifically intended to be made in, or *directed or targeted at*, a particular jurisdiction then there is likely to be a use in that jurisdiction when the mark is downloaded. Of course, once the website intends to make and makes a specific use of the mark in relation to a particular person or persons in a jurisdiction there will be little difficulty in concluding that the website proprietor used the mark in that jurisdiction when the mark is downloaded.[79]

In what began as an April Fools Day joke, Dow Jones, through its publication the *Wall Street Journal*, published an article about the floating of shares by Harrods, with the headline 'The Enron of Britain?' Action was commenced by Harrods in England, claiming that the article caused serious damage to the company's reputation. Dow Jones filed suit in the Southern District of New York, seeking to block the action. In the US case Judge Marrero said that Dow Jones was eager to avoid a British trial because Britain's libel laws were far more friendly to plaintiffs.[80] In a declaratory judgment, Judge Marrero expressed concern that any relief granted by the New York Court would nevertheless be subject to recognition and enforcement in the United Kingdom. Dow Jones argued that a declaratory judgment was appropriate in that it would free them from 'vexatious and oppressive' litigation abroad. In refusing relief, the judge held that granting authority to enjoin foreign lawsuits should be done sparingly and only with care and restraint.

Single publication rule

Most US states have adopted a single publication rule which treats all sales of a defamatory book or newspaper as a single publication. The time of the single publication (download) is fixed as the time of the first publication. In the United Kingdom and Australia each publication gives rise to a separate cause of action. That principle was confirmed by the English Court of Appeal in *Loutchansky v Times Newspapers Ltd (Nos 2–5)*.[81]

Samuels JA's observation in *Australian Broadcasting Corporation v Waterhouse*,[82] that a single publication rule could only be introduced throughout Australia by statute, was approved by Kirby J in *Dow Jones v Gutnick*.[83]

In the same case Gaudron J described the single publication rule as 'a legal fiction which deems a widely disseminated communication . . . to be a single communication regardless of the number of people to whom, or the number of states in which, it is circulated'.[84] Dow Jones described the Australian position as 'primitive'.

79 [2005] FCA 471, para 43, emphasis added.
80 *Dow Jones & Co. Inc. v Harrods Ltd*, 237 F. Supp. 2d 394 (SD NY 2002).
81 [2002] QB 783.
82 (1991) 25 NSWLR 519.
83 [2002] HCA 56, para 127.
84 [2002] HCA 56, para 57. See Debra Cohen, 'The single publication rule: One action, not one law', (1966) 62 *Brooklyn Law Review* 921.

Callinan J rejected Dow Jones' submission that publication occurs at one place, such as the place where the matter is provided, or first published, on the grounds that that view 'cannot withstand any reasonable test of certainty and fairness'. However, he expressed concern that publishers would set up in a 'defamation free jurisdiction' or one in which the defamation laws are tilted towards defendants. Why, he asked, 'would publishers, owing duties to their shareholders to maximise profits, do otherwise?'[85]

Gleeson CJ, McHugh, Gummow and Hayne JJ noted that in *Firth v State of New York*[86] the New York Court of Appeals decided that the one-year statute of limitation in New York runs from the first posting of defamatory matter upon an internet site and that the single publication rule applies to that first posting. The Australian position on the position of the limitation period is yet to be determined.

In *Harrods Limited v Dow Jones & Co. Inc.*[87] Justice Eady noted the limited amount of publication in England (only 10 copies of the *Wall Street Journal*) and that there were very few English hits to the *Wall Street Journal* website. Nevertheless jurisdiction was granted to protect a reputation in England. His Honour rejected the single publication rule, which provides that a defendant should only be sued in the place of the first publication of the defamatory material, finding instead that if a target of a publication is based in England, then an online or offline publisher can be brought before the UK courts.

Substantial publication

Jurisdiction was not taken by the English courts in *Jameel v Dow Jones & Co. Inc.*,[88] where Dow Jones published the article 'WAR ON TERROR, List of Early al Qaeda Donors Points to Saudi Elite, Charities' in the *Wall Street Journal*, allegedly defaming Yousef Jameel, a Saudi businessman. Jameel was unable to prove 'substantial harm'. Although there were 6000 online subscribers to the *Wall Street Journal*, technical evidence showed that only five subscribers had actually accessed the article, and that three of those five were associated with Jameel. The other two subscribers had no knowledge of or connection to Jameel. The court ruled that without a 'substantial' publication in England, there could have been no damage to Jameel's reputation.

Uniform defamation legislation – choice of law

From 1 January 2006, in all states and territories in Australia except the Northern Territory, legislation for uniform defamation law commenced:[89]

85 [2002] HCA 56, para 199.
86 775 NE 2d 463 (2002).
87 [2003] EWHC 1162.
88 [2005] EWCA Civ 75.
89 The Northern Territory's uniform legislation commenced on 26 April 2006. See Chapter 12.

Choice of law for defamation proceedings

(1) If a matter is published wholly within a particular Australian jurisdictional area, the substantive law that is applicable in that area must be applied in this jurisdiction to determine any cause of action for defamation based on the publication.

(2) If there is a multiple publication of matter in more than one Australian jurisdictional area, the substantive law applicable in the Australian jurisdictional area with which the harm occasioned by the publication as a whole has its closest connection must be applied in this jurisdiction to determine each cause of action for defamation based on the publication.

(3) In determining the Australian jurisdictional area with which the harm occasioned by a publication of matter has its closest connection, a court may take into account:
  (a) the place at the time of publication where the plaintiff was ordinarily resident or, in the case of a corporation that may assert a cause of action for defamation, the place where the corporation had its principal place of business at that time, and
  (b) the extent of publication in each relevant Australian jurisdictional area, and
  (c) the extent of harm sustained by the plaintiff in each relevant Australian jurisdictional area, and
  (d) any other matter that the court considers relevant.[90]

Subsection (2) in particular assist in determining jurisdiction within Australia. This provision is in part a codification of the 'effects test' postulated in the US case of *Calder v Jones*.[91]

Conclusion

The application of traditional rules has proved appropriate for cases such as Bensusan and Macquarie Bank. Where the internet is used peripherally – for example, where conduct remains in a single, specific, geographical location – the mere use of the internet should not complicate the application of the jurisdictional principles.

In the absence of an unusual feature involving the internet as the medium of choice, courts should apply the usual principles for jurisdictional questions. Quite often there are a number of factors involved in a given dispute, including the use of the telephone and mail. Any peripheral effect of the internet should be disregarded unless it is central to the dispute. Where the internet is central to a dispute, consideration should be given to the effect of non-internet factors. The mere possibility of access to a website should be insufficient grounds for jurisdiction. Similarly, the mere viewing of a site, as opposed to engaging in some further interaction, should be disregarded. The courts should adopt the distinction between passive and active sites. The greater the degree of interactivity the greater the likelihood that the site's owner is engaging, or should be

90 *Defamation Act 2005* (NSW) s11, *Defamation Act 2005* (Qld) s11, *Defamation Act 2005* (Tas) s11, *Defamation Act 2005* (Vic) s11, *Defamation Act 2005* (WA) s11, *Defamation Act 2005* (SA) s11; *Defamation Act 2006* (NT) s10; and *Civil Law (Wrongs) Act 2002* (ACT) s123.

91 See *Effects test*, above.

aware that he or she is engaging, in commerce with the user. Correspondingly, where the site owner purposefully avails himself or herself of the benefits of commerce from that jurisdiction, he or she should be more likely to be subject to that jurisdiction. If courts take jurisdiction too broadly or recklessly there should be legislative intervention to impose a fairness requirement similar to the approach taken in the US cases.[92] The effect of action by the defendant and its likelihood to result in harm to a known target will also affect the courts' reasoning in taking jurisdiction.

The development of the internet permits various uses.[93] Distinctions will continue to be made between the intended, expected and foreseeable consequences of those uses. The typical website is open to the world, and evidence of use and access within a court's jurisdiction is necessary before that court will grant jurisdiction. The same cannot be said of subscription websites, where it is reasonable to assume use and access by subscribers. Discussion forums or bulletin boards may similarly be open or closed. Proof of use and of impact are necessary for jurisdiction to be granted. Electronic mail is less of an issue as it is typically targeted. Nevertheless, the location of the recipient, the computer and server become complicating factors.

Further reading

Anonymous, 'A "category-specific" legislative approach to the internet personal jurisdiction problem in US law', (2004) 117 *Harv L Rev* 1617.

Cohen, Debra, 'The single publication rule: One action, not one law', (1966) 62 *Brooklyn Law Review* 921.

92 In relation to due process and the 14th Amendment; *International Shoe Co. v Washington* 326 US 310.

93 For example, *Australian Football League v Age Company Ltd* [2006] VSC 308 involved a discussion forum. The Victorian Supreme Court described the discussion forum as enabling 'opinions, gossip, trivia, rumour and speculation to be published as an assertion of fact by anonymous contributors' (para 55).

12

Defamation in cyberspace

From the beginning, the boundaries of appropriate and acceptable behaviour on the internet have been challenged. The notion of the internet as the last bastion of free speech has produced a general mindset that the laws that bind and regulate social behaviour should not apply in that space. Arguably, the laws that have been most challenged relate to intellectual property rights. However, there has developed a sense that anything written in an electronic forum should somehow be immune from oversight and censure. Courts internationally have disagreed with this view, particularly in the realm of defamation.

There is no universal definition of defamation, and so no one set of common elements that need to be satisfied in all jurisdictions. Articles 17 and 19 of the UN International Covenant on Civil and Political Rights (ICCPR) provide for both freedom of expression and the right to hold opinions, but balance these rights and freedoms with considerations of unlawful interference, privacy and the protection of honour and reputation.[1]

These articles were considered and applied by Kirby J in the internet defamation case of *Dow Jones v Gutnick*.[2] (The case concerned the reputation of a Melbourne-based businessman, Joseph Gutnick, who alleged that the Dow Jones corporation defamed him by publishing an article titled 'Unholy Gains' in *Barrons*, an international financial magazine. *Barrons* could be purchased in hardcopy or readers could subscribe online.) According to Kirby J:

> any development of the common law of Australia, consistent with such principles, should provide effective legal protection for the honour, reputation and personal privacy of individuals. To the extent that our law does not do so, Australia, like

1 See Chapter 11.
2 [2002] HCA 56. See Chapter 11.

> other nations so obliged, is rendered accountable to the relevant treaty body for such default.[3]

The laws dealing with defamation have always struggled, in a philosophical sense, with freedom of speech, freedom of the press, the framing of political debate, and the public disclosure of malfeasance, as these have all in some ways always inherently contradicted the rights of protection afforded by the principles of defamation. While the 'truth' of defamatory remarks has generally been held to be a defence to defamation suits,[4] the burden of proof for establishing truth has occasionally been shown to be so onerous as to destroy any protection afforded by this defence.

The opportunity for a significant body of 'e-defamation' case law to develop has been limited. Many jurisdictions are struggling with content-specific issues of cyberspace and have yet to address the implications of content, such as its effect on reputation. Defamation law in cyberspace will be tested in the courts and the interface between the law and the internet will continue to be built.

Any discussion of the law of defamation in cyberspace must be conducted in conjunction with a discussion of jurisdictional issues in cyberspace.[5] This chapter will provide an overview of the legal position regarding online defamation in Australia and, to some extent, in the north American and UK jurisdictions.

Defamation principles

Kirby J in *Dow Jones v Gutnick*[6] restated defamation principles in the following terms:

> First, a starting point for the consideration of the submission must be an acceptance that the principles of defamation law invoked by the respondent are settled and of long standing. Those principles are:
> (1) that damage to reputation is essential for the existence of the tort of defamation;
> (2) that mere composition and writing of words is not enough to constitute the tort; those words must be communicated to a third party who comprehends them;
> (3) that each time there is such a communication, the plaintiff has a new cause of action; and
> (4) that a publisher is liable for publication in a particular jurisdiction where that is the intended or natural and probable consequence of its acts.[7]

3 [2002] HCA 56, para 116. See also *Chakravarti v Advertiser Newspapers Ltd* [1998] HCA 37; (1998) 193 CLR 519, 575.
4 *Defamation Act 2005* (Qld) s26, *Defamation Act 2005* (NSW) s26, *Defamation Act 2005* (Vic) s26, *Defamation Act 2005* (Tas) s26, *Defamation Act 2005* (WA) s26; *Defamation Act 2005* (SA) s24; *Defamation Act 2006* (NT) s23; *Civil Law (Wrongs) Act 2002* (ACT) s136.
5 See Chapter 11.
6 [2002] HCA 56.
7 [2002] HCA 56, para 124.

Defamation reform

The law of defamation in Australia is provided by state and territory legislation. Until recently it has differed quite significantly between jurisdictions, particularly in terms of the elements that need to be satisfied to establish an action for defamation. The term 'defamation' has historically been a composite tort expression encompassing both slander and libel, each of these being separate torts for verbal and written defamation respectively, and this distinction persists in some jurisdictions. All state and territory jurisdictions recently reformed their statutes in order 'to ensure the law of defamation within the State(s) . . . is uniform in substance with the law of defamation in all Australian jurisdictions'.[8]

One of the reforms provides for the abolition of the distinction between spoken and published defamation. None dealt specifically with online defamation. The Working Group of State and Territory Officers that created the reform proposals did, however, highlight the need for the defence of 'innocent dissemination' to be drafted in such a way as to 'ensure that the defence takes proper account of modern means of mass communication and the problems associated with its monitoring'.[9] This has been reflected in the legislation by provisions that exempt from the definition of 'primary distributor' those who provide services for 'the processing, copying, distributing or selling of any electronic medium in or on which the matter is recorded or the operation of, or the provision of, any equipment, system or service, by means of which the matter is retrieved, copied, distributed or made available in electronic form'.[10]

These reforms were specifically aimed at standardising the legislative response to the case law across all state and territory jurisdictions, not at codifying the existing case law. The revised statutes do not affect the operation of the precedents already in law except to the extent that the statutes specifically provide.

Defamation in cyberspace – actions and issues

It has been established by both Australian courts and courts in other jurisdictions that a person may be defamed in an online forum and that damages may be awarded. The decision of the Supreme Court of Western Australia in 1993 in

8 Standing Committee of Attorneys-General (SCAG) Working Group of State and Territory Officers, Legislation and Policy Division of the NSW Attorney General's Department, (2004) *Proposal for Uniform Defamation Laws*.

9 Recommendation 12: The common law defence of innocent dissemination should apply, with careful drafting to ensure that the defence takes proper account of modern means of mass communication and the problems associated with its monitoring.

10 *Defamation Act 2005* (Qld) s32(3)(f)(i) and (ii); *Defamation Act 2005* (NSW) s32(3)(f)(i) and (ii); *Defamation Act 2005* (Vic) s32(3)(f)(i) and (ii); *Defamation Act 2005* (Tas) s32(3)(f)(i) and (ii); *Defamation Act 2005* (WA) s32(3)(f)(i) and (ii); *Defamation Act 2005* (SA) s30(3)(f)(i) and (ii); *Defamation Act 2006* (NT) s29(3)(f)(i) and (ii); *Civil Law (Wrongs) Act 2002* (ACT) s139C(3)(f)(i) and (ii).

Rindos v Hardwick[11] was the first internet-based defamation determination in any jurisdiction. The decision did not advance principles of online defamation but the award of $40 000 in damages for untrue allegations of paedophilia demonstrated that a person can be held liable for statements made in the public space of the internet.

There have been few defamation actions related to online publication to date.[12] In the Australian context, this is likely to remain so given that the defamation reforms of 2006 specifically provide for an 'offer of amends' process that must precede formal litigation.[13] Historically, the introduction of such alternative dispute resolution mechanisms has resulted in a reduction in litigation; the impact on defamation actions is likely to be the same.

Electronic publications using the internet increase the scope of defamation actions in three ways. First, it is an additional mode of communication: many newspapers, magazines and media outlets republish their material online, and radio and television stations also place videos, sound files or transcripts online. Second, online publication of new material, including material that is otherwise intended to be relatively private, has a broad reach. Third, the broader reach can affect the amount of damages. Publication can occur through:

- the sending and redistribution of emails;
- postings to bulletin boards, news groups or discussion lists;
- chat rooms;
- information placed on web pages, including text, sound and video; and
- files that are made available for downloading.

These actions can occur easily and unintentionally, and can reach a large and indeterminate audience. In Australia, a separate cause of action arises each time defamatory material is published (see Single publication rule, below). This means that when defamatory material on the internet is downloaded and read, a new cause of action arises. If a person merely repeats another's defamatory statement, a new and separate cause of action arises,[14] so employees who forward defamatory emails may also be liable.

It is often the case that the authors of defamatory statements are protected from legal proceedings by their impecuniosity; those damaged may then seek out a party more likely to be able to meet any damages awards. This is often unsuccessful because the law in many jurisdictions allows a publisher or disseminator protection through the principle of 'innocent dissemination'. Such 'innocence' is greatly qualified, however, and defendants using this defence have a high evidentiary burden in many jurisdictions. In Australia, the innocent dissemination defence appears in every state and territory jurisdiction:

11 (1993) Unreported decision of the Supreme Court of Western Australia, case 1994 of 1993.

12 See *The Buddhist Society of Western Australia Inc. v Bristile Ltd* [2000] WASCA 210.

13 *Defamation Act 2005* (Qld) Part 3; *Defamation Act 2005* (NSW) Part 3; *Defamation Act 2005* (Vic) Part 3; *Defamation Act 2005* (Tas) Part 3; *Defamation Act 2005* (WA) Part 3; *Defamation Act 2005* (SA) Part 3; *Defamation Act 2006* (NT) Part 3; *Civil Law (Wrongs) Act 2002* (ACT) Part 9.3.

14 See *McLean v David Syme & Co. Limited* (1970) 72 SR (NSW) 513.

Defence of innocent dissemination

(1) It is a defence to the publication of defamatory matter if the defendant proves that:
  (a) the defendant published the matter merely in the capacity, or as an employee or agent, of a subordinate distributor, and
  (b) the defendant neither knew, nor ought reasonably to have known, that the matter was defamatory, and
  (c) the defendant's lack of knowledge was not due to any negligence on the part of the defendant.

(2) For the purposes of subsection (1), a person is a 'subordinate distributor' of defamatory matter if the person:
  (a) was not the first or primary distributor of the matter, and
  (b) was not the author or originator of the matter, and
  (c) did not have any capacity to exercise editorial control over the content of the matter (or over the publication of the matter) before it was first published.

(3) Without limiting subsection (2)(a), a person is not the first or primary distributor of matter merely because the person was involved in the publication of the matter in the capacity of:
  (a) a bookseller, newsagent or news-vendor, or
  (b) a librarian, or
  (c) a wholesaler or retailer of the matter, or
  (d) a provider of postal or similar services by means of which the matter is published, or
  (e) a broadcaster of a live programme (whether on television, radio or otherwise) containing the matter in circumstances in which the broadcaster has no effective control over the person who makes the statements that comprise the matter, or
  (f) a provider of services consisting of:
    (i) the processing, copying, distributing or selling of any electronic medium in or on which the matter is recorded, or
    (ii) the operation of, or the provision of any equipment, system or service, by means of which the matter is retrieved, copied, distributed or made available in electronic form, or
  (g) an operator of, or a provider of access to, a communications system by means of which the matter is transmitted, or made available, by another person over whom the operator or provider has no effective control, or
  (h) a person who, on the instructions or at the direction of another person, prints or produces, reprints or reproduces or distributes the matter for or on behalf of that other person.[15]

In *Stratton Oakmont Inc. v Prodigy Services Co.*,[16] the innocent dissemination defence was disallowed in the Supreme Court of New York. Prodigy Services Co. (Prodigy) was an online provider of a website that offered internet forum services to subscribers. Unlike most providers, Prodigy sought to differentiate itself in the market by providing a degree of censorship on the site. This censorship was

15 *Defamation Act 2005* (Qld) s32; *Defamation Act 2005* (NSW) s32; *Defamation Act 2005* (Vic) s32; *Defamation Act 2005* (Tas) s32; *Defamation Act 2005* (WA) s32; *Defamation Act 2005* (SA) s30; *Defamation Act 2006* (NT) s29; *Civil Law (Wrongs) Act 2002* (ACT) s139C; *Defamation Act* (NZ) s8.
16 1995 WL 323710 (NY Sup Ct 1995).

minimal, comprising only a software-driven review of content for a preset range of specific words, mostly profanities. Prodigy did not advertise the limitations of its censorship role. When one of Prodigy's site users posted defamatory material in a forum about Stratton Oakmont Inc., the court was asked to determine the extent to which Prodigy's limited censorship program constituted editorial control. The court determined that while Prodigy in fact had no specific knowledge of the defamatory content, the fact that they practised some editorial control and informed users of that meant that they were liable for the defamation and could not claim innocent dissemination.

There was some initial misinformation regarding the Prodigy decision. One interpretation appeared to be that any control over the content to be disseminated would make the administrator liable. Administrators and Internet Service Providers (ISPs) thus often chose not to censor sites to avoid liability. In the United States, section 230(c)(1) of the *Communications Decency Act 1996* (US) effectively overturned this restrictive interpretation of the Prodigy decision:

> No provider or user of an interactive computer service shall be treated as the publisher or speaker of any information provided by another information content provider.

This section excludes administrators, ISPs and other service providers from liability for defamatory statements made by users of public systems unless they know of defamatory material and fail to take action on it, or are themselves involved in either the creation or development of the material.[17] Critics of the section claim that it is too broad, but it is yet to be repealed or modified. In *Barrett v Rosenthal* the court stated (in *obiter*):

> We conclude that section 230 prohibits 'distributor' liability for Internet publications. We further hold that section 230(c)(1) immunizes individual 'users' of interactive computer services, and that no practical or principled distinction can be drawn between active and passive use. We acknowledge that recognizing broad immunity for defamatory republications on the Internet has some troubling consequences. Until Congress chooses to revise the settled law in this area, however, plaintiffs who contend they were defamed in an Internet posting may only seek recovery from the original source of the statement.[18]

In *Cubby v CompuServe*,[19] CompuServe was the administrator for an online forum. However, unlike Prodigy, CompuServe did not exercise any editorial control over the online forum content. The court intimated, but did not specifically find, that electronic publishers ought not be found liable for content on sites hosted on their servers in these circumstances:

17 Ken S Myers, 'Wikimmunity: Fitting the Communications Decency Act to Wikipedia', (2006) 20 *Harvard Journal of Law and Technology* 163.
18 (2006) 40 Cal 4th 33; 146 P. 3d 510.
19 776 F. Supp. 135 (SD NY 1991).

> [the] requirement that a distributor must have knowledge of the contents of a publication before liability can be imposed for distributing that publication is deeply rooted in the First Amendment, made applicable to the states through the Fourteenth Amendment.[20]

CompuServe's defence was aided by the fact that the system operators for their online forums were independent entrepreneurs who were contracted to 'manage, review, create, delete, edit and otherwise control the contents'[21] of the forum on which the offensive material appeared. Cubby also sought to hold CompuServe vicariously liable for the libel, but the court held there to be no agency between the defendant and the other parties to the libel.

Statute of limitations

The question of precisely when a defamatory tort occurs (for the purposes of deciding when a Statute of Limitation commences to operate) was addressed by the New York Court of Appeals in *Firth v State of New York*[22] and considered by the High Court of Australia in *Dow Jones v Gutnick*.[23] The difficulty in relation to online publication is that, unlike other forms of publication, internet publication is continuous: 24 hours a day, seven days a week. Also, material placed online two years ago may only come to the attention of the general public after some form of publicity or report in the news. In Australia the limitation for commencing legal proceedings is one year from publication, with extensions of up to three years in limited circumstances.[24]

The US court held that the statute runs from the first posting of the defamatory article onto an internet site. In Australia the question has not definitively been decided.

Single publication rule

Strictly speaking, a defamatory utterance or publication arises each time a newspaper or book is sold or read, or separately in the homes of each person who hears or views a radio or television broadcast. A rule of convenience has developed to treat these multiple actions as one. The rule has been described as 'a legal fiction which deems a widely disseminated communication . . . to be a single

20 776 F. Supp. 135 (SD NY 1991), 136.
21 Ibid.
22 706 NYS 2d 835.
23 [2002] 210 CLR 575 at 602–603.
24 *Limitation Act 1969* (NSW) ss14B and 56A; *Limitation of Actions Act 1974* (Qld) ss10AA and 32A; *Limitation of Actions Act 1936* (SA) s37(1) and (2); *Defamation Act 2005* (Tas) s20A(1) and (2); *Limitation of Actions Act 1958* (Vic) s23B; *Limitation Act 2005* (WA) ss15 and 40(3); *Limitation Act 2005* (NT) ss12(2)(b) and 44A(2); *Limitation Act 1985* (ACT) s21B(1) and (2). *Limitation Act 1950* (NZ) s4 provides for a two year limitation period with a maximum extension to four years.

communication regardless of the number of people to whom, or the number of states in which, it is circulated'.[25]

Applying this principle to internet web pages is a minefield. As a continuous publication, the defamatory statement may be read, heard or viewed at completely different times, perhaps over a number of years, in locations all over the world.

In their joint judgment in *Dow Jones v Gutnick*,[26] Gleeson CJ, McHugh, Gummow and Hayne JJ considered the single publication rule and its development in relation to the place of publication.

> To trace, comprehensively, the origins of the so-called single publication rule . . . may neither be possible nor productive. It is, however, useful to notice some of the more important steps that have been taken in its development. Treating each sale of a defamatory book or newspaper as a separate publication giving rise to a separate cause of action might be thought to present difficulties of pleading and proof. Following early English authority holding that separate counts alleging each sale need not be pleaded in the declaration, American courts accepted that, where the defamatory matter was published in a book or newspaper, each publication need not be pleaded separately. Similarly, proof of general distribution of a newspaper was accepted as sufficient proof of there having been a number of separate publications. It was against this background that there emerged, at least in some American States by the late nineteenth century, the rule that a plaintiff could bring only one action against a defendant to recover damages for all the publications that had by then been made of an offending publication. The expression 'one publication' or, later, 'single publication' was first commonly used in this context.
>
> In the early decades of the twentieth century, the single publication rule came to be coupled with statements to the effect that the place of that single publication was the place where the newspaper or magazine was published. The source of this added proposition was given as a case of prosecution for criminal libel where the question was that raised by the Sixth Amendment to the United States Constitution and its reference to the 'state or district wherein the crime shall have been committed.' Despite this difference in the context in which the question of location arose, the statement that the place of publication was where the newspaper or magazine was published was sometimes taken as stating an element of (or at least a consequence of) the single publication rule applied to civil defamation suits.[27]

In *Firth v State of New York*[28] the court noted that the rationale behind the single publication rule was even more applicable to the internet, as there is greater potential there for multiple triggerings of the statute of limitations as well as for multiple suits and vexatious actions. The court regarded unrelated modifications to the website as insufficient to restart the one year limitation period. This is to be distinguished from the issue of a new edition of a book, which is sold to a new and different readership.

25 Debra Cohen, 'The single publication rule: One action, not one law', (1966) 62 *Brooklyn Law Review* 921, 924. See Chapter 11.
26 [2002] 210 CLR 575.
27 [2002] 210 CLR 575, paras 33–34.
28 706 NYS 2d 835.

The principle of the single publication rule is codified in legislation only in the United States. It has not found favour in United Kingdom or Australasian jurisdictions. The *Uniform Single Publication Act* (US) provides that multistate defamation must be dealt with in one jurisdiction and that the findings of the court in that jurisdiction are binding on the plaintiff in all other jurisdictions for any single defamatory publication. The Uniform Single Publication Act is legislation that becomes law when a state enacts it. Its aim is to create a template. It has been widely adopted across the United States with very similar wording to this:

> No person shall have more than one (1) cause of action for damages for libel, slander, invasion of privacy or any other tort founded upon a single publication, exhibition or utterance, such as any one (1) edition of a newspaper, book or magazine, any one presentation to an audience, any one (1) broadcast over radio or television or any one (1) exhibition of a motion picture. Recovery in any action shall include all damages for any such tort suffered by the plaintiff in all jurisdictions.[29]

In the United Kingdom the House of Lords was called upon to deal with the single publication rule in the defamation case of *Berezovsky v Michaels.*[30] An article published online was alleged to have defamed two Russian businessmen residing in Britain. The article was published by the Forbes Corporation in hardcopy and on their website. Their Lordships held that 'The Uniform Single Publication Act does not assist in selecting the most suitable court for the trial: it merely prevents a multiplicity of suits'. The court stated that English law could find 'no support for this argument' but then reiterated their support for the long-established legal principle that 'each publication is a separate tort'.[31]

In the Dow Jones case, Dow Jones sought to persuade the court that as they were incorporated in the United States, and the server upon which the defamatory material was published was also physically located in that country, the most appropriate forum was the United States. The High Court disagreed on a number of grounds, including that Victoria was not a 'clearly inappropriate forum' to hear the action. The court expressed concerned that applying a single publication rule for internet-based defamation (in Australia) would in effect differentiate such defamation from all other types of defamation.

Kirby J gave a number of reasons for 'declining an Internet-specific single publication rule'. First, he concluded that the court could not justify changing the rules of Australian common law merely as a response to the Dow Jones submission. Second, he added:

> Where rules such as these are deeply entrenched in the common law and relate to the basic features of the cause of action propounded, their alteration risks taking the judge beyond the proper limits of the judicial function.[32]

29 Idaho Single Publication Act s6-702.
30 [2000] UKHL 28.
31 [2000] UKHL 28.
32 [2002] HCA 56, para 124.

Third, he expressed concern about a separate rule for the internet, stating that 'Rules should be technology-neutral.'[33] Fourth, he noted that there would be 'special difficulties' in achieving judicial reform of the multiple publication rule in Australian law, even if such reform were warranted.[34]

In the absence of a specific single publication rule, Australian courts have estopped parties from raising the same issue in subsequent litigation. It was said in *Henderson v Henderson* that:

> where a given matter becomes the subject of litigation in, and of adjudication by, a Court of competent jurisdiction, the Court requires the parties to that litigation to bring forward their whole case, and will not (except under special circumstances) permit the same parties to open the same subject of litigation in respect of matter which might have been brought forward as part of the subject in contest, but which was not brought forward, only because they have, from negligence, inadvertence, or even accident, omitted part of their case.[35]

Single controversy principle

Australian courts have *de facto* adopted a 'single controversy' principle: where there are multiple publications of the same defamatory matter there is only a 'single controversy' to be litigated. That controversy must include all separate utterances of the defamatory allegation, in order to properly determine breach and damages. The cases to date have not adequately addressed the issues of timing and place, particularly where internet publication is concerned.

Single cause rule

The single publication rule is not to be confused with the 'single cause' provision in the uniform Defamation Acts across Australia, which provides that:

> A person has a single cause of action for defamation in relation to the publication of defamatory matter about the person even if more than one defamatory imputation about the person is carried by the matter.[36]

Adventitious or opportunistic conduct

In Dow Jones, both parties raised the issue of adventitious or opportunistic conduct.[37] Dow Jones claimed that permitting jurisdiction in Victoria – that is,

33 [2002] HCA 56, para 125.
34 See *Australian Broadcasting Corporation v Waterhouse* (1991) 25 NSWLR 519 at 537. See also Australian Law Reform Commission, *Unfair Publication: Defamation and Privacy*, Report No. 11, (1979), 60–61, para; Australian Law Reform Commission, *Choice of Law*, Report No. 58, (1992), 57 paras 6.53–54.
35 [1843] 3 Hare 100, 115; 67 ER 313, 319.
36 *Defamation Act 2005* (Qld) s8; *Defamation Act 2005* (NSW) s8; *Defamation Act 2005* (Vic) s8; *Defamation Act 2005* (Tas) s8; *Defamation Act 2005* (WA) s8; *Defamation Act 2005* (SA) s8; *Defamation Act 2006* (NT) s7; *Civil Law (Wrongs) Act 2002* (ACT) s120; *Defamation Act* (NZ) s7 is a similar provision.
37 See para 131ff. See also Chapter 11.

at the place of download rather than the place of upload – would encourage plaintiffs to commence actions in jurisdictions with favourable defamation laws. Gutnick's response was that should a plaintiff improperly attempt to take such an advantage, the courts should consider rejecting jurisdiction. He claimed that his choice of venue was not adventitious or opportunistic: Victoria was the place of his residence and his prime business interests.

Gutnick claimed that permitting jurisdiction in New Jersey – that is, at the place of upload rather than the place of download – would encourage defendants to move their servers to jurisdictions with weak defamation laws. Dow Jones' response was a copy of Gutnick's: should a defendant improperly attempt to take such an advantage, the courts should consider rejecting jurisdiction. Equally, Dow Jones claimed that its choice of location for the server was not adventitious or opportunistic.

Jurisdiction for defamatory statement

To date the definitive online defamation case in Australian jurisprudence remains *Dow Jones v Gutnick*.[38]

The High Court held firm to the Australian legal doctrine that *forum non conveniens* can only be applied by determining whether or not the Australian forum is 'clearly inappropriate'.[39] The burden under this doctrine rests with the applicant. To succeed in this, and thereby have the matter removed to another forum, the applicant must demonstrate under the 'Spiliada principles' that their preferred forum, in this case the United States, was the 'natural forum' in which to hear the matter. In *Spiliada Maritime Corp. v Cansulex Ltd*,[40] Lord Goff described the 'natural forum' as being 'that with which the action had the most real and substantial connexion'. In the Dow Jones case, the court declined to refuse the exercise of its jurisdiction, holding that when 'Victorian law was the *lex loci delicti*, Victoria is the natural forum [and] . . . the appropriate and the convenient forum'.[41]

Conclusion

While much of the law dealing with online defamation and other torts committed on the internet is still in its earliest days of development and interpretation, some

38 [2002] 210 CLR 575. For further details see Chapter 11.

39 *Oceanic Sun Line Special Shipping Co. v Fay* (1988) 62 ALJR 389; *Voth v Manildra Flour Mills Pty Ltd* (1990) 65 ALJR 83; *Henry v Henry* (1996) 185 CLR 571. The High Court reaffirmed its position in *Regie National des Usines Renault SA v Zhang* (2002) 76 ALJR 551: '[a]n Australian court cannot be a clearly inappropriate forum merely by virtue of the circumstance that the choice of law rules which apply in the forum require its courts to apply foreign law as the *lex causae*'. See David Rolph, 'Before the High Court – The massage, not the medium: Defamation, publication and the internet in *Dow Jones & Co. Inc. v Gutnick*', (2002) 24 *Sydney Law Review* 263.

40 [1987] AC 460, 478, Lord Goff of Chieveley.

41 *Dow Jones and Co. Inc. v Gutnick* [2002] 210 CLR 575 at 590. See also *Macquarie Bank v Berg* [1999] NSWSC 526.

general principles have already emerged. The first is that it would appear that for the purposes of mitigating risk, it is wiser for forum providers to not provide any editorial control whatsoever – and to advertise this fact widely.

Next, employers would be well advised to ensure that there are rigorous and effective systems in place to prevent employees from putting the firm into a defensive position. These might include a staff training régime, a compulsory Code of Conduct, and clear guidelines on the roles and responsibilities of every employee who is exposed to the internet as part of their normal duties.

As the law develops, and both governments and industry groups better understand their rights and responsibilities in this domain, the law will be developed and amended to provide direction and protection for those most at risk.

Further reading

Australian Law Reform Commission, (1992) *Choice of law*, Report No. 58.

Australian Law Reform Commission, (1979) *Unfair publication: Defamation and privacy*, Report No. 11.

Debra Cohen, 'The single publication rule: One action, not one law', (1966) 62 *Brooklyn Law Review* 921.

Uta Kohl, 'Defamation on the internet – A duty free zone after all? *Macquarie Bank Ltd & Anor v Berg*', (2000) 22 *Sydney Law Review*.

Lyrissa Barnett Lidsky, 'Silencing John Doe: Defamation & discourse in cyberspace', (2000) 49 *Duke L J* 855.

Ken S Myers, 'Wikimmunity: Fitting the Communications Decency Act to Wikipedia', (2006) 20 *Harvard Journal of Law and Technology* 163.

David Rolph, 'Before the High Court – the massage, not the medium: Defamation, publication and the internet in *Dow Jones & Co Inc. v Gutnick*', (2002) 24 *Sydney Law Review* 263.

Standing Committee of Attorneys-General (SCAG) Working Group of State and Territory Officers, Legislation and Policy Division of the NSW Attorney General's Department, (2004), *Proposal for Uniform Defamation Laws*, www.ag.gov.au/www/agd/rwpattach.nsf/VAP/(03995EABC73F94816C2AF4AA2645824B)~Uniform+Defamation+States+Territories.pdf/$file/Uniform+Defamation+States+Territories.pdf.

13

Privacy and data protection in cyberspace

'Privacy' has numerous meanings, and its importance varies greatly among individuals, communities, organisations and governments. It is an aspect of freedom and human rights. Civil libertarians may believe that our actions and behaviour should not be subject to public or governmental scrutiny; protectionists may accept such erosions for the greater good in the name of law and order. The development of technology, together with social, economic and political factors, has raised the antennas of those concerned with interference from governments, enterprises and others on personal freedoms.

Historically, governments seem to have pursued increasing and systemic invasions of privacy in the name of law and order, fighting crime and terrorism. However, their role is in fact to ensure and balance security issues and the proper protection of the privacy rights of the individual. In the 1990s the Clipper Chip was proposed by the US government, ostensibly for the purpose of allowing the government to override individual encryption to protect society from 'gangsters, terrorists and drug users'.[1] Such a process would have allowed the government to access and decipher all encrypted files. The proposal was unsuccessful. In Australia in 1984, an attempt to pass a Privacy Act failed because it set in place an anti-privacy provision: a central national identification card.[2]

Search engines such as Google use sophisticated programs called spiders, robots and wanderers to trawl the internet gathering data on several billion websites, creating an index that handles several million enquiries per day. Individuals are often surprised by the digital persona visible when their name is searched. This challenges notions of fundamental human rights, including privacy. False,

1 See Mark Berthold, 'Regulating surveillance: Hong Kong's proposals', (1996) *PLPR* 44 and Graham Greenleaf, 'OECD searches for crypto-consensus', (1996) *PLPR* 22.
2 Michael Kirby, 'Privacy in cyberspace', (1998) 21(2) *UNSW LJ*.

distorted and damaging information may be compiled. A consequence of this massive index of the internet creates dataveillance. Privacy is only one attribute of the internet in which challenges arise for established values. Organised crime, terrorism, intellectual property rights, pornography, the integrity of financial markets and tax systems, and cultural and sovereign diversity are others.

In 2001 the NSW Supreme Court[3] expressed its view on the expectation of privacy or confidentiality when information is placed on the internet:

> It must be said at the outset that part of the information that the defendants have used is in the public domain. I regard everything which is accessible through resort to the internet as being in the public domain. It is true that someone can obtain that information only if they have access to a computer which has a modem which connects to an internet service provider who, for a fee, provides a connection to the internet. But those barriers are, in my view, no more challenging or significant in today's Australia, complete with internet cafes, than those involved in access to a newspaper or television content, both of which should, according to precedent, be seen as involving the public domain.[4]

Information wants to be free

This is the catchcry of the digital age.[5] The absence of a controlling and enforceable law facilitates free expression, the communication of ideas and notions of individual liberty (which are themselves important human rights). Such values are not the only human rights: see the Universal Declaration of Human Rights.[6] There are other fundamental human rights which compete, or conflict, with the right to free expression. The right to privacy and to reputation and honour, and the confidentiality of communications must also be protected. In the world of the internet, technological capacity tends to favour the spread of information.

Electronic communication and storage of data have a real impact on issues of privacy and censorship. Previously, many of the safeguards of our privacy were in fact only the costs of retrieving personal information. Retrieving data stored in hardcopy form involved time and expense. Data could be lost, destroyed or misfiled. Hardcopies could deteriorate. Many data collections were incompatible. Considerable time could be spent gaining access. Our privacy was protected only by the fact that methods of data storage were impractical and inconvenient.

These so-called safeguards have evaporated in the digital age. One body, less cognisant of privacy issues and more concerned about costs and time, may respond to another body's request for information by transmitting its entire

3 *EPP v Levy* [2001] NSWSC 482, per Barrett J.

4 *EPP v Levy* [2001] NSWSC 482, para 22. See also *Duchess of Argyle v Duke of Argyle* [1967] Ch 302 and *G v Day* [1982] 1 NSWLR 24.

5 An aphorism attributed to Stewart Brand from the First Hackers Conference in 1984. The full quote is: 'On the one hand, information wants to be expensive, because it's so valuable. The right information in the right place just changes your life. On the other hand, information wants to be free because the cost of getting it out is getting lower and lower all the time. So you have these two fighting against each other.'

6 Adopted and proclaimed by General Assembly Resolution 217A (III) of 10 December 1948.

database, the password or the entire record of an individual rather than a response to the specific request. Data is now stored in compatible operating systems. Due to the ease of storage, bodies both public and private are collecting and amassing a greater amount of personal material than ever before. This is a side of globalisation that is both irreversible and inevitable. Privacy dilemmas in the digital world are a Pandora's Box. Individuals now have a virtual existence in cyberspace, a digital persona made up of a collection of otherwise unconnected and previously unconnectable data, and the quantity of personal information in cyberspace is likely to increase.

Privacy and regulation

Privacy can be divided into two broad categories: information privacy and personal privacy. Information privacy refers to the ways in which information is gathered, recorded, accessed and released. In the digital age there are multitudes of records of individuals on databases. Only since the advent of electronic recording has information privacy become a serious matter for regulation and control. The most significant step towards regulation began with the drafting of privacy principles by the Organisation for Economic Co-operation and Development (OECD) in the 1970s.

Personal privacy relates to privacy of the person, of an individual's personal space: it can be 'invaded' by those seeking to photograph, film and record in public and private places. Media outlets pay substantial amounts for even blurred long-distance photographs of celebrities. Australian laws permit surveillance cameras in shops, malls, service stations, railway stations and many other places.

Information privacy

As law and society evolved, pressing issues such as lawful behaviour – in both criminal and civil senses – and the enforcement of commercial agreements and promises were the priority of lawmakers. Philosophical concepts such as privacy were insignificant in comparison with survival with warring neighbours and the enforcement of property rights. The courts and legislatures failed to recognise inherent concepts of isolation, seclusion and the protection of personal information. In the Middle Ages only the literate could meaningfully access written information. With literacy came dissemination. With government came secrets: national security secrets as well as personal ones. Although crude methods of encryption were developed during World War II, it was the Cold War and the perceived need to protect state secrets that pushed cryptography into the limelight as a science.

Modern concepts of information privacy emerged in the late 1970s, and the OECD set up an international Experts Group to draft Guidelines on the Protection of Privacy and Transborder Flows of Personal Data (the OECD Guidelines). The Experts Group was chaired by the Hon Mr Justice Michael Kirby,[7] then Chairman of the Australian Law Reform Commission (ALRC). The OECD Guidelines were formally adopted and disseminated in September 1980. They included these recommendations:

1 That Member countries take into account in their domestic legislation the principles concerning the protection of privacy and individual liberties set forth in the Guidelines contained in the Annex to this Recommendation which is an integral part thereof;
2 That Member countries endeavour to remove or avoid creating, in the name of privacy protection, unjustified obstacles to transborder flows of personal data;
3 That Member countries co-operate in the implementation of the Guidelines set forth in the Annex;
4 That Member countries agree as soon as possible on specific procedures of consultation and co-operation for the application of these Guidelines.

The OECD supplemented these Guidelines with the Declaration on Transborder Data Flows (1985) and the Ministerial Declaration on the Protection of Privacy on Global Networks (1998).

Electronic commerce law is about the need to find solutions for novel circumstances. The OECD Guidelines noted that it was the increase in international data transmission that had made it 'necessary' to address privacy protection in relation to personal data. The OECD observed that privacy protection laws were at that time being contemplated or introduced in OECD countries such as Austria, Belgium, Canada, Denmark, France, Germany, Iceland, Luxembourg, Norway, the Netherlands, Spain, Sweden, Switzerland and the United States, and expressed its concern about new 'violation[s] of fundamental human rights',[8] including the unlawful storage of personal data, the storage of inaccurate personal data, and the abuse or unauthorised disclosure of such data.

Even in the late 1970s the rapid growth of digital communication was foreseen: automatic data processing that would result in the transmission of vast quantities of data across national boundaries was already being developed. The OECD's major concern was that disparities in national legislation could lead to interruptions in international flows of data and hamper the free flow of personal data across frontiers; restrictions on these flows, the OECD noted, 'could cause serious disruption in important sectors of the economy, such as banking and insurance'.[9]

7 Former Justice of the High Court of Australia.
8 Preface to the OECD's Guidelines on the Protection of Privacy and Transborder Flows of Personal Data; www.oecd.org.
9 Ibid.

The OECD Guidelines require that personal information not be collected unless the person concerned either consents to its collection or is informed of why it is being collected, and is informed of who will use it and how the person may access and correct it. The preamble to the Guidelines states, in part:

> Member countries have a common interest in protecting privacy and individual liberties, and in reconciling fundamental but competing values such as privacy and the free flow of information; that automatic processing and transborder flows of personal data create new forms of relationships among countries and require the development of compatible rules and practices; . . . [that] transborder flows of personal data contribute to economic and social development; . . . [that] domestic legislation concerning privacy protection and transborder flows of personal data may hinder such transborder flows.

Part Two of the Guidelines sets out the key principles:

Part Two – Basic Principles of National Application

Collection Limitation Principle

7. There should be limits to the collection of personal data and any such data should be obtained by lawful and fair means and, where appropriate, with the knowledge or consent of the data subject.

Data Quality Principle

8. Personal data should be relevant to the purposes for which they are to be used, and, to the extent necessary for those purposes, should be accurate, complete and kept up-to-date.

Purpose Specification Principle

9. The purposes for which personal data are collected should be specified not later than at the time of data collection and the subsequent use limited to the fulfilment of those purposes or such others as are not incompatible with those purposes and as are specified on each occasion of change of purpose.

Use Limitation Principle

10. Personal data should not be disclosed, made available or otherwise used for purposes other than those specified in accordance with Paragraph 9 except:
    a) with the consent of the data subject; or
    b) by the authority of law.

Security Safeguards Principle

11. Personal data should be protected by reasonable security safeguards against such risks as loss or unauthorised access, destruction, use, modification or disclosure of data.

Openness Principle

12. There should be a general policy of openness about developments, practices and policies with respect to personal data. Means should be readily available of establishing the existence and nature of personal data, and the main purposes of their use, as well as the identity and usual residence of the data controller.

Individual Participation Principle

13. An individual should have the right:
    a) to obtain from a data controller, or otherwise, confirmation of whether or not the data controller has data relating to him;
    b) to have communicated to him, data relating to him
        - within a reasonable time;
        - at a charge, if any, that is not excessive;
        - in a reasonable manner; and
        - in a form that is readily intelligible to him;
    c) to be given reasons if a request made under subparagraphs(a) and (b) is denied, and to be able to challenge such denial; and
    d) to challenge data relating to him and, if the challenge is successful to have the data erased, rectified, completed or amended.

Accountability Principle

14. A data controller should be accountable for complying with measures which give effect to the principles stated above.[10]

Australia

In Australia there is no specific right to privacy at common law.[11] Privacy was not a significant issue for legislators in the immediate postwar period. However, it began to attract attention in Europe and North America, and in 1969 Zelman Cowen (later Governor General), in his ABC Boyer Lecture Series, discussed the real privacy concerns faced by Australians. In 1972, the NSW Attorney-General, John Madison, commissioned a report into the law of privacy by Professor Morison of the University of Sydney. Morison concluded that privacy was an interest, not a right. Concerned at possible legislative overreaction, the Australian Computer Society lobbied the government to prevent the perceived imposition of excessive or inappropriate regulation.

While the Australian Constitution makes no direct reference to a privacy right or power, it contains sufficient powers, in section 51, to justify such legislation. The relevant other heads of power in that section include:

- trade and commerce [s51(i)];
- postal, telegraphic and telephonic [s51(v)];
- banking [s51(xiii)];
- insurance [s51(xiv)];
- foreign corporations and trading or financial corporations [s51(xx)];
- marriage [s51(xxi)];
- divorce and matrimonial causes [s51(xxii)];

10 Other Parts are: Part One – General Definitions; Part Three – Basic Principles of International Application: Free Flow and Legitimate Restrictions; Part Four – National Implementation; Part Five – International Co-Operation.

11 See *Personal privacy*, below.

- invalid and old-age pensions [s51(xxiii)];
- social security and other allowances [s51(xxiiiA)];
- external affairs [s51(xxix)];
- matters referred by states [s51(xxxvii)];
- incidental matters [s51(xxxix)]; and
- public service [s52(ii)].
- financial assistance to the states [s96].

New South Wales enacted the *Privacy Committee Act 1975*, pre-dating the OECD Guidelines. The Act created a complaints-driven investigative process and a research organisation. The work of the latter contributed to establishing information privacy principles for organisations using computers.

In April 1976, the Commonwealth government gave the ALRC a reference to study interferences with privacy arising under the laws of the Commonwealth. The report, which was handed down in 1983, took into account the OECD Guidelines. In 1986 the government introduced a Privacy Bill which included a controversial national identification card. Many argued that this was in fact an anti-privacy move. The Bill was defeated by a hostile Senate. In 1988 the government set out to enhance the Tax File Number (TFN) scheme used by the Australian Tax Office. This was intended in part as a replacement of the defeated Australia Card scheme.

The *Privacy Act 1988* (Cth) was passed in December 1988. Australia's first Privacy Commissioner was immediately appointed. The Act applied to the public sector, not to the populace at large. In 1989 the Privacy Act was amended to protect consumers from adverse consumer credit reporting. Subsequent amendments extended coverage to 'spent' criminal convictions[12] and to data matching.[13] In 1994 New South Wales introduced the Privacy and Data Protection Bill, but it was withdrawn after heavy criticism.

In 1996, the NSW Health Commission issued a consolidation of privacy law, policy and practice for health care workers. In 1997, the Asia Pacific Smart Cards Forum Code of Conduct included privacy principles for members. In 1998, the Australian Privacy Commissioner released a document entitled *National Principles for the Fair Handling of Personal Information* (FHPI). This was an application of the OECD Guidelines intended for the wider community.

The *Privacy Act 1988* (Cth) currently includes 11 Information Privacy Principles[14] (IPP) which Commonwealth government departments and agencies are obliged to follow when handling personal information. The Act imposes restrictions on how credit providers and credit reporting agencies may handle personal information, together with rules applicable to the entire community in relation to the handling of tax file number information. The Act implements the OECD

12 By way of an amendment to Part VIIC of the *Crimes Act 1914* (Cth).
13 *Data-Matching Program (Assistance and Tax) Act 1990* (Cth).
14 *Privacy Act 1988* (Cth) s14.

Privacy Principles and observes Australia's obligations under Article 17 of the ICCPR.[15] The Privacy Act provides that the IPP be treated as law.[16]

All firms and businesses need to monitor the changing Australian and international legal environment to ensure that the risk of data protection breaches is minimised.

National Privacy Principles (NPP)

In December 2001 the federal government introduced a co-regulatory scheme for information privacy protection for the private sector. The *Privacy Act 1988* (Cth) was significantly extended, with the National Privacy Principles (NPP) made applicable. The scheme allows certain organisations to comply with either the NPP or approved industry-developed privacy codes, provided they maintain an equivalent legislative standard. The 10 NPP are set out in Schedule 3 of the Act and are generally based on the OECD standard. The NPP cover collection (NPP 1), use and disclosure (NPP 2), data quality (NPP 3), data security (NPP 4), openness (NPP 5), access and correction (NPP 6), identifiers (NPP 7), anonymity (NPP 8), transborder data flows (NPP 9) and sensitive information (NPP 10). The IPP continue to apply to the public sector.

Under the *Privacy Act 1988* (Cth), individuals have the right to know why an organisation is collecting their personal information, what information it holds about them, how the information will be used and who else may receive the information. Generally, individuals have the right to access this information and correct it if it is wrong. Individuals can also make a complaint to the Privacy Commissioner if they believe their information is not being handled properly. Alternatively, they can apply to the Federal Court or the Federal Magistrates' Court for an order to restrain an organisation from engaging in conduct that breaches the NPP. The individual's rights arise where an 'act or practice' results in an 'interference with privacy'. Section 7 comprehensively defines an 'act or practice'; section 13A provides that an act or practice of an organisation is an 'interference with the privacy of an individual' if the act or practice breaches the NPP.

Schedule 3 of the Act sets out the NPP. They state that an organisation should:

- only collect personal information it actually needs;
- collect information fairly and lawfully;
- let people know what it intends to do with the information it collects;
- collect information directly from the individual, rather than from other sources;
- only use information in ways that:

15 1. No one shall be subjected to arbitrary or unlawful interference with his privacy, family, home or correspondence, nor to unlawful attacks on his honour and reputation. 2. Everyone has the right to the protection of the law against such interference or attacks.
16 *Privacy Act 1988* (Cth) s13.

  - the individual would expect;
  - the individual has consented to; or
  - are required by the public interest;
- ensure the accuracy of the information;
- keep information securely;
- be open about how it handles personal information;
- give people access to information about themselves;
- limit its use of government identification numbers (like the Medicare number);
- let people operate anonymously where it reasonably can;
- only send personal information overseas if it will be properly protected there; and
- limit its collection of information about people's race, religion, sex life or political beliefs.

In a cyberspace context, these provisions require, for example, a website operator that is classified as an organisation and that collects personal information online to take reasonable steps to ensure that internet users know who is collecting their information and how it is used, stored and disclosed.[17] A breach of the NPP amounts to an interference with the privacy of an individual, and gives rise to a right to complain to the Privacy Commissioner and a right to seek compensation.

The Act broadly defines an organisation as an individual, a body corporate, a partnership, any other unincorporated association or a trust, that is not a small business operator, a registered political party, or a government agency or instrumentality.[18] A small business operator is defined as a business which:

- has an annual turnover of three million dollars or less;
- is not related to a business with an annual turnover of greater than three million dollars;
- does not provide a health service and hold health records;
- does not disclose personal information about an individual for a benefit, service or advantage;
- does not provide a benefit, service or advantage to collect personal information; and
- is not a contracted service provider for a Commonwealth contract (even if the entity is not a party to the contract).[19]

Employee records are exempt if the organisation is or has been an employer of the individual and the act or practice is directly related to the employment relationship.[20] Acts and practices engaged in by a media organisation in the course of journalism are exempt.[21] Registered political parties are exempted by

17 See the Attorney General's Department, (2000) 'Factsheet: Privacy and the electronic environment', www.law.gov.au/privacy/newfacts/Electronic.htm.
18 *Privacy Act 1988* (Cth) s6C.
19 *Privacy Act 1988* (Cth) s6D.
20 *Privacy Act 1988* (Cth) s7B(3).
21 *Privacy Act 1988* (Cth) s7B(4).

being excluded from the definition of organisation.[22] Certain acts or practices of members of parliament, local government councillors and their contractors, for any purpose in connection with an election under an electoral law, a referendum or in connection with participation of the member or councillor in an aspect of the political process, are exempt.[23]

Data protection

In relation to data protection in Australia, a limited form of privacy protection was introduced by the *Telecommunications Act 1997* (Cth). Part 13 prohibits the disclosure by carriers, carriage service providers and others, of certain information acquired as a result of their normal business activities. However, the same Act requires carriers to have wiretapping capabilities in place for 'lawful' surveillance. The *Crimes Act 1914* (Cth), Part VIIC, Div 5 affords some protection for past criminal convictions. The *Data-Matching Program (Assistance and Tax) Act 1990* (Cth) provides protection for specific major data matching.

Victoria

There are several Victorian Acts related to privacy. The *Surveillance Devices Act 1999* (Vic) aims to regulate the use of surveillance devices, to restrict the communication and publication of records of private conversations and to establish procedures for law enforcement officers to obtain warrants or emergency authorisations. The Act imposes requirements for the secure storage and destruction of records obtained by police through the use of surveillance devices.

The *Information Privacy Act 2000* (Vic) aims to:

- establish a regime for the responsible collection and handling of personal information;
- provide individuals with rights of access to information about them held by organisations; and
- provide individuals with the right to require an organisation to correct errors.

The Act covers the handling of all personal information except health information in the public sector in Victoria. It adopts ten Information Privacy Principles which are similar to the NPP set out in the federal Privacy Act.[24]

The *Health Records Act 2001* (Vic) covers the handling of all personal information held by health service providers in the state's public sector and also seeks to govern acts or practices in the Victorian private health sector. The Act contains a set of principles adapted from the NPP.[25]

22 *Privacy Act 1988* (Cth) s6C.
23 *Privacy Act 1988* (Cth) s7C.
24 See the Office of the Victorian Privacy Commissioner: www.privacy.vic.gov.au.
25 See the Office of the Health Services Commissioner: www.health.vic.gov.au/hsc/.

Other relevant Victorian laws include: *Information Privacy Act 2000*; *Health Records Act 2001*; *Freedom of Information Act 1982*; *Public Records Act 1973*; *Surveillance Devices Act 1999*; and *Telecommunications (Interception) (State Provisions) Act 1988*.

New South Wales

The *Privacy Committee Act 1975* (NSW) established the NSW Privacy Committee. The committee has made significant local progress .[26] The *Privacy and Personal Information Protection Act 1998* (NSW) regulates the management of personal information within NSW public sector agencies. It also sets out the role of the Office of the NSW Commissioner. New South Wales has also developed statutory guidelines in the form of legally binding documents that define the scope of particular exemptions in the health privacy principles.

Other relevant NSW laws include: *Health Records and Information Privacy Act 2002*; *Freedom of Information Act 1989*; *State Records Act 1998*; *Criminal Records Act 1991 (Spent Convictions)*; *Listening Devices Act 1984*; *Workplace Surveillance Act 2005*; *Telecommunications (Interception) (New South Wales) Act 1987*; and *Access to Neighbouring Land Act 2000* (note sections 16 and 26).

Queensland

In 2001 the Queensland government implemented a privacy scheme based on the federal IPP. The scheme applies to state government agencies and corporations. The scheme requires Queensland Health to comply with the NPP. The Queensland Health Rights Commission provides an enquiry service and a health complaint system which covers privacy-related complaints involving the state public health sector.[27]

Other relevant Queensland laws include: State Government Standards Nos 42 (Information Privacy, September 2001) and 42A (Information Privacy for the Queensland Department of Health, September 2001); *Freedom of Information Act 1992*; *Public Records Act 2002*; *Criminal Law (Rehabilitation of Offenders) Act 1986* (spent convictions); *Invasion of Privacy Act 1971* (listening devices, invasion of privacy of the home); *Invasion of Privacy Regulations* (1998); *Whistleblowers Protection Act 1994*; and *Police Powers and Responsibilities Act 2000* (Chapter 4 deals with covert evidence-gathering powers).

Western Australia

The public sector in Western Australia does not have a privacy regime. However, the *Freedom of Information Act 1992* (WA) includes some privacy principles and its confidentiality provisions cover government.

26 See also *Health Records and Information Privacy Act 2002* (Cth).
27 See Queensland government website: www.privacy.qld.gov.au.

Other relevant WA laws include: *State Records Act 2000*; *Spent Convictions Act 1988*; *Surveillance Devices Act 1998*; and *Telecommunications (Interception) Western Australia Act 1996*.

South Australia

South Australia has issued Information Privacy Principle Instructions that government agencies must comply with. South Australia has a Code of Fair Information Practice, based on the NPP, which is applicable to the SA Department of Health, its funded service providers and others with access to the Department's personal information.[28]

Other relevant SA laws include: *Freedom of Information Act 1991*; *State Records Act 1997*; *Listening and Surveillance Devices Act 1972*; and *Telecommunications (Interception) Act 1988*.

Tasmania

In Tasmania the *Personal Information and Protection Act 2004* (Tas) implements Personal Information Protection Principles (PIPP) applicable to 'personal information custodians' such as public and local government sectors and the University of Tasmania. The PIPP are based on the federal NPP. The Act is administered by the Department of Justice and complaints may be made to the Tasmanian Ombudsman.

Other relevant Tasmanian laws include: *Personal Information Protection Act 2004*; *Freedom of Information Act 1991*; *Archives Act 1983*; *Annulled Convictions Act 2003* (spent convictions); *Listening Devices Act 1991*; and *Telecommunications (Interception) Tasmania Act 1999*.

Northern Territory

The Information Commissioner for the Northern Territory is the independent authority responsible for overseeing the freedom of information and privacy provisions of the *Information Act 2002* (NT). The Act deals with the protection of personal information, record keeping and archive management in the public sector and incorporates freedom of information and privacy principles. The Health Information Privacy website provides information and links to health privacy-related matters in the Territory, and includes a code of conduct. A discussion paper on protecting the privacy of health information in the Territory was issued in March 2002.[29]

Other relevant laws include: *Information Act 2002* (NT) (privacy, FOI and public records); *Criminal Records (Spent Convictions) Act 1992* (NT); *Surveillance*

28 See SA Information Privacy Principles Instruction: www.premcab.sa.gov.au/pdf/circulars/Privacy.pdf.
29 Office of the Information Commissioner Northern Territory: www.privacy.nt.gov.au.

Devices Act 2007 (NT); *Telecommunications (Interception) Northern Territory Act 2001* (NT).

Australian Capital Territory

The federal Privacy Act[30] applies to ACT government agencies and is administered by the federal Privacy Commissioner on behalf of the ACT government. The *Health Records (Privacy and Access) Act 1997* (Health Records Act) covers health records held in the public sector in the ACT and also seeks to apply to acts or practices in the private sector that are not covered by the Privacy Act. It contains privacy principles based on the federal legislation but modified to suit the requirements of health records. The ACT Community and Health Services Complaints Commissioner handles health record privacy complaints.

Abuses

There have already been many early warning signs of the potential for privacy abuse. Naval officer Timothy R McVeigh was discharged from the US Navy after he came under investigation following details of his use of the internet which revealed the use of the word 'gay' on his Internet America Online (AOL) profile. AOL agreed to pay damages to McVeigh for having improperly disclosed his identity.[31] While the US Senate considered the nomination of Judge Robert Bork to the Supreme Court, a journalist retrieved and reported the record of the judge's 146 video rentals as itemised by computer from his local video store.

The increase in the amount of data being collected and kept also leads to an increase in mistakes. A brother who had made a few payments of rent for his sister was blacklisted when the sister later defaulted. Records that show that a particular computer or network was used to access or download 'undesirable material' do not identify the actual individual using the facility.

Cookies

Normal use of the internet typically causes information from personal computers to be sent to the hosts of sites visited. One such process uses 'cookies'. A cookie, in an internet context, is a small text file placed on the user's hard drive, often unknown to the user, by the host of the website visited. The cookie can act like an identification card, but cannot be executed as code or deliver viruses. It can only

30 As amended by the *Australian Capital Territory Government Service (Consequential Provisions) Act 1994* (Cth).

31 *Timothy R McVeigh v William Cohen* 983 F. Supp. 215 (DDC January 1998). (Not to be confused with the infamous criminal Timothy J McVeigh.)

be read by the host that presented it to the user. Internet browsers are designed to facilitate cookies.

A positive example of a cookie may be accessing a website of a particular cinema chain.[32] The user typically must click through several links and pages, choosing the state, city and suburb of the local cinema. However, such a website may allow the user to customise the page so that the web page showing the local cinema appears immediately on access. A cookie is created and is stored on the user's computer. It is read by the cinema website and automatically responds by customising the information to be displayed.

Bill Gates, in his book *The Road Ahead*,[33] laid out his vision of an interconnected world built around the internet. Gates theorises that life will be transformed by the convergence of inexpensive computing and inexpensive communications. He describes the ideal house, where on entering, rooms illuminate to the user's preset level, favourite music is played, the temperature is adjusted, the favourite style of art is projected on to the walls and the telephone nearest the user is activated. Visitors are given electronic tags so that the house computer knows and keeps records of the location of all occupants, including previous visitors. The house computer does not need to know names, just the presets. Cookies work in a similar manner to this electronic tag.

Cookies are automatically created by such sites as Wikipedia (the internet encyclopaedia), the *New York Times*, and Google. Many websites allow the user to set preferences for return visits – Wikipedia has a link 'My Preferences' to do this.[34]

Cookies can be used to identify users and track their movements through the site. A particular website may comprise several web pages. One web page may include graphics and links from a range of outside sources. Some of these may be advertising banners. Indeed the owner of the page may not necessarily know which advertisement will be displayed. A user in Europe may call up a web page based in Canada, but the advertising banner, text and material may be automatically accessed from a location in the United States. It may be that both the host in Canada and the advertising owner place cookies on the user's page. Their purposes may be entirely different: the web host may want information in order to provide a more functional site, and the advertiser may be collecting information for marketing purposes.[35]

Web hosts may systematically share information for identification or tracking. Typically, the main purpose of sharing is identification. Where the user voluntarily supplies these details, they can be placed in a cookie that lets the host take a shortcut to supplying unique information. The host may register each visit, and the time and number of pages visited within its site. The host may also track the visit: this means using cookies to trace usage and status from page to page. For

32 For example, www.greaterunion.com.au.
33 Bill Gates, *The Road Ahead*, Penguin, London, 1996.
34 en.wikipedia.org.
35 Sharon Nye, 'Internet privacy – Regulating cookies and web bugs', (2002) *PLPR* 26.

example, where a user accesses an online shopping basket, each time the basket is filled with items to purchase from separate web pages within a service, the cookies verify that the user is the person requesting those items. When a user is at the 'check-out', the details of all the items can be retrieved using the cookie file.[36]

A website cannot uncover your email address, for example, without its being specifically disclosed. However, many sites require an email address as part of their operation; when it is provided, it may be placed or referenced in a cookie. This information can then be shared. As a result of cookie technology, a user can be tracked.

In the most usual and typical situation, when a user visits a particular web page, the web browser usually makes a request for http code to be displayed and for the graphics that appear on the screen. The graphics may come from completely different locations and the user will be unaware of the origin. There may be, for example, two banner advertisements from different companies. Any information provided by the user can go to each of these companies. With this information the advertising companies may develop a database for future marketing. Many advertising banners on search engine web pages can record the search engine enquiries. Depending on the information in the cookie, next time you visit a site with that advertiser, a personal banner that reflects your profile may appear. For many people, all this raises serious privacy concerns.[37]

One view is that a cookie is a unique identifier, like a serial number, that is used to retrieve your records from their databases. Cookies may have an expiry date, but many last for years. Cookies are stored on users' hard drives and may be viewed. Both Netscape and Microsoft Internet Explorer[38] have dedicated directories for this purpose. Each individual cookie may be viewed or deleted.

Cookies operate in a surreptitious manner. The majority of users are unaware of their existence and of those that do few appreciate the extent of their operation. The cookies make use of the user's computer without express consent to both write cookies and read them. The default setting for most web browsers is to permit the use of cookies. The latest version of the browser Microsoft Internet Explorer has six optional settings in relation to cookies, from accept all to reject all.

Cookies provide significant benefits for users of the internet. However, their operation has been largely clandestine, and in certain situations they will stretch privacy rules to the limit, if not break them. Users must take care with the type

36 See Morris Averill, 'The spiders stratagem on the Web: Hunting and collecting web users', (2004) *DTLJ* 1.

37 For example, see www.junkbusters.com/cookies.html, with subheadings such as: 'The pages you read tell marketers what junk to push on you'; 'What your browser tells them'; and 'Cookies tell them it's you every time you click'.

38 Microsoft Internet Explorer, version 7.

of response they make to questions asked of them on websites; but this applies to all sites, not just those with cookies.[39]

Web bugs

Web bugs are objects embedded in web pages or email, and they are used to monitor use. Originally web bugs were actually tiny invisible or virtually invisible images – one pixel in size – on a computer screen. A web bug is sometimes referred to as 'clear gif': it is invisible and is usually a 'gif' file (image). The image need not be invisible and can be any size, but invisibility of course assists surreptitious use.

Web bugs may initiate contact with another server.[40] When a web bug is loaded unknowingly by a user, it will yield the IP address of the computer that fetched the web bug, the URL of the web page with the web bug, the URL of the web bug image, the time and date the web bug was viewed, the type and version of the browser that fetched the web bug image and cookie information related to the website the bug is on.

As with cookies, web bugs may be used nefariously by advertising agencies to gather information for marketing purposes. This includes information on the personal profile of the users. This information may be used in conjunction with cookies to provide statistics, such as the number of visitors to a website, and other information.

In emails, web bugs can be used to determine whether the email was read, and when it was read, the IP address of the recipient and how often a message is being forwarded and read. This can give an indication of the success of, for example, junk email and thus on whether or not the person should be sent future emails. Web bug programmers can synchronise the web bug with a cookie to a particular email address. This connects the identity of the user with future access to websites. Web bug detectors are freely available, but web bugs cannot be removed.

Most privacy laws require disclosure of when information is being collected. This is a standard privacy principle and appears in the OECD principles, the IPP and the NPP. Anecdotally, the majority of web-based privacy policies do not disclose the use of web bugs. This is either in ignorance of the applicable law or is deliberate, so as not to alert potential users. The minimum disclosure standard typically requires there to be a privacy policy on the web page's site. For email, either disclosure within the email or a link to the privacy policy would be required.[41]

39 See Australia Law Reform Commission (ALRC), (2007) 'Review of Australian Privacy Law', Discussion Paper 72, paras 11.9–11.11.
40 See Kaman Tsoi, 'Web bugs and internet advertising', (2001) *PLPR* 21.
41 See ALRC, 'Review of Australian Privacy Law', paras 11.12–11.13.

International Covenant on Civil and Political Rights (ICCPR)

The ICCPR, which is based on the Universal Declaration of Human Rights, provides for privacy protection. More than 170 state parties have ratified it, and several pending ratification. Article 17 provides:

> 1 No one shall be subjected to arbitrary or unlawful interference with his privacy, family, home or correspondence, nor to unlawful attacks on his honour and reputation.
> 2 Everyone has the right to the protection of the law against such interference or attacks.

The Universal Declaration of Human Rights, in Article 12, states:

> No one shall be subjected to arbitrary interference with his privacy, family, home or correspondence, nor to attacks upon his honour and reputation. Everyone has the right to the protection of the law against such interference or attacks.

Many nations have enshrined the right of privacy into their constitutions and laws. The Constitution of France, for instance, includes the 'Declaration of the Rights of Man and of the Citizen'. However, in many there is an absence of a controlling and enforceable law to facilitate free expression, privacy, to reputation and honour, and the confidentiality of communications. The United Nations has a role to play in enunciating such issues and encouraging adoption and enforcement.

The US Senate has declared that acceptance of the ICCPR 'will not create a private cause of action in US Courts'.[42] The treaty binds the United States as a matter of international law, but does not form part of its domestic law. The US Supreme Court has stated that the Constitution of the United States includes 'penumbras' that guarantee the right to privacy against government intrusion. For example in *Griswold v Connecticut*[43] the state legislature in Connecticut prohibited the use of contraceptives. The US Supreme Court, by a majority of seven to two, invalidated the law on the grounds that it violated the 'right to marital privacy'. The court used the first nine amendments to the Bill of Rights, in particular the ninth amendment, in reaching its decision. The US Supreme Court has used the right to privacy in several decisions since Griswold's case, particularly in relation to healthcare. In *Roe v Wade*[44] the Supreme Court stated that a woman's choice to have an abortion was protected as a private decision between her and her medical practitioner.

Data protection

Data privacy (or data protection) are issues in electronic commerce. 'Data' refers to electronic symbols typically stored on some form of disk for computers or

42 S Exec. Rep., No. 102-23, 15 (1992).
43 381 US 479 (1965).
44 410 US 113 (1973).

transmitted from computer to computer. The Freedom of Information legislation[45] deals with documents, data and records, but not the much larger concept of information. There is a view that data has now become so freely available and accessible that privacy has largely disappeared – every individual has some personal data recorded on computer databases, and so has a kind of digital persona. Some of this data is freely available online. The technology has, in a sense, released data, making it now relatively easy to locate, copy and transfer.

Internationally, legislation has focused on the concept of data protection. This means that the laws protect the data directly rather than the people who are the subject of the data.

At common law, the tort of passing off and the principles relating to confidentiality and trade secrets provide indirect protection. The main protection for privacy in the past arose from the sheer costs of retrieving personal information, themselves a result of the tangible nature of the forms in which that information was stored and the inconvenience experienced in procuring access (assuming that its existence was known). Privacy was further protected by the incompatibility of collections with available indexes and the difficulty in ascertaining the existence of most personal data. These practical safeguards for privacy have largely disappeared.

Review

As cyberspace and its use continue to grow at an enormous rate, so does the potential for abuse of privacy rights. This has led many commentators to consider a review of pre-cyberspace privacy principles. Since the OECD Guidelines were released, the internet has become such a significant component in privacy issues that a review of the underlying principles would seem to be not just be prudent, but critical. All nations should examine their laws and policies and adapt them to the new technologies. Standards that reflect cybermanners and protocols should be formulated. Internet standards and the initiatives of bodies such as the Global Internet Liberty Campaign have assisted in some nations in doing this.

In 2007 the ALRC released its 'Review of Australian Privacy Law' Discussion Paper.[46] The discussion paper was produced following the largest public consultation process in the ALRC's history: more than 300 submissions were received and 170 meetings were held.

The ALRC proposes a new set of principles entitled Unified Privacy Principles (UPP). The UPP, set out in full in the discussion paper, uses the NPP as a template. The UPP is designed to minimise costs to businesses by allowing them to retain

45 *Freedom of Information Act 1982* (Cth); *Freedom of Information Act 1989* (NSW); *Freedom of Information Act 1982* (Vic); *Freedom of Information Act 1992* (Qld); *Freedom of Information Act 1991* (SA); *Freedom of Information Act 1992* (WA); *Freedom of Information Act 1991* (Tas); *Freedom of Information Act 1989* (ACT); *Right to Information Act 2005* (India); *Official Information Act 1982* (NZ); *Freedom of Information Act 2000* (UK) ; *Freedom of Information (Scotland) Act 2002*; *Freedom of Information Act 1966* (US) and *Privacy Act 1974* (US).

46 Discussion Paper 72, 2007.

policies and processes they have already implemented. The ALRC acknowledges the need to co-ordinate and unify the IPP and NPP.

The ALRC proposes changes to the definition of 'personal information' and 'record' so that they would include email addresses and IP addresses where these can be used to contact, target or affect an individual. Many cyberspace individuals have developed digital personas, often with complex and detailed identities. The ALRC is further proposing that the Australian Privacy Commissioner be able to issue website operators with take-down notices for inappropriate information placed online: information that may constitute an invasion of an individual's privacy, for example. Unfortunately, the discussion paper does not specify a trigger or standard on the basis of which the Privacy Commissioner could make such a determination. Such a determination must balance freedom of speech and freedom of expression issues and lawful behaviour; this may involve complex issues of law.

The ALRC Discussion Paper also considers expanding the meaning of 'identifiers' as used in NPP 7, biometric information, children, young people and the internet, developing technology, spam and telemarketing.[47]

The 'Review of Australian Privacy Law' provides a detailed basis for discussion by legislators, and demonstrates many deficiencies of legislation in the face of galloping advances in technology.

In the 1990s in the United Kingdom, Rosemary West was accused of involvement in notorious serial killings. Efforts to control traditional media outlets were well established and effective. However, they were ineffective for the internet. As Michael Kirby[48] wrote:

> The case illustrated the effective powerlessness of most national courts to enforce, in a truly effective way, local norms and values affecting global information... Governments and legislatures are not wholly powerless... The force of the technology (and the vast audiences which it gathers up) suggest that common global standards will tend to swamp local susceptibilities. In most countries, there will be little which they can do to influence the information flow except to enact laws enforceable in their courts in the comparatively rare instances in which they can catch those who offend against such laws within their jurisdiction.[49]

He has also reflected on his role as the Chair of the OECD Expert Group on the Protection of Privacy, noting that the OECD Guidelines have become largely obsolete and raising the following potential rights:

- a right not to be indexed – if a 'rogue' robot indexer ignores existing or new contemporary standards which exclude indexing;

47 See Alan Davidson, 'Privacy in a brave new world: ALRC proposals for privacy and technology', (2007) *Privacy Law Bulletin* 61 and Alan Davidson, 'Privacy reforms: Technological considerations in the age of the internet', (2008) *Internet Law Bulletin* 21.

48 The Hon Justice Michael Kirby AC CMG, President of the International Commission of Jurists, former Justice of the High Court of Australia and one-time Chairman of the OECD Expert Group on the Protection of Privacy (1978–80) and on Security of Information Systems (1991–92).

49 M Kirby, 'International dimensions of cyberspace law: Protection of privacy and human rights in the digital age', (1999) 30(2) *Library Automated System Exchange*, State Library of New South Wales, 12.

- a right to encrypt personal communications effectively;
- a right to fair treatment in public key infrastructures, so that no person is unfairly excluded in a way that would prejudice that person's ability to protect their privacy;
- a right to human checking of adverse automated decisions and a right to understand such decisions; and
- a right, going beyond the aspiration of the OECD openness principle, of disclosure of the collections to which others will have access and which might affect the projection of the profile of the individual concerned.[50]

All data collection practices, Kirby claims, must be open and transparent to individuals prior to operation and permit the individual the ability to disengage the feature if desired.[51]

Kirby's first suggestion deals with the process by which internet search engines search for websites and index them for all to see. Many websites are intended for private or limited use. Yet many search engines cannot discriminate. The underlying code in websites, hypertext transfer protocol (http), includes a 'no robot' code as an option for such index systems to take note of. When in place it signals to search engines a request to ignore the website and not to index it. Kirby suggests that this request should be elevated to a right.

The second and third suggestions relate to the use of high-level encryption systems for protecting private information. Public key cryptography currently provides the best protection. However, some governments have made attempts to circumvent or restrict public key cryptography. In the post-September 11 era, the right to such privacy has been eroded by legislation such as the PATRIOT Act of the United States, as well as by public opinion (to some degree). Nevertheless, the argument remains that lawful use of such privacy protection should be a right.

Personal privacy

Personal privacy involves what could be described as 'invasion of the person'. This may take the form of surveillance cameras, stalking, and the media following people, such as politicians and celebrities.

In 1937, in *Victoria Park Racing v Taylor*,[52] the High Court of Australia determined, in a majority verdict, that there was no right to privacy at common law. The case was about a prominent Sydney radio station setting up a platform adjacent to a horseracing track for the purpose of calling the races live on the radio. They did not have the permission of the racetrack authorities. The High Court ruled that the racetrack authorities could not prevent a party viewing or broadcasting from an adjacent private property.

50 Kirby, 'Privacy in cyberspace'.
51 Ibid.
52 (1937) 58 CLR 479.

Whether or not Australian law recognises a tort of invasion of privacy was raised in the High Court case *Australian Broadcasting Corporation v Lenah Game Meats Pty Ltd.*[53] In particular, the question was whether or not such a right might attach to a corporation. The facts were that unidentified trespassers gained access to the abattoirs and filmed the slaughtering and processing of possum meat for the export market. The ABC intended to broadcast excerpts on a television current affairs program in Tasmania.

Most of the High Court held that whether or not a tort of 'invasion of privacy' might develop under Australian law was still an open issue. Callinan J gave tentative approval for such a development in several paragraphs of his judgment. He did so first by arguing for a property right in a spectacle:

> It may be that the time is approaching, indeed it may already have arrived, for the recognition of a form of property in a spectacle. There is no reason why the law[s] should not, as they emerge, or their value becomes evident, recognise new forms of property.[54]

His Honour went on:

> It seems to me that, having regard to current conditions in this country, and developments of the law in other common law jurisdictions, the time is ripe for consideration whether a tort of invasion of privacy should be recognised in this country, or whether the legislatures should be left to determine whether provisions for a remedy for it should be made. Any consideration of that matter should be undertaken with regard to the separation of the roles of the judiciary and the legislature . . . [55]

The High Court's decision in Lenah Game Meats was based on the law of trespass, but the observations and statements made in *obiter* revived the issue of a possible tort of invasion of privacy to the person.

Gummow and Hayne JJ referred to the case of *Church of Scientology v Woodward*,[56] in which Murphy J identified 'unjustified invasion of privacy' as one of the 'developing torts'.[57] Kirby J stated that 'cheque-book journalism', intrusive telephoto lenses, surreptitious surveillance, gross invasions of personal privacy, deliberately deceptive 'stings' and trespass onto land 'with cameras rolling' are mainly phenomena of recent times.[58] Such phenomena have produced applications to the courts for relief, including injunctive relief. Adapting the words of Cardozo J used in another context, '[t]he cry of distress is the summons to relief'.[59] In Australia, generally, 'courts exercising equitable jurisdiction have upheld the entitlement to relief where to turn their backs would be seriously offensive to conscience'.[60]

53 [2001] HCA 63.
54 [2001] HCA 63, para 316.
55 [2001] HCA 63, para 335.
56 [1982] HCA 78.
57 [1982] HCA 78, para 13.
58 *Australian Broadcasting Corporation v Lenah Game Meats Pty Ltd* [2001] HCA 63, para 172.
59 *Wagner v International Ry Co.* 133 NE 437 at 437 (NY 1921).
60 [2001] HCA 63, para 172.

While the Supreme Court of New South Wales,[61] Supreme Court of Victoria[62] and the Federal Court of Australia have questioned this development, both the District Court of New South Wales[63] and the County Court of Victoria[64] have availed themselves of opportunities to develop the principles of invasion of personal privacy.

In *Grosse v Purvis*[65] the District Court of Queensland recognised a separate common law cause of action for invasion of privacy. The plaintiff, Ms Grosse, a prominent Queensland political figure, sued the defendant for, amongst other things, damages for invasion of privacy. The conduct of the defendant included instances of loitering, spying, unlawful entry to her home, unwelcomed physical contact, repetitious phone calls and use of offensive language and behaviour towards the plaintiff over an extended period of time. The judge was satisfied on the evidence that the defendant developed an extraordinary and active infatuation with the plaintiff.

As a result of the defendant's stalking behaviour, the plaintiff suffered post-traumatic stress disorder (PTSD). The plaintiff's condition was held to seriously and adversely affect her enjoyment of life and ability to function, including in her elected position. The plaintiff was successful and was awarded $178 000 in compensatory, aggravated and exemplary damages – for invasion of privacy and other causes of action.

Skoien SDCJ stated:

> It is not my task nor my intent to state the limits of the cause of action nor any special defences other than [are] necessary for the purposes of this case. In my view the essential elements would be:
> (a) a willed act by the defendant,
> (b) which intrudes upon the privacy or seclusion of the plaintiff,
> (c) in a manner which would be considered highly offensive to a reasonable person of ordinary sensibilities, [and]
> (d) . . . which causes the plaintiff detriment in the form of mental psychological or emotional harm or distress or which prevents or hinders the plaintiff from doing an act which she is lawfully entitled to do.[66]

In *Jane Doe v Australian Broadcasting Commission*,[67] Hampel J of the Victorian County Court awarded a rape victim $234 190 in damages based in part on invasion of her privacy. Despite a court order, the ABC broadcast in its news bulletins the name of the rapist, who was the husband of the victim, the name of the victim and her suburb. Without formulating an exhaustive definition, her Honour held that it was wrong to publish this personal information where

61 *NRMA v John Fairfax* [2002] NSWSC 563.
62 *Giller v Procipets* [2004] VSC 113.
63 *Grosse v Purvis* [2003] QDC 151.
64 *Jane Doe v ABC* [2007] VCC 281.
65 [2003] QDC 151
66 [2003] QDC 151, para 444.
67 [2007] VCC 281.

there was no corresponding public interest and where there was a prohibition on publishing.

Surveillance cameras have become a part of modern life. They are largely taken for granted. In petrol stations, supermarkets, hotels and malls surveillance cameras record our movements. The community as a whole has come to accept the underlying rationale: the need to protect personal safety and to reduce theft.

The cost of surveillance equipment has reduced dramatically, making increased digital scrutiny within the reach of nearly all businesses. Walls of TV screens allow authorities in the control rooms in major cities to monitor major roads. Concerned parents have set up cameras which record actions by their babysitters. Some link the cameras to the internet for remote surveillance. Internet cameras are commonplace, and are used for a variety of purposes: there are now more than 40 million cameras linked online.

More employers are using surveillance cameras and connecting them to the internet for remote surveillance. Cameras are being placed not only in places the public may frequent, but also in employee-only areas. The areas in which employees have the greatest sensitivity in relation to the use of video surveillance are toilets, showers and areas where employees may change clothing. However, employees also express concern about surveillance in areas set aside for employees when they not engaged in actual employment, such as dining, recreational areas, practising regular religious observances; and also personal behaviour such as flirting and other general behaviour where there is an assumption of privacy. The employers' concerns may also relate to areas with expensive furnishings or recreational equipment and other 'vulnerable' assets (such as vending machines).

It is clear that employers have a legitimate interest in placing cameras in positions frequented by members of the public. However, this has the secondary effect of keeping an eye on employees. Misuse is not unknown. One US advocate has documented police parties where 'best of' videos are shown. In call centres, where there are high levels of monitoring of employees, studies have shown that workers experience greater health problems such as depression, tension, anxiety and lower productivity levels. Employers must consider these effects of surveillance on productivity. The employer has the right to manage the workplace, but also an obligation to protect employees and the company from unlawful activities.

In an effort to address these competing interests, the NSW government introduced the *Workplace Video Surveillance Act 1998* (now replaced by the *Workplace Surveillance Act 2005*), which was designed to restrict and regulate the use of video surveillance equipment in the workplace. Under the 2005 Act, a court may order surveillance only where 'reasonable grounds exist to justify its issue . . . [having] regard to the seriousness of the unlawful activity with which the application is concerned'[68] and only under the following conditions:

68 *Workplace Surveillance Act 2005* (NSW) s25.

> The notice must be given at least 14 days before the surveillance commences [and] . . . must indicate:
> (a) the kind of surveillance to be carried out (camera, computer or tracking), and
> (b) how the surveillance will be carried out, and
> (c) when the surveillance will start, and
> (d) whether the surveillance will be continuous or intermittent, and
> (e) whether the surveillance will be for a specified limited period or ongoing.[69]

The two Acts prohibit 'covert' surveillance unless an 'authority' has been issued by a magistrate. When the 1998 Act came into effect, the then Attorney-General and Minister for Industrial Relations, Jeff Shaw, stated: 'the secret filming of workers in the workplace will be illegal unless there are reasonable grounds to suspect an employee is committing an unlawful act and a court authority is obtained'. There is still no specific legislation dealing with workplace surveillance in other Australian states. However, there are provisions which relate to general surveillance activities by 'authorities', which provide some protection.

In July 2003 ACI Glass Packaging was fined $500 under the 1998 NSW Act for installing a secret camera in a room where employees undressed. ACI claimed that it was installed in a first-aid room to monitor oxygen tanks, which had been stolen in the past. The room was frequently used by employees to change their clothing as well as to receive treatment for ailments. A Sydney lawyer breached the Act when he installed secret cameras in the female toilets of his law firm, allegedly because he suspected employees were using drugs in the toilet. However, a box of videotapes with the ratings 'best', 'beauty' and 'great' was found.[70]

In the recent unfair dismissal case in the Australian Industrial Relations Commission, Commissioner Larkin considered videotape of an employee removing a bag of coins from a weighing scale and placing the bag between the scales and her till.[71] The commissioner expressed concern that the applicant did not have access to the surveillance videotape and had not been shown the videotape at any point during the investigation. However, the commissioner still accepted the surveillance, stating: 'I am not satisfied that the applicant made an honest mistake on the night in question. In my view, *the videotape is clear on its face*.'[72] The commissioner ruled that the termination was not harsh, unjust or unreasonable.

Disclosure in the contract of employment clarifies the position for all parties. New employees have the 'choice' of refusing employment under such conditions. However, imposing conditions on existing employees may prove problematic. It is advisable for employers to disclose the use of surveillance cameras. Indeed, one may argue that the surveillance, if its purpose is to deter thieving, is ineffective unless its use is known. The employer needs to decide whether the priority is to prevent theft or to catch a thief.

69 *Workplace Surveillance Act 2005* (NSW) s10.
70 Belinda Harding, 'Surveillance in the work place', findlawaustralia.com.au/articles/.
71 *Y Liu v Star City Pty Limited* PR903625 [2001] AIRC 394.
72 *Y Liu v Star City Pty Limited* PR903625 [2001] AIRC 394, para 30.

Formulating a policy may prove useful. The policy would presumably note that public areas are subject to surveillance and that personal areas such as toilets and changing areas are not. As for the middle ground, a more circumspect policy may state that surveillance will only be undertaken if a legitimate concern regarding an employee's actions has arisen. Even if the policy is not initially disclosed to the employees, its release in the event of a complaint or other action would demonstrate a considered and judicious approach to the predicament.

Cell phone cameras provide a clandestine method for taking photographs and videos. If someone photographs a bather in a change room and posts the image on the internet, clearly the bather's privacy has been violated. The YMCA and the Royal Life Saving Society of Australia have banned camera phones because of their concerns about inappropriate pictures being taken surreptitiously.

In *Grosse v Purvis*[73] Senior Judge Skoien stated, 'there can be a civil action for damages based on the actionable right of an individual person to privacy'.[74] The elements of the cause of action proposed for personal privacy are most appropriate for camera surveillance situations.[75]

Reasonable use surveillance would be unlikely to be regarded as an invasion of privacy. However, misuse, such as for voyeurism or to eavesdrop for negotiation purposes may cross the line.

New Zealand

The *Privacy Act 1993* (NZ) came into effect on 1 July 1993. It followed the *Privacy Commissioner Act 1991* (NZ) which had established the office of Privacy Commissioner.

The *Privacy Act 1993* makes use of the 1980 OECD Guidelines on the Protection of Privacy and Transborder Flows of Personal Data as its template. Generally the Act applies to both the public and private sectors. Its underlying aim is the promotion and protection of individual privacy. The Act contains 12 IPP, dealing with collecting, holding, use and disclosure of personal information and assigning unique identifiers. The privacy principles give individuals the right to access personal information and to request correction of it. Like the Australian Act, the New Zealand Privacy Act gives the Privacy Commissioner the power to issue codes of practice that become part of the law. In certain circumstances the Privacy Commissioner may authorise agencies to collect, use or disclose information even though that would otherwise breach information privacy principles.[76]

73 [2003] QDC 151 (based on the High Court Justices comments in *ABC v Lenah Game Meats* (2001) HCA 63).
74 [2003] QDC 151, para 442.
75 [2003] QDC 151, para 444.
76 See, generally, the website of the NZ Privacy Commissioner: www.privacy.org.nz/. For a judicial consideration of the *Privacy Act 1993* (NZ) see *Harris v Selectrix Appliances* (Complaints Review Tribunal, Decision

United States

In the United States, the right of freedom of speech granted in the First Amendment has limited the effects of lawsuits for breach of privacy.

The US *Privacy Act 1974* was passed during the administration of President Nixon. The Act provides:

> no agency shall disclose any record which is contained in a system of records by any means of communication to any person, or to another agency, except pursuant to a written request by, or with the prior written consent of, the individual to whom the record pertains.[77]

There are a number of specific exceptions. They include for statistical purposes by the Census Bureau and the Bureau of Labor Statistics, archival purposes, law enforcement purposes, congressional investigations and other administrative purposes. Every US federal government agency must put in place an administrative and physical security system to prevent the unauthorised release of personal records.[78]

All government agencies are required to maintain a system permitting access upon request by any individual to permit the review and copying of the record, and to allow the individual to request amendments of their record.

The US *Computer Matching and Privacy Protection Act 1988* amended the Privacy Act to include protections for the subjects of Privacy Act records whose records are used in automated matching programs. These protections are designed to ensure procedural uniformity in carrying out matching programs, due process, and oversight of matching programs through the establishment of Data Integrity Boards at each agency engaging in matching to monitor the agency's matching activity.

The full name of the USA PATRIOT Act is the *Uniting and Strengthening America by Providing Appropriate Tools Required to Intercept and Obstruct Terrorism Act 2001*.[79] This Act was in response to the attacks of September 11, 2001 and was conceived, written, debated and passed within 45 days.

The Act significantly expands the authority of US law enforcement agencies for the purpose of fighting terrorism. It has many provisions that affect the lives of Americans in America. It affects their privacy by arming authorities with a range of investigative rights to access and gather information about communications, health and finances. The Act permits 'sneak and peek' searches where law enforcement officers may search a home or business without the owner's or the occupant's permission or knowledge. 'National Security Letters' allow the Federal Bureau of Investigation to search telephone, email and financial records without a court order. At a time when the right to privacy was solidifying in the

No. 12/2001, 2001), *L v L* (Complaints Review Tribunal, Decision No. 15/2001, 2001), *Commissioner of Inland Revenue v B* [2001] 2 NZLR 566.

77 Privacy Act 1974, 5 USC § 552a.

78 See also the US *Computer Matching and Privacy Protection Act*.

79 (Public Law 107-56), also known as the USA PATRIOT Act.

US courts and Congress, this Act single-handedly sets back information privacy rights by decades.

Final comment

Law makers and business need to monitor the changing national and international legal environments to ensure that the risk of privacy breaches in the future is minimised. Regulations in cyberspace must evolve in a way that includes fundamental human rights and national governance, and reflects global values and human diversity. The rule of cyberspace encompasses law, practice, procedure and presence in cyberspace. The internet should develop in a way that demonstrates respect for fundamental and universal human rights and democratic governance. Its expansion should reflect global values and human diversity.

Further reading

Australia Law Reform Commission (ALRC), 'Review of Australian Privacy Law', Discussion Paper 72, 2007.

Alan Davidson, 'Privacy in a brave new world: ALRC proposals for privacy and technology', (2007) *Privacy Law Bulletin* 61.

Alan Davidson, 'Privacy reforms: Technological considerations in the age of the internet', (2008) *Internet Law Bulletin* 21.

Mark Davison, *The legal protection of databases*, Cambridge University Press, Cambridge, 2003.

Sharon Nye, 'Internet privacy – Regulating cookies and web bugs', (2002) *PLPR* 26.

14

Electronic mail and online presence

The availability of electronic mail (email) goes beyond the introduction of postal systems, the telephone, telexes and facsimile. In the near future verbal communication will take place with greater ease and clarity than the telephone. Additionally, users will have the advantage of digitally saving conversations in a similar way to the way in which they can already save email messages. The tools to access and navigate between systems are steadily becoming cheaper, more powerful and easier to use. The internet provides tools to maximise the use of time to engage in discourse, study and recreation in a manner previously unimaginable. Email adds a further dimension to communication. Email will be required by many government departments and courts for the lodging of materials and to facilitate correspondence. Email is a fast and efficient worldwide communication portal. This chapter deals with legal and practical issues relating to the use of electronic mail and maintaining an online presence.

Email

Email is one of the most popular applications of the internet. It allows users to send messages created on the computer to any other internet computer in a matter of seconds. Data files such as photographs, sound clips and, more usually, word-processed document files, can be attached to an email message. In fact any file which may be stored digitally may be transferred by email. It is usually cheaper and quicker than a telex or facsimile and more reliable than the ordinary post. Email is becoming integrated with other communication technologies such as facsimiles, pagers and personal digital assistants (PDAs). Email messages can be accessed on the screens of the newer models of mobile phone.

The internet's use as a recreational facility and business tool is relatively new. The integration of the telephone into the office, now almost before living memory, had a significant effect on business and communication. Many can recall similar effects when the telex, the facsimile and the photocopier were introduced to offices.

In cyberspace, the address for each person, each computer and each internet page is unique. Unlike names, no two email addresses may be the same. Many individuals possess several unique email addresses.

When a person writes a standard letter, puts it in a stamped, addressed envelope and drops it in a letter box, they do not have to understand the processes conducted by the mail carrier to know that the letter will be delivered. Email is the same. Users do not need to understand that the email facility creates text, typically using standard protocol (ASCII or HTML files), for transmission, or that the protocol used on the internet to standardise the transfer of mail is called SMTP (Simple Mail Transfer Protocol). SMTP is activated when mail is sent.

The interface with the internet mail protocol allows incoming messages to be read, deleted or saved, and new messages to be created. Users can have more than one email program on their computer. Several companies have produced software applications: these aim to maximise ease of use and offer attractive features such as folders for filing mail, feature buttons, checks on delivery and receipt of mail sent, encryption for security, preparation of group lists, forwarding on to others and so forth.

Attachments

In common with normal mail, emails are generally text. Although it is impossible to email hard copies of photographs, plans, documents for signature and so forth, it is usually possible to attach data files to email.

Depending upon the intention of the sender, the attachment is an electronic record, proof of which may be determined by inherent meta-data. It may also be the intention of the sender that the attachment be printed and the printout be the original document. In this manner the email itself is regarded as a courier, much like an envelope. Depending upon the requirements of the sender, it is prudent to make an express statement of this intention either within the accompanying email, or in the attachment itself.[1]

Authentication

Authentication of electronic messages will become increasingly important for evidentiary purposes. Australian Evidence Acts do not address all aspects of email communication. Sections 160–2 of the Evidence Acts of the Commonwealth,

1 See Chapter 17.

New South Wales, Tasmania and the ACT make presumptions regarding the sending and receipt for postal articles, telexes, lettergrams and telegrams. There is no similar presumption regarding email. However, they do provide that:

> The hearsay rule does not apply to a representation contained in a document recording a message that has been transmitted by electronic mail or by a fax, telegram, lettergram or telex so far as the representation is a representation as to:
> (a) the identity of the person from whom or on whose behalf the message was sent, or
> (b) the date on which or the time at which the message was sent, or
> (c) the message's destination or the identity of the person to whom the message was addressed.

Recently section 161, dealing with the facilitation of proof, was expanded to apply to electronic communications – this includes all modern electronic technologies as well as more outmoded ones such as facsimile and telex. The exception to the hearsay rule in section 71 was similarly expanded.[2] Courts increasingly need to be satisfied regarding the authenticity of transmissions. The Electronic Transactions Acts in the Commonwealth, states and territories were enacted to facilitate admission of electronic communications where they are functionally equivalent to traditional paper-based communications.[3]

Language

Several texts that discuss electronic communications emphasise the need to keep language formal. Grammar, syntax and an appreciation of the form of the medium involve customs which many of us take for granted. The use of language without careful and appropriate consideration may have unintended and potentially disastrous consequences. However, email can be handled with such relative ease and speed that communications tend to become less formal, perhaps more akin to speaking in person or on the telephone. The parties may soon be involved in a conversation that involves an exchange of a dozen or so emails within a single hour. The chance increases for informal and glib comments. Irony and sarcasm do not translate well on email, or on any form of written communications. Some people take care by writing the obvious '(just kidding)'. Others use smiling faces :-) or ;-) (Turn your head sideways to the left.) Smiley is used to communicate subtle nuances which may otherwise be absent. The use of Smiley has led others to use the unhappy face to express displeasure:

> :-(or ;-(

In any event, when writing, it is important to be aware of the context and to consider how the written communication will appear if read by other parties, including a court.

2 See *Evidence Amendment Act 2008* (Cth).
3 See Chapter 3 and Chapter 17.

Viruses

Many myths and rumours have been circulated regarding the dangers of catching a computer virus from email, particularly unsolicited email. Several of these rumours have been hoaxes. Most users have a rudimentary understanding as to how data is transmitted by email and how it is read, interpreted and converted into meaningful text and files. Many have nonetheless decided to err on the side of caution.

The reality is that a virus cannot be contracted by simply opening an email message. However, they can be transmitted by attachments. As a general rule, users should only open attachments from persons they know, or sources which are regarded are secure.

Disclaimers

Disclaimers often appear at the end of emails. Many are ignored by recipients. High numbers of mail defamation incidents, unintended contract formations and misdirected emails demonstrate the lack of appeal of email disclaimers. The value of disclaimers, however, is well known.[4]

Disclaimers are much more routinely included in emails than in standard mail. The reasons suggested for this vary. One is that the nature of email means that writers are often less formal and more unguarded than they would be with hardcopy mail. They reply and send without taking sufficient time to reflect, and they may not have as good a system of checks and balances as they have with standard mail. Another could be the fact that Norwich Union paid out £450 000 to a competitor, Western Provident, for defamatory emails circulated by an employee.

Defamation, unintended contract formation, misdirected emails, confidentiality, legal privilege, infringement of copyright, plus viruses, sexual and racial discrimination and harassment are just a few issues of concern addressed in disclaimers.

However, the value of these disclaimers is questionable. First, the courts will typically attach more weight to the substantive content of the email. There are occasions where a standard disclaimer is clearly inappropriate in relation to the actual content of the email. This occurs where the sender includes a standard, all-purpose disclaimer but does not address the reasons for its inclusion in the email in question.

Second, courts will look to the surrounding circumstances. This may include such factors as prior communications, the nature of the relationship (for example, whether or not it is contractual) and how prior disagreements were resolved.

Importantly, the disclaimer may ward off legal action. A person contemplating legal action may think twice if an appropriately worded disclaimer was included

4 *Hedley Byrne v Heller* [1964] AC 465 is a precedent on negligent misstatement. However, the judgment on that issue is strictly *obiter* as the defendant succeeded because of the inclusion of an effective disclaimer.

in the transmission. The disclaimer may provide a useful argument in negotiations to resolve a dispute early and for a lower sum.

Generally, if in doubt, a disclaimer should be included, even though one of the greatest problems with disclaimers is the use of inappropriate ones. The writer should consider the purpose for which the email is sent and which areas require protection. There is a vast difference, for example, between commercial and personal emails. In a commercial context, there is a variety of concerns regarding legal liability: contractual or defamatory concerns, concerns regarding confidentiality and regarding accidental delivery to the wrong address.

Confidentiality

An express statement that a communication is confidential may well make the difference between its being treated as confidential or not. It could be argued that the notice may be ineffectual if it is in small print or placed after the sign-off at the end of the message. Prepending such notices (putting them at the start of a message) rather than appending them (at the end) may be useful.

Viruses

Many email writers include a disclaimer that they take no responsibility for checking for viruses: that responsibility falls on the recipient. Whether these disclaimers would be effective in court is questionable, but they may discourage disgruntled recipients from initiating a dispute.

Defamation

Employers have been held liable for defamatory statements made in emails by employees. A disclaimer will most likely not excuse the act where it is made in the ordinary course of business. Where it is not made in the ordinary course of business the disclaimer would not be needed. Nevertheless, adding a disclaimer may be a useful negotiation tactic in appropriate circumstances.

Copyright

A disclaimer would be unlikely to serve as a defence to a charge of breach of copyright. However, an appropriately worded disclaimer may indicate the level of care taken and the intention behind the transmission. Additionally, the disclaimer may resolve internal responsibilities and liabilities between employee and employer.

Negligent misstatement

By law, a person is obliged to take care when giving advice that a third party relies on. If an employee were to give professional advice in an email, the company will be liable for the effect of the advice that the recipient, or even a third party, reasonably relies upon. A suitably worded disclaimer could protect the sender and the sender's organisation from this kind of liability.

Accidental contract formation

A disclaimer could clearly set out the extent to which personnel have actual authority to bind the company or employer. The company should establish procedures to guard against such situations. However, the nature of email is that an immediate reply is more likely. Such a disclaimer may state: 'No employee or agent is authorised to conclude any binding agreement on behalf of this firm/company without the express written confirmation by a partner/director of the firm/company.'

Sexual and racial discrimination and harassment

Internal emails may give rise to claims of discrimination or harassment. Employees should be informed of the employer's policy and expected practice. Liability will depend upon whether the act was or was not in the course of business, and on factors such as the level of supervision and position and authority of the offending employee. A disclaimer on internal emails may put all parties on notice about the employer's policy and concerns.

Risk assessment

Each organisation and individual should consider undertaking a risk audit on its email procedures and policies. The audit should determine who sends emails, their authority, how often emails are sent, whether there is a supervisory process or checking process, and how much time is allowed or available for reflection, thought and consideration before replying. Many employers may be surprised and disturbed at the use and misuse of email by employees. Once a risk analysis has been completed, the organisation is in a much better position to prepare a policy or code of conduct and to determine the extent to which disclaimers ought to be used in their email system.

Service of documents by email

In the NSW Supreme Court case of *Macquarie Bank v Berg*[5] the plaintiff sought an order restraining the defendant, Berg, from publishing certain material on the internet. The summons first came before B M James J, who abridged the time for service and directed that service be effected by delivering a copy of the documentation on the solicitors acting for the defendant in other proceedings between the parties, and by sending copies of the documents, marked for the attention of the defendant, to a specified email address.

Since that case New South Wales has amended its *Electronic Transactions Act 2000* by inserting Part 2A Court Administration. This Part established an Electronic Case Management (ECM) system to enable documents with respect to legal proceedings to be created, filed, issued, used and served in electronic form. For example, section 14M provides that any document filed in or issued by

5 [1999] NSWSC 526, 1.

an ECM court by means of the ECM system may be served electronically. Other states and territories have made similar provisions in their equivalent statutes and regulations.

Time and place of dispatch and receipt

All nine jurisdictions in Australia have enacted an Electronic Transactions Act that includes a provision regarding the time and place of dispatch and receipt of electronic communications.[6] Electronic communication includes emails, facsimiles, SMS and instant messaging.[7]

Time of dispatch

If an electronic communication enters a single information system outside the control of the sender, then, unless otherwise agreed, the dispatch occurs when it enters that information system.[8]

Time of receipt

If the recipient of an electronic communication has designated an information system for the purpose of receiving electronic communications, then, unless otherwise agreed, the time of receipt is the time when the electronic communication enters that information system. The designation of an information system may be by prior conduct, prior email use, or by including the email address in correspondence, such as a letterhead or a business card. If the recipient has not designated such an information system then, unless otherwise agreed, the time of receipt is the time when the electronic communication comes to the attention of the addressee.[9]

Place of dispatch and receipt

Unless otherwise agreed, the electronic communication is taken to have been dispatched at the place where the sender has its place of business, and is taken to have been received at the place where the recipient has its place of business. If the sender or recipient has more than one place of business, and one of those places has a closer relationship to the underlying transaction, it is to be assumed that that place of business is only place of business. If the sender or recipient has more than one place of business, but it cannot be determined which has a closer relationship to the underlying transaction, then it is assumed that the principal place of business is the only place of business. If the sender or recipient does not have a place of business, it is assumed that the place of business is the place where the sender or recipient ordinarily resides.[10]

6 The provisions were examined in detail in Chapter 3.
7 Electronic Transactions Acts – NSW s12; Cth – s14; Qld – s23; SA – s13; Tas – s11; Vic – s12; WA – s13; ACT – s13; NT – s13.
8 Ibid.
9 Ibid.
10 Ibid.

Web page presence

All forms of business have taken hold of internet technology, many just in order to have simpler and speedier communication. Others have created web pages. Some use the pages in a simple static way. That is, they provide a simple electronic brochure. This form of advertising is cheap and continuous. For a modest fee, often included in the price of connection, the user may have allocated internet space to place a number of web pages with text, graphics and links. The pages may be updated at any time, and the brochure is available to all 24 hours a day.

Alternatively, the business may produce dynamic web pages. These are pages where potential consumers may interact with the host by placing orders and making enquiries; the pages can act as an online storefront.

Some issues faced by the host in the design of the website are technical and practical. However, there are many legal concerns as well.

As most hosts lack the expertise to develop professional web pages, they generally seek the services of a web developer. There are many fly-by-night operators arising as demand has grown. Many of these are small businesses juggling the joint considerations of design, implementation, programming, multimedia artwork and graphics. A contract with a web designer should include provisions dealing with:

- intellectual property aspects of the content, structure, graphics, domain names and computer and web code;
- the product delivered – this aspect should include details of not only content but also compatibility, and functionality and performance criteria;
- the degree of subcontracting, including rights between the head contractor and the subcontractors;
- marketing strategy – such as the steps to be taken to actively place the pages on the most appropriate search engines and the strategy for indexing by search engines, such as the placement of key terms in the web pages and in meta tags (HTML searched and indexed by the search engines – called Search Engine Optimisation);
- maintenance and future amendments;
- indemnities.

Liability for online material

Misleading and deceptive conduct

Online businesses are liable for material and information that is misleading and deceptive within the meaning of laws such as the *Trade Practices Act 1974* (Cth) and the Fair Trading Acts of the states and territories. These laws apply to statements and representations made in web pages just as they apply to hardcopy advertisements and brochures. The nature of a web page is that it is a continuing publication.[11] Steps should be taken to ensure that information remains current, accurate and correct. In particular, note the application of *Trade Practices Act 1974* (Cth) (TPA) section 52, and the criminal provisions of section 53:

> Section 52(1) A corporation shall not, in trade or commerce, engage in conduct that is misleading or deceptive or is likely to mislead or deceive.
>
> **Section 53 False or misleading representations**
>
> A corporation shall not, in trade or commerce, in connexion with the supply or possible supply of goods or services or in connexion with the promotion by any means of the supply or use of goods or services:
>
> (a) falsely represent that goods are of a particular standard, quality, value, grade, composition, style or model or have had a particular history or particular previous use;
> (aa) falsely represent that services are of a particular standard, quality, value or grade;
> (b) falsely represent that goods are new;
> (bb) falsely represent that a particular person has agreed to acquire goods or services;
> (c) represent that goods or services have sponsorship, approval, performance characteristics, accessories, uses or benefits they do not have;
> (d) represent that the corporation has a sponsorship, approval or affiliation it does not have;
> (e) make a false or misleading representation with respect to the price of goods or services;
> (ea) make a false or misleading representation concerning the availability of facilities for the repair of goods or of spare parts for goods;
> (eb) make a false or misleading representation concerning the place of origin of goods;
> (f) make a false or misleading representation concerning the need for any goods or services; or
> (g) make a false or misleading representation concerning the existence, exclusion or effect of any condition, warranty, guarantee, right or remedy.[12]

Defamation

Statements made online – by web pages or email – are of course subject to defamation laws. Indeed, web pages often have a greater potential for being viewed, as by their very nature they are continuing publications.[13]

Disclaimers – conditions of use

Before proceeding to the main pages, some sites require the user to click a button titled 'I agree' which follows 'terms and conditions of use'. The courts have

11 For example, in 2000 the Dow Chemical Co. audited their employees' usage. This led to the dismissal of some employees for engaging in inappropriate email messaging: www.cnn.com/2000/TECH/computing/09/19/dow.firing.idg/index.html.

12 *Trade Practices Act 1974* (Cth). Substantially similar provisions appear in the Fair Trading Acts of the states and territories.

13 See Chapter 12.

held that users are bound by such 'agreements'.[14] Some sites place less obvious links to the site's terms and disclaimers.[15] All users (including website designers and administrators) need to understand the significance of such notices and to balance legal considerations with practical and aesthetic considerations.

Linking and framing

Some US actions have been taken where the host site includes an 'unauthorised' link to another website. The plaintiffs' concerns have included the fact that links sometimes bypass the 'home page' thus avoiding security, implementation of cookies, tracking techniques and stated terms and disclaimers. Some sites place a 'frame' around other sites, giving the appearance that the 'framed' site is the creation of the host site.[16] The liability for such framing is yet to be ironed out by the courts. However, misleading and deceptive conduct provisions such as TPA section 52 have been applied broadly to sections of the community. Many websites are directed to the general public, which the courts have described as the knowledgeable, and those who are not, the profound, the gullible as well as the cautious.[17] Actions may also be based on the tort of passing off.

Information to be placed on pages for practical and legal purposes

Other sites may link not to the host's opening page but to some interesting feature in the middle of the pages. Users need to find their way out of this situation. Consideration should be given to placing certain essential information on every web page, as part of the page template. Design considerations should be not only aesthetically pleasing for marketing purposes, but also practical in terms of liability. The use of headers, footers and frames is a way to provide fundamental information, such as a privacy policy and disclaimers, on every page of a website. Including the date of revision in such information tells users how current the information on the web pages is. Although one of the advantages of websites is that they can be updated quickly, one of the dangers is that users rely on them for current information when the page may not in fact have been altered recently. Many legal researchers cite the date they visited web pages for this very reason. Many law firms provide online legal bulletins to the general public. A statement placed online regarding the law may quickly be obsolete. With a time stamp, members of the public will be sure that they may rely on the advice; in the end, this will be an advantage to the host.

14 See Chapter 4.
15 See *Disclaimers*, above.
16 See Chapters 6 and 7.
17 See *World Series Cricket Pty Ltd v Parish* (1977) 16 ALR 181.

Newsgroups and mailing lists

Newsgroup and mailing list administrators need to consider the same issues regarding liability as other online providers and email writers. Newsgroups and mailing lists allow participants to post messages to a central computer file for access by anyone. There are thousands of newsgroups where people discuss matters of common interest, from rugby to aliens, recipes to *Star Trek*. A mailing list permits a participant to send one message to a central computer and have the message then forwarded automatically to every email address on the list. Issues of control and responsibility should be brought to the attention of users, preferably contractually, using clickwrap techniques.[18]

The professional office and email

Just as each office maintains its own set of procedures in relation to the sending and receipt of mail and facsimiles, so each office must appropriately determine its approach to email.

Backup copies

Before photocopying, carbon copies were often kept. For electronic documents and records, legal offices must have in place a regimented system of creating and maintaining backup copies. Some offices print duplicates and retain 'hard copies' in a physical file. Whatever the form of record-keeping set up, consideration must be given to methods of backing up electronic mail. Some have incoming mail on disk and backup daily or print out that mail for retention and filing while retaining the incoming mail on disk. In many jurisdictions it is vital to retain the electronic version: evidentiary rules may exclude a printout of an email, or a scanned copy of paper documents, as evidence.[19]

Maintain supervisory checks

If the organisation includes procedures for the checking by supervisors of mail or facsimiles before sending, then an appropriate gateway should be put in place to ensure that the checking process is maintained for email as well. It is too easy to create emails without the assistance of the secretarial staff and then press the send button. Electronic gateways can be put in place by systems programmers.

18 See Chapter 4.
19 For example the secondary evidence rule applies in Queensland, Victoria, South Australia and Western Australia; see Chapter 17.

Records and costing

Many offices used to have secretarial staff to keep records of the production of letters and other documents for costing purposes. If the professionals now bypass such a system, because of the ease and speed of email, an appropriate procedure must be put in place to ensure that such work is costed.

Confidentiality

Email has a range of potential security problems. Confidentiality is often most important. As a parallel, when facsimiles are used, it is usually prudent to ask the recipient in advance if sensitive information may be sent by that medium. In certain cases the recipient may wish to know when the facsimile is being sent so that he or she may be on hand to receive it to ensure privacy and confidentiality. It would be prudent to prepare a standard communiqué regarding the risks of email and obtain the recipient's written consent before sending sensitive information electronically. Where appropriate, the recipient may agree to the use of an encryption program as added security.[20]

Internal trial

There are organisations which have severely restricted access to email facilities because they fear abuse by staff. Organisations that do not yet use email, or use it only in restricted ways, could consider undertaking a month's trial of sending inter-office memos by email. Informing staff that there will be no more hard copies of memos will ensure that they open their email daily. This form of familiarity will lead to office acceptance and use. Staff acceptance and morale are important.

Confirmation of sending

One continuing problem with email is determining whether a particular email has been delivered and read. This problem existed with standard mail and facsimiles as well. Some email programs claim to have a facility that will send a message to the sender when the email was received, and again when it has been read. However, such facilities are dependent on the mail systems and protocols used by the other party or the mail path. The most certain method of reassurance is to ask the recipient – even insist, in appropriate situations – to send a reply that is as short as 'Received', and that includes a copy of the original email. Most email systems allow the user to reply either with or without the body of the received message included in the reply. It may be worthwhile to adopt a policy of sending confirmations and requesting confirmations from recipients.

20 For a common public example, see PGP: www.pgpi.com.

Access to files

A procedure that monitors and controls access to electronically saved files, regardless of whether or not hard copies have been created and maintained, should be put in place. Sensitive hardcopy files may be kept locked in cabinets with limited access. Just as a duplicate key may be entrusted to another person, passwords to electronic data records should be entrusted in a similar way, with similar security constraints. Using different passwords for different levels of sensitivity increases the security of documents.

A new form of expression

There are reports from a number of organisations that they are having trouble with employees sending other employees attachments which are questionable movie files. Sometimes, the sending of these files could be regarded as constituting sexual harassment. While the form of communication may now be electronic, such communications have always been undesirable in any form. Email is merely a new form of communication, and like all other forms, it may be misused.

Conclusion

Just as each organisation maintains procedures in relation to the dispatch and receipt of mail and facsimiles, each organisation must also determine its policies on email; this is preferable to defending legal proceedings because of a failure to do so.

Email and other forms of electronic communication are replacing the traditional formal letter. Commercial expediency will win out. With this comes several challenges. Many commercial parties have jumped into electronic commerce with little consideration of the legal consequences. This too is not new. Commercial parties forged ahead in centuries past without law that dealt with their actions, and the result was the *lex mercatoria*, a reactive body of law. Electronic communicators must be aware of the current laws that affect their actions, and they must validate their documents in terms of the place of dispatch and receipt, the timing of dispatch and receipt, the admissibility of the email and attachments sent. Disclaimers can be effective, but should be drafted with care; they may need to be specific to particular uses.

15

National electronic surveillance

Electronic surveillance methods have become the subject of several pieces of federal legislation since the events of September 11, 2001. Anti-terrorism legislation internationally has introduced powers and regulations which (among other things) have an impact on cyberlaw, communications, the internet and security (confidentiality and safe computer systems). The most recent proposal, the *Surveillance Devices Act 2004* (Cth), includes significant increases in powers of investigation and surveillance.

Prior to September 11, 2001, there were no federal or state laws relating to terrorism. The Charter of the United Nations (Anti-Terrorism Measures) Regulations 2001 implemented aspects of UN Security Council Resolution 1373 of 28 September 2001, which called on all nation states to prevent and suppress the financing of terrorist acts. However, beginning in March 2002, the Australian government introduced a number anti-terrorism Bills. In 2003, the government announced the creation of an Ambassador for Counter-Terrorism; the government has since signed numerous memorandums of understanding on counter-terrorism.

Many of these measures affect cyberlaw and electronic commerce. The extent to which law enforcement and security agencies can now intercept, search and seize information, electronically and otherwise, has been dramatically enhanced. Some anti-terrorism provisions permit Australian law enforcement and security agencies to intercept unread emails in routine investigations. This chapter briefly examines these electronic surveillance measures.

The USA PATRIOT Act

In response to the attacks of September 11, 2001 the US Congress drafted, debated and passed the *Uniting and Strengthening America by Providing*

Appropriate Tools Required to Intercept and Obstruct Terrorism Act 2001 (commonly known as the PATRIOT Act), all within seven weeks.

The Act contains features which previously would have made civil libertarians cringe. Yet in the changed environment, the US Senate passed the Act by a vote of 98 to 1. The Act gives US federal officials greater authority to track and intercept communications, both for law enforcement and foreign intelligence-gathering purposes. It includes powers to combat the use of US financial institutions for foreign money-laundering purposes. It tightens requirements for entry into the US and contains powers to detain and remove a wider class of persons. It creates new crimes, new penalties, and new procedural efficiencies, all for use against domestic and international terrorists. The Act contains a number of safeguards, but critics contend that these are inadequate, and that many of the Act's provisions go too far. Others are concerned that it does not go far enough.

The Act prohibits electronic eavesdropping on telephone conversations, face-to-face conversations, and computer and other forms of electronic communications in most instances. It mandates a specifically defined process for authorities that wish to use electronic surveillance, and specifies that such surveillance is to be used only as a last resort in serious criminal cases. The court notifies the parties to any conversations seized under the order after the order expires. The Act authorises nation-wide execution of court orders for pen registers, trap and trace devices, and access to stored email or communication records. The Act permits authorities to intercept communications to and from a trespasser within a computer system.

New computer crimes are created. Electronic cooperation between law enforcement and foreign intelligence investigators is encouraged.

Australian response

Criminal Code Amendment (Anti-Hoax and Other Measures) Act 2002 (Cth)

This Act creates new offences relating to the use of postal services to perpetrate hoaxes, the making of threats and the sending of dangerous articles. A person is guilty of an offence if the person causes or threatens to cause an article to be carried by a postal or similar service, and does so with the intention of inducing a false belief that the article encloses or contains an explosive, dangerous or harmful substance, or that an explosive, dangerous or harmful substance has been left in any place.[1] The maximum penalty is seven years' imprisonment. It is an offence to perpetrate a menace, harass or cause offence in these terms.[2]

1 *Criminal Code 1995* (Cth) ss471.10, 471.11.
2 *Criminal Code 1995* (Cth) s471.12.

Security Legislation Amendment (Terrorism) Act 2002 (Cth)

This Act creates a new offence of 'engaging in a terrorist act'. The effect is to modernise the offence of treason,[3] making it an offence to have certain links with terrorist organisations,[4] to train with a terrorist organisation,[5] to receive funds from a terrorist organisation,[6] and to provide support to a terrorist organisation.[7] The Act enhances powers to deal with terrorist-related offences and creates a new regulations in relation to identifying terrorist organisations: six organisations have been specified under this new process.

Suppression of the Financing of Terrorism Act 2002 (Cth)

This Act is designed to prevent the movement of funds for terrorist purposes. It implements international obligations, creates the offence of the collection of funds used to facilitate a terrorist act, requires financial institutions and cash dealers to report suspected terrorist-related transactions, and introduces a penalty for using the assets of those involved in terrorist activities. This makes it an offence to knowingly or recklessly fund a terrorist act.[8]

Criminal Code Amendment (Suppression of Terrorist Bombings) Act 2002 (Cth)

This Act makes it an offence to place or detonate bombs or other lethal devices in prescribed places, such as government facilities or places of public use.[9] The offences do not apply to defence force members. The Attorney-General must consent to proceedings for offences under this Act. For example, within the terms of the Act, someone who murders an Australian outside Australia may face imprisonment for life.[10]

Telecommunications Interception Legislation Amendment Act 2002 (Cth)

The Act allows the use of telecommunications interception in the investigation of the following offences: conduct involving acts of terrorism,[11] child pornography,[12] serious arson offences, and police officer conduct leading to a dismissal decision.

3 *Criminal Code 1995* (Cth) s80.1.
4 *Criminal Code 1995* (Cth) s102.1.
5 *Criminal Code 1995* (Cth) s102.5.
6 *Criminal Code 1995* (Cth) s102.6.
7 *Criminal Code 1995* (Cth) s102.7.
8 *Criminal Code 1995* (Cth) s103.1.
9 *Criminal Code 1995* (Cth) s72.1.
10 *Criminal Code 1995* (Cth) s115.1.
11 *Telecommunications Interception Legislation Amendment Act 2002* (Cth) s5(1)(c).
12 *Telecommunications (Interception and Access) Act 1979* (Cth) s5D(2)(d) .

Criminal Code Amendment (Offences Against Australians) Act 2002 (Cth)

The Act amends the federal Criminal Code by inserting new provisions to make it an offence to murder, commit manslaughter or intentionally or recklessly cause serious harm to an Australian outside Australia.

Australian Security Intelligence Organisation Legislation Amendment (Terrorism) Act 2003 (Cth)

The Act amends the *Australian Security Intelligence Organisation Act 1979* (Cth) by reincorporating terrorism within the definition of 'politically motivated violence',[13] thus extending the investigative powers of ASIO in relation to terrorism permitting personal searches to be authorised in conjunction with search warrants and providing a power to detain and question people who are believed to have information about planned terrorist attacks for a period of seven days.

In relation to warrants to access and search computers and electronic equipment, the minister may allow the following:

(a) where there is reasonable cause to believe that data relevant to the security matter may be accessible by using a computer or other electronic equipment, or a data storage device, brought to or found on the subject premises – using the computer, equipment or device for the purpose of obtaining access to any such data and, if necessary to achieve that purpose, adding, deleting or altering other data in the computer, equipment or device;

(b) using the computer, equipment or device to do any of the following:
   - (i) inspecting and examining any data to which access has been obtained;
   - (ii) converting any data to which access has been obtained, that appears to be relevant to the collection of intelligence by the Organisation in accordance with this Act, into documentary form and removing any such document;
   - (iii) copying any data to which access has been obtained, that appears to be relevant to the collection of intelligence by the Organisation in accordance with this Act, to any data storage device and removing the device.[14]

'Terrorism act'

'Terrorism act' is defined by the federal Criminal Code in three parts.

First there must be an action or threat of an action that causes serious harm that is physical harm to a person; causes serious damage to property; causes a person's death; endangers a person's life, other than the life of the person taking the action; or creates a serious risk to the health or safety of the public or a section of the public.[15]

13 *Australian Security Intelligence Organisation Act 1979* (Cth) s4.
14 *Australian Security Intelligence Organisation Act 1979* (Cth) s25(5).
15 *Criminal Code 1995* (Cth) s100.1.

In relation to electronic systems, it is a terrorist act to:

> seriously interfere with, seriously disrupt, or destroy, an electronic system, including, but not limited to:
> (i) an information system; or
> (ii) a telecommunications system; or
> (iii) a financial system; or
> (iv) a system used for the delivery of essential government services; or
> (v) a system used for, or by, an essential public utility; or
> (vi) a system used for, or by, a transport system.[16]

Second, the action must be done or the threat be made with the intention of advancing a political, religious or ideological cause. Third, the action must be done or the threat be made with the intention of coercing, or influencing by intimidation, the government of the Commonwealth or a state, territory or foreign country, or of part of a state, territory or foreign country or intimidating the public or a section of the public.[17]

A terrorist act does not include an action if it:

> (a) is advocacy, protest, dissent or industrial action; and
> (b) is not intended:
> (i) to cause serious harm that is physical harm to a person; or
> (ii) to cause a person's death; or
> (iii) to endanger the life of a person, other than the person taking the action; or
> (iv) to create a serious risk to the health or safety of the public or a section of the public.[18]

Anti-Terrorism Act 2004 (Cth)

This Act, which commenced on 30 June 2004, doubles the questioning time for terrorist suspects to 24 hours. Compared with other nations, this is quite modest. In Britain, the police can detain suspected terrorists for 48 hours, extendable for a further five days. However, the Australian Act provides for time when the clock does not run. For example, time waiting for a reply from international agencies. This cannot 'exceed the amount of the time zone difference', but it can still add an extra 23 hours.

Surveillance Devices Act 2004 (Cth)

This Act establishes procedures for enforcement agencies to obtain warrants, emergency authorisations and authorisations for the installation and use of surveillance devices in Australia and overseas. Additionally, it regulates the use,

16 *Criminal Code 1995* (Cth) s100.1(2)e.
17 *Criminal Code 1995* (Cth) ss100.1(1), 100.1(2).
18 *Criminal Code 1995* (Cth) s100.1(3).

communication, publication, storage, destruction and making of records in connection with surveillance device operations. The then Attorney-General, Philip Ruddock, told the Federal Parliament:

> One increasingly important tool is the use of surveillance devices, which can range from a pair of binoculars, a tiny microphone or camera hidden in a suspect's vehicle to a piece of software to capture the input of information to a computer . . . as criminal and terrorist groups make use of sophisticated technology, our police must be able to match and better them.[19]

The Act began as an initiative of the Leaders' Summit on Terrorism and Multi-Jurisdictional Crime held in April 2002. A joint working group of Commonwealth, state and territory officials was established. The joint working group has developed model laws to improve cross-border criminal investigations in the areas of controlled operations, assumed identities, protection of witness identity and electronic surveillance. The federal government provided $14.5 million over four years, (from 2004) to the Australian Federal Police and the Australian Crime Commission, to enable them to use the new surveillance device powers provided in the Act. In addition, funds are provided to the Commonwealth Ombudsman and the Administrative Appeals Tribunal in support of these new powers.

Electronic surveillance is regarded as a crucial tool for effective and efficient law enforcement, especially in pursuing serious drug traffickers, terrorists, paedophiles and criminals involved in national and transnational crime. Complementary legislation, the *Telecommunications (Interception) Amendment (Stored Communications) Act 2004* (Cth), removes the need for a telecommunications interception warrant under the *Telecommunications (Interception and Access) Act 1979* (Cth) in order to access the content of communications temporarily delayed and stored on a telecommunication service provider's equipment during passage. This includes access to undelivered email, SMS and voice mail messages.

Victoria has passed the *Surveillance Devices Act 1999*. The purpose is to regulate the use of surveillance devices, to restrict the communication and publication of records of private conversations and to establish procedures for law enforcement officers to obtain warrants or emergency authorisations. The Act imposes requirements for the secure storage and destruction of records obtained by police through the use of surveillance devices.

Anti-Terrorism Act (No. 2) 2005 (Cth)

This Act amends the law relating to terrorist acts to bring it in line with international developments. The Act extends the definition of 'terrorist organisation' to enable the listing of organisations that advocate terrorism. A regime was

19 Second Reading Speech, Surveillance Devices Bill 2004.

established in relation to 'control orders', which allow for the overt close monitoring of terrorist suspects who are considered to pose a risk to the community. Police may detain a person without charge for up to 48 hours as a 'preventative detention' measure, where they believe it is 'reasonably necessary to prevent a terrorist act' or to preserve evidence of such an act.

The Act updated sedition offences so that they now cover those who urge violence or assistance to Australia's enemies. It tightened up provisions regarding offences of financing of terrorism: there is now better coverage of the collection of funds for terrorist activity. There is a new regime of stop, question, search and seize powers that will be exercisable at airports and other Commonwealth places to prevent or respond to terrorism. The Act gives law enforcement and intelligence agencies access to airline passenger information and creates a legal basis for the use of video surveillance at Australia's major airports and on aircraft. The powers may also be exercised *carte blanche* in a 'prescribed security zone'.

Some states have enacted mirror legislation to provide for similar stop and search powers for state police in relation to events or areas, but the police still need to apply to the Supreme Court for an authorisation in relation to a specific event or area.[20]

Ambassador for Counter-Terrorism

On 7 March 2003 the position of Ambassador for Counter-Terrorism was created.[21] It was one of the government's actions taken in response to the events of September 11, 2001. The role of the Ambassador is to facilitate cooperation within Australian agencies and between Australian agencies and their international counterparts. In the 2006/07 Budget, the Australian government provided an additional $92.6 million over four years to 'boost Australia's capacity to work with regional partners to combat the continuing and evolving threat of terrorism'.[22]

In announcing the appointment of a new ambassador in 2006 the Minister for Foreign Affairs described the responsibility of the Ambassador in the following terms:

> The Ambassador for Counter-Terrorism is responsible for the development and implementation of Australia's international counter-terrorism policy. The Ambassador has a key role in ensuring a coherent and effective approach to Australia's counter-terrorism cooperation with both regional and extra-regional partners. This entails identifying needs and opportunities for practical action to counter the terrorist threat and maximising Australia's capacity to respond to these internationally.[23]

20 For example, the *Terrorism (Community Protection) Act 2003* (Vic).
21 See www.dfat.gov.au/homs/auct.html.
22 See Minister for Foreign Affairs, press release, July 2006: www.foreignminister.gov.au/releases/2006/fa079_06.html.
23 Ibid.

Memorandums of Understanding on counter-terrorism

Australia has signed a number of memorandums of understanding on counter-terrorism: with Indonesia, Malaysia, Thailand, the Philippines, Fiji, Cambodia, East Timor, India, Papua New Guinea, Brunei, Pakistan, Afghanistan and the United States. They are intended to establish a bureaucracy to coordinate security, intelligence, law enforcement and defence issues. They also make provision for training, education and technical assistance programs in counter-terrorism.

For example, in December 2005 Australia signed an MOU with Afghanistan. The Minister for Foreign Affairs stated:

> Afghanistan is an important ally for Australia in the war on terror and has played a pivotal role in international efforts to dismantle global terrorist networks, such as Al Qaida. Strong international cooperation is crucial in combating the global terrorist threat. This CT MOU will enhance the security of both Australia and Afghanistan through exchanges of information and intelligence, joint training activities and capability building initiatives.[24]

International Conventions

Australia is party to 11 of the 12 anti-terrorism Conventions. They have been implemented into Australia's domestic legislation as follows:

Convention on Offences and Certain Other Acts Committed on Board Aircraft 1963 (Tokyo Convention)

Australia deposited an Instrument of Accession in 1970. Implemented under the *Crimes (Aviation) Act 1991* (Cth).

Convention for the Suppression of Unlawful Seizure of Aircraft 1970 (Hague Convention)

Australia signed the Hague Convention on 15 June 1971 and ratified it on 9 November 1972. Implemented under the *Crimes (Aviation) Act 1991* (Cth), which covers the hijacking of a civilian aircraft.

Convention for the Suppression of Unlawful Acts against the Safety of Civil Aviation 1971 (Montreal Convention)

Australia signed the Convention on 12 October 1972 and ratified it on 12 July 1973. Implemented under the *Crimes (Aviation) Act 1991* (Cth).

24 See Minister for Foreign Affairs, press release, December 2005: www.foreignminister.gov.au/releases/2005/fa161_05.html.

Supplementary Protocol for the Suppression of Unlawful Acts of Violence at Airports Serving International Civil Aviation 1988 (Montreal)

Australia deposited an Instrument of Accession on 23 October 1990, effective 22 November 1990. Implemented under the *Crimes (Aviation) Act 1991* (Cth).

Convention for the Suppression of Unlawful Acts against the Safety of Maritime Navigation 1988 (Rome)

This Convention makes it an offence to hijack a civilian ship, or attack a person on board a civilian ship or the ship itself, which might endanger the safe navigation of the ship, or deliberately endanger shipping through sabotage or misinformation. The responsible body is the International Maritime Organisation (IMO). The Convention came into force 1 March 1992. Australia deposited an Instrument of Accession on 19 February 1993, effective 20 May 1993. Implemented under the *Crimes (Ships and Fixed Platforms) Act 1992* (Cth).

Protocol for the Suppression of Unlawful Acts Against the Safety of Fixed Platforms Located on the Continental Shelf 1988 (Rome) (Supplementary to the Convention for the Suppression of Unlawful Acts Against the Safety of Maritime Navigation 1988 (Rome))

This Protocol makes it an offence to hijack a fixed platform, attack a person on board a fixed platform or the platform itself, which may endanger its safety. The Protocol entered into force on 1 March 1992. Australia deposited an Instrument of Accession on 19 February 1993, effective 20 May 1993. Implemented under the *Crimes (Ships and Fixed Platforms) Act 1992* (Cth).

Convention on the Prevention and Punishment of Crimes Against Internationally Protected Persons 1973

Australia signed the Convention on 30 December 1974 and ratified it on 20 June 1977. Implemented under the *Crimes (Internationally Protected Persons) Act 1976* (Cth).

International Convention against the Taking of Hostages 1979 (Hostages Convention)

Australia deposited an Instrument of Accession on 21 May 1990. Implemented under the *Crimes (Hostages) Act 1989* (Cth).

Convention on the Physical Protection of Nuclear Material 1980 (Nuclear Materials Convention)

Australia signed the Convention on 22 February 1984 and ratified it on 22 September 1987. Implemented under the *Nuclear Non-Proliferation (Safeguards) Act 1987* (Cth).

International Convention for the Suppression of Terrorist Bombings 1997 (New York)

This Convention makes it an offence to unlawfully or intentionally deliver, place, discharge or detonate an explosive or other lethal device into or against a public place, a state or government facility, public transport or infrastructure, with intent to cause death or serious injury, or extensive damage to the place, facility or system, where the destruction results in, or is likely to result in, major economic loss. The Convention entered into force on 23 May 2001, and was deposited with the United Nations. Australia deposited an Instrument of Accession on 8 September 2002. Implemented under the *Criminal Code Amendment (Suppression of Terrorist Bombings) Act 2002* (Cth).

International Convention for the Suppression of the Financing of Terrorism 1999 (New York)

This Convention makes it an offence to unlawfully and wilfully provide or collect funds with the intention of carrying out one of the offences listed in the above Conventions, and any other act intended to cause death or serious bodily injury to a civilian, or to any other person not taking an active part in the hostilities in a situation of armed conflict, when the purpose of such an act is to intimidate a population, or to compel a government or an international organisation to do or abstain from doing any act. The Convention entered into force on 10 April 2002, and was deposited with the United Nations. Australia signed the Convention on 15 October 2001 and ratified it on 26 October 2002. Implemented under the *Suppression of the Financing of Terrorism Act 2002* (Cth).

International Convention for the Marking of Plastic Explosives for the Purposes of Detection 1991 (Montreal)

This Convention makes it an offence to fail to properly mark plastic explosives for the purposes of detection. The Convention entered into force on 21 June 1998, and was deposited with the International Civil Aviation Organisation (ICAO). Australia is not a party to this Convention.

Conclusion

The debate continues in relation to the powers and measures that authorities should have so that they can undertake appropriate action for the prevention of terrorism, and about a perceived lack of transparency. Civil libertarians express outrage at the curtailment of fundamental human rights and freedoms, citing conflicts with commitments to international human rights conventions and guarantees under the rule of law.

These new powers affect the 'right to silence', the right to confidential legal advice, and freedoms of expression and association. Decisions made under these powers are not subject to judicial review, and there remain questions about the nexus between the alleged breach and a terrorist act or a terrorist organisation. In the future, this period of time may be viewed as a dark age. However, it may be the beginning of permanent laws and tools to counter terrorism.

The powers given to law enforcement agencies allow for extensive access to electronic systems of all kinds, via interceptions, 'wire tapping' and surveillance.

16

Cybercrime

The advance of information technology and computer technology has led to a corresponding increase in computer crime. There is no accepted definition of computer crime. Computers may be the subject of a crime, such as theft, or a computer may be used to commit a crime.[1] Typically a computer is used, or misused, to elicit or manipulate data or processing. Simple unlawful access to a computer system can be regarded as an offence. Telecommunications may be involved. The result may be a transfer of funds or of confidential information. Sending an email to place a virus can be unlawful. The free flow of information has generated undesirable and abhorrent material. The public has expressed concern about pornography, information on how to make bombs and information about suicide techniques. New offences have been created in response to these concerns. Persons gaining unlawful access to computers for these purposes are typically referred to as crackers: hackers with malicious intent. Many crimes that are not specifically related to computers can be substantially facilitated by the use of computers. Crimes involving electronic commerce typically involve the use of computers and telecommunications. This area is often referred to as cybercrime.

In surveys undertaking by the Australian Computer Emergency Response Team (AusCERT),[2] the most common breaches involved viruses (programs or codes that can replicate themselves and infect computers without the owner's consent and often without the owner's knowledge), worms (programs or codes which replicate themselves using a network system, and that are designed to

1 The US has many unique examples of computer crime – stealing an ATM with a front-end loader or shooting an ATM with a handgun.

2 AusCERT, *2006 Australian computer crime and security survey*. AusCERT is an independent, not-for-profit team of IT security professionals, based at the University of Queensland.

cause malicious damage) and Trojan infections (code incorporated into an existing program which gives the appearance of performing a desirable function but which performs malicious functions).[3] In 2003 and 2004, 80 per cent of organisations experienced this type of breach, but in 2005 and 2006 the proportion dropped to 60 per cent. The second most common breach was laptop theft. Insider abuse of computer systems increased: 26 per cent of organisations reported this in 2003, 32 per cent in 2006. Security breaches – in particular, viruses, laptop theft and financial fraud – generated the highest cost to the surveyed organisations. The percentage of organisations reporting electronic attacks on organisations that harmed confidentiality, integrity or availability of network data or systems decreased from 42 per cent in 2003 to 22 per cent in 2006.

This chapter examines cybercrime: unlawful access, malicious damage, spam, cyberstalking, internet gambling, child pornography and child grooming online and more. Just as fraudsters use the latest technology to gain an advantage, so cybercriminals have also crossed the technological boundaries. Such criminal activities cost individuals and organisations internationally several trillion dollars per annum.

The Commonwealth Criminal Code and computer crime

In Australia cybercrime is generally dealt with by federal law, because the Australian Constitution includes 'postal, telegraphic, telephonic, and other like services' as legislative powers of the federal Parliament.[4] Although the drafters of the Constitution did not envisage the internet, the power has been interpreted as including modern telecommunications.[5] The majority of computer crime is committed using telecommunication networks such as the telephone systems and the internet.

Amendments to the *Crimes Act 1914* (Cth) in 1989 introduced a range of limited computer offences. These have been superseded by the amendments to the federal *Criminal Code 1995*, by the *Cybercrime Act 2001* (Cth)[6] and by the *Crimes Legislation Amendment (Telecommunications Offences and Other Measures) Act (No. 2) 2004* (Cth). The former Act was based on a draft of the Council of Europe Convention on Cybercrime,[7] specifically the 25th draft, released on

3 The organisations surveyed represent the manufacturing, information technology, federal and state government, utilities, finance, and education sectors, and the number of organisations included differed from year to year.

4 *Australian Constitution* s51(v).

5 Geraldine Chin, 'Technological change and the Australian Constitution', (2000) *MULR* 25. See *R v Brislan; Ex parte Williams* (1935) 54 CLR 262 and *Jones v Commonwealth [No. 2]* (1965) 112 CLR 206.

6 See Second reading speech of Attorney-General Daryl Williams, www.aph.gov.au/legis.htm.

7 See Cybercrime Bill 2001 (Cth) Explanatory Memorandum.

22 December 2000.[8] The Council of Europe released the Final Draft on 29 June 2001.[9]

The Cybercrime Act inserted seven computer offences into the *Criminal Code 1995* (Cth).

It is an offence to access or modify computer data or impair electronic communications to or from a computer without authority, with the intention of committing a serious offence.[10] A 'serious offence' is defined as an offence punishable by five or more years' imprisonment. The maximum penalty for this offence is equal to the maximum penalty for the serious offence. For example, if a person hacked into a bank computer and accessed credit card details with the intention of using the details to obtain money, the penalty would be equivalent to that for the fraud offence, which is 10 years' imprisonment. 'Data' includes 'information in any form . . . or any program (or part of a program)'.[11] 'Electronic communication' is defined as 'communication by means of guided or unguided electromagnetic energy or both'.[12]

Proving intent is always problematic. Hackers often experiment with computer code unsure of the consequences and with no specific intent. It is an offence for a person to cause unauthorised modification of data in a computer where that person is reckless as to whether that modification will impair data. The maximum penalty for this offence is 10 years' imprisonment. The offence covers a range of situations, including a hacker who obtains unauthorised access to a computer system and impairs data and a person who circulates a disk containing a computer virus which infects a Commonwealth computer.[13]

Another offence is causing an unauthorised impairment of electronic communications to or from a computer knowing that such impairment is unauthorised. This prohibits strategies such as 'denial of service attacks', where, for example, a service provider is swamped by useless messages causing the service to be inoperative. This offence recognises the importance of internet communications. The maximum penalty for this offence is 10 years' imprisonment.[14]

It is an offence for an unauthorised person to access or modify data that is protected by a password or some other security feature. This is referred to as restricted data. The offence targets hackers attempting to circumvent password-protected computer systems.[15] 'Restricted data' is defined as '(a) data held in a computer; and (b) to which access is restricted by an access control system associated with a function of the computer'.[16] The maximum penalty for this offence is two years' imprisonment.

8 See Council of Europe website: conventions.coe.int/.
9 Ibid.
10 *Criminal Code 1995* (Cth) s477.1.
11 *Criminal Code 1995* (Cth) Dictionary.
12 *Criminal Code 1995* (Cth) Dictionary.
13 *Criminal Code 1995* (Cth) s477.2.
14 *Criminal Code 1995* (Cth) s477.3.
15 *Criminal Code 1995* (Cth) s478.1.
16 *Criminal Code 1995* (Cth) s478.1.

The Criminal Code targets destructive actions such as passing a magnet over a credit card or cutting a computer disk in half. Specifically, it is an offence to cause unauthorised impairment of the reliability, security or operation of any data held on a Commonwealth computer disk or credit card or other device. The maximum penalty for this offence is two years' imprisonment.[17]

Finding evidence of these computer offences is problematic. They may be difficult to trace, and have originated within another jurisdiction. In an attempt to address such concerns, the Criminal Code also makes it an offence to possess, control or supply data or programs which are intended for use in the commission of a computer offence. The maximum penalty for each of these offences is three years' imprisonment. The offences cover persons who possess, create or trade in programs and technology designed to hack or damage other computer systems.[18]

Telecommunications services

The *Criminal Code 1995* (Cth) Part 10.6 contains comprehensive provisions dealing with telecommunications services.[19] Division 473 of Part 10.6 includes a broad list of definitions which are technology specific. Division 474 contains the substantive provisions dealing with telecommunication offences. These include criminal offences that involve use of the internet and other telecommunications and carriage services: internet child pornography and child abuse and assisting suicide.[20]

Part 10.6 applies to the generic medium of 'carriage service', which includes the internet, email and SMS messages. Service carriage is given the same meaning as defined in the *Telecommunications Act 1997* (Cth): 'a service for carrying communications by means of guided and/or unguided electromagnetic energy'. Guided electromagnetic energy utilises a physical means such as a wire, cables and optical fibre: unguided electromagnetic energy includes radio and infra-red waves.[21] It is a broad definition intended to cover future technologies.

Division 474 begins with general dishonesty provisions. A person is guilty of an offence if the person does anything with the intention of dishonestly obtaining a gain from, or causing a loss to, a carriage service provider by way of the supply of a carriage service. A person similarly commits an offence for knowingly risking such a loss. Dishonesty is determined by reference to the standards of ordinary people. The maximum penalty is five years' imprisonment.[22]

17 *Criminal Code 1995* (Cth) s478.2.

18 *Criminal Code 1995* (Cth) ss478.3, 478.4.

19 The *Crimes Legislation Amendment (Telecommunications Offences and Other Measures) Act (No 2) 2004* (Cth) replaced the previous Part 10.6, which was notable for its brevity and lack of detail.

20 For applicable decisions on the prior legislation see *R v Idolo* [1998] VICSC 57; *DPP (Cth) v Rogers* [1998] VICSC 48 and *R v Stevens* [1999] NSWCCA 69.

21 *Criminal Code 1995* (Cth) Dictionary.

22 See *Criminal Code 1995* (Cth) ss474.1, 474.2.

The focus of Division 474 is the misuse of carriage services and telecommunications including International Mobile Equipment Identity (IMEI) (the actual phone number) and mobile phone Subscriber Identity Module (SIM) card data. The early legislative approach was to simply deal with concepts, such as access and manipulation of computer systems. Division 474 goes beyond this, referring to specific mediums and intent, and in particular covering acts such as possessing, producing, supplying and obtaining. In their attempt to be comprehensive, the legislature has created many overlapping offences. A prosecutor will have quite an arsenal.

It is an offence if a person causes a communication to be received by a person or carriage service other than the person or service to whom it is directed. It is also an offence to tamper with, or interfere with, a carriage service facility.[23] Modifying or interfering with a telecommunications device identifier is an offence. Possessing or controlling data or a device with intent to modify a telecommunications device identifier is an offence, though there are a range of defences for manufacturers, certain employees, and law enforcement agencies and intelligence and security officers in the course of their duties.[24]

Producing, supplying or obtaining data or a device with intent to modify a telecommunications device identifier is an offence; possessing or controlling such data or devices with the same intent is also an offence.[25] Copying subscription-specific secure data from an existing account or onto a new account identifier is an offence.[26]

Using a telecommunications network with intention to commit a serious offence is an offence.[27] Notably, using a carriage service to make a threat, a hoax threat or to menace, harass or cause offence are offences.[28] Improper use of emergency call service is also an offence.[29]

Child pornography

Prior to 2005, child pornography offences were dealt with by the states and territories; the approaches taken varied. A federal approach was warranted given the extensive use made of the internet to upload and download child pornography and the use of internet communications, such as email and chat rooms, to contact vulnerable children. The *Criminal Code 1995* (Cth) now provides a consistent nationwide approach.[30] The new offences target online 'grooming' activities by offenders, where adults use the internet to forge relationships with children as a first step in luring them into sexual abuse.

23 *Criminal Code 1995* (Cth) ss474.5, 474.6 respectively.
24 *Criminal Code 1995* (Cth) ss474.7, 474.8 respectively.
25 *Criminal Code 1995* (Cth) ss474.9, 474.11 respectively.
26 *Criminal Code 1995* (Cth) s474.10.
27 *Criminal Code 1995* (Cth) s474.14.
28 *Criminal Code 1995* (Cth) ss474.15, 474.16, 474.17.
29 *Criminal Code 1995* (Cth) s474.18.
30 *Criminal Code 1995* (Cth) ss474.19–474.29.

'Child pornography' and 'child abuse' are defined in detail.[31] However, each specific provision concludes with the expression: 'and does this in a way that reasonable persons would regard as being, in all the circumstances, offensive'. The term 'offensive' can vary from person to person and community to community. The inability by the US Supreme Court in *Reno v American Civil Liberties Union*[32] to determine a unified standard of 'offensive' within the US community was a significant part of the rationale for invalidating much of the US *Communications Decency Act 1996*.[33] The Criminal Code includes a guide for determining whether or not material is offensive:

> **473.4 Determining whether material is offensive**
> The matters to be taken into account in deciding for the purposes of this Part whether reasonable persons would regard particular material, or a particular use of a carriage service, as being, in all the circumstances, offensive, include:
> (a) the standards of morality, decency and propriety generally accepted by reasonable adults; and
> (b) the literary, artistic or educational merit (if any) of the material; and
> (c) the general character of the material (including whether it is of a medical, legal or scientific character).

The expression 'reasonable adult' provides flexibility; it is therefore also imprecise.

A person is guilty of an offence if the person uses a carriage service to: access material; cause material to be transmitted; transmit material; make material available online; or publish or otherwise distribute material, and the material is child pornography or child abuse material.[34] The penalty is imprisonment for 10 years.[35] Possessing, controlling, producing, supplying or obtaining child pornography or child abuse material for use through a carriage service is an offence punishable by up to 10 years' imprisonment.[36]

It is a defence to such prosecutions to find that the conduct in question is 'for the public benefit'. Conduct is of public benefit if, and only if, the conduct is necessary for or of assistance in enforcing the law; monitoring compliance with, or investigating a contravention of the law; the administration of justice; or in conducting scientific, medical or educational research that has been approved by the relevant minister in writing.[37] 'In determining whether the person is . . . not criminally responsible for the offence, the question whether the conduct is of public benefit is a question of fact and the person's motives in engaging in the conduct are irrelevant.'[38] Action by law enforcement authorities in accessing and

31 *Criminal Code 1995* (Cth) s473.1.
32 521 US 844 (1997). Janet Reno was the Attorney General for the Clinton Administration.
33 This case permits online pornography under the guise of freedom of speech and freedom of expression and is the foundation of the escalation of online pornography in the modern era.
34 *Criminal Code 1995* (Cth) ss474.19, 474.22.
35 *Criminal Code 1995* (Cth) ss474.19, 474.22.
36 *Criminal Code 1995* (Cth) ss474.20, 474.23.
37 *Criminal Code 1995* (Cth) s474.24.
38 *Criminal Code 1995* (Cth) s474.24.

downloading the material in the performance of their duty is not an offence.[39] Acts in good faith for the sole purpose of assisting the Australian Communications Media Authority (ACMA) to detect prohibited content (within the meaning of *Broadcasting Services Act 1992* (Cth) Schedule 7),[40] or to manufacture, develop or update certain content filtering technology (including software), are also exempt.[41]

The term 'grooming' is not defined. It refers to enticing a child with the intent of engaging in sex. A person (the sender) commits an offence if he or she uses a carriage service to transmit a communication to another person (the recipient) and:

- the communication includes material that is indecent;
- the sender does this with the intention of making it easier to procure the recipient to engage in, or submit to, sexual activity with the sender or another person (the third party);
- the recipient is someone who is, or who the sender believes to be, under 16 years of age; and
- the sender or the third party is at least 18 years of age.[42]

The maximum penalty is 12 years. 'Indecent' means 'indecent according to the standards of ordinary people', another flexible and imprecise term.[43]

Assisting suicide

The ease with which information can be accessed online has prompted a response by lawmakers. A person is guilty of an offence if that person uses a carriage service to access, distribute, transmit or make available material (such as on an internet web page) that directly or indirectly counsels or incites suicide. An element of the offence is that the person intended the material to be used for counselling or inciting suicide. It is also an offence to use a carriage service to promote a particular method of committing suicide.[44]

The offence also covers persons who possess, control, produce, supply or obtain suicide-related material for use through a carriage service. A person may be found guilty of an offence even if the suicide is impossible.

Police and security powers

Law enforcement and security agencies are able to monitor and eavesdrop on suspected terrorists' and criminals' emails, SMS messages and voicemail. The

39 *Criminal Code 1995* (Cth) s474.24.
40 *Criminal Code 1995* (Cth) s474.24. On the role of ACMA and detecting 'prohibited content', see Chapter 18.
41 *Criminal Code 1995* (Cth) ss474.21, 474.24.
42 *Criminal Code 1995* (Cth) s474.27.
43 *Criminal Code 1995* (Cth) s474.27.
44 *Criminal Code 1995* (Cth) ss474.29A, 474.29B, inserted by *Criminal Code Amendment (Suicide Related Material Offences) 2004* (Cth).

federal Parliament passed problematic telecommunications interception legislation which had been sought by police, regulatory and security agencies to aid investigations.[45]

Authorities are able to apply for search warrants to inspect 'stored communications'.[46] A 'stored communication' is a communication that is held on equipment that is operated by a carrier and cannot be accessed on that equipment, by a person who is not a party to the communication, without the assistance of an employee of the carrier, but does not pass over a telex system.[47]

Recent amendments substantially improved the original telecommunications interception legislation, which was drafted at a time when telecommunications was largely land-based and involved live telephone conversations.[48]

Investigative powers

The *Cybercrime Act 2001* (Cth) enhanced the criminal investigation powers under the *Crimes Act 1914* (Cth) and the *Customs Act 1901* (Cth) relating to the search, seizure and copying of electronically stored data. Police may analyse the computer on site or seize computer equipment, including data storage disks, to be copied and analysed elsewhere.

The use of security measures such as encryption and passwords present particular problems for law enforcement agencies. Encrypted data can be impossible to decipher. A magistrate may order a person with knowledge of a computer system to provide information or assistance.[49] This power extends to the compulsory disclosure of passwords, keys, codes, cryptographic and steganographic methods (the art and science of writing hidden messages in such a way that no one, apart from the sender and intended recipient, suspects the existence of the message – it is a form of security through obscurity) used to protect information 'as is necessary and reasonable':

> 3LA(1) The executing officer may apply to a magistrate for an order requiring a specified person to provide any information or assistance that is reasonable and necessary to allow the officer to do one or more of the following:
> (a) access data held in, or accessible from, a computer that is on warrant premises;
> (b) copy the data to a data storage device;
> (c) convert the data into documentary form.
>
> (2) The magistrate may grant the order if the magistrate is satisfied that:
> (a) there are reasonable grounds for suspecting that evidential material is held in, or is accessible from, the computer; and

45 *Telecommunications (Interception and Access) Act 1979* (Cth).
46 *Telecommunications (Interception and Access) Act 1979* (Cth) s117.
47 'Carrier' means a carrier and a carriage service provider within the meaning of the *Telecommunications Act 1997* (Cth), such as Internet Service Providers (ISPs).
48 For more on the express powers see Chapter 15.
49 *Crimes Act 1914* (Cth) s3LA.

(b) the specified person is:
 (i) reasonably suspected of having committed the offence stated in the relevant warrant; or
 (ii) the owner or lessee of the computer; or
 (iii) an employee of the owner or lessee of the computer; and
(c) the specified person has relevant knowledge of:
 (i) the computer or a computer network of which the computer forms a part; or
 (ii) measures applied to protect data held in, or accessible from, the computer.

Failing to comply with the magistrate's order is punishable by up to six months' imprisonment.[50] Some commentators have described the new investigative powers as draconian and dangerous. Their argument is that the provision contravenes the privilege against self-incrimination, often colloquially referred to as the right to silence. Supporters of the provision claim that it is a valuable tool to fight organised crime, paedophilia, terrorist activities and any criminal who hides material in an encrypted form. The provision is akin to requiring a person to hand over a key to a filing cabinet. In reality the police can break open a filing cabinet if necessary; but they need the 'key' to access encrypted computer data. In fact, the provision provides law enforcement agencies with a necessary tool to fight crime. There has also been exaggerated criticism of the provisions which make it offence to possess hacker toolkits, scanners and virus code, on the basis that these are tools of the trade for security vendors.

The *Crimes Act 1914* (Cth) permits both the Defence Signals Directorate, the Australian Security Intelligence Organisation (ASIO) and the Australian Secret Intelligence Service (ASIS) to hack: it is seen as part of their role of providing national security.[51]

State legislative offences relating to computers

There has been no uniform approach taken in relation to offences relating to the use of computers in electronic commerce or otherwise. Every Australian state and territory legislature has prohibited unlawful access to a computer.[52] Only New South Wales, Queensland, Tasmania and the ACT have specific legislation dealing with damage to computer data.[53] In addition, there are a range of offences relating to the misuse of computers. 'Misuse' includes the falsification of documents, dishonest use of computers, fraudulent use of computers, obtaining

50 Some accused may prefer to accept imprisonment rather than reveal the contents.

51 See Part IAA – Search, Information Gathering, Arrest and Related Powers.

52 *Crimes Act 1900* (NSW) s308C; *Crimes Act 1958* (Vic) s247B; *Criminal Code 1899* (Qld) ss408D, 408E; *Summary Offences Act 1953* (SA) s44; *Criminal Code Act 1913* (WA) s440A; *Criminal Code 1924* (Tas) s257D; *Criminal Code 2002* (ACT) s415; *Criminal Code Act* 1983 (NT) s276B.

53 *Crimes Act 1900* (NSW) ss308, 308I; *Criminal Code 1899* (Qld) s408D(2), (3) ; *Criminal Code 1924* (Tas) s257C; *Criminal Code 2002* (ACT) s416.

property by deception and child pornography.[54] The Northern Territory has a specific provision dealing with unlawful appropriation of access time.[55]

New Zealand

The official approach by the New Zealand government has been to pursue a common cybercrime policy by adopting the Council of Europe's Cybercrime Convention. The aim is the protection of society against cybercrime through appropriate legislation and cooperation between countries and private industry. Countries which have adopted the Convention legislate standards guiding the definition of and response to cybercrime. New Zealand legislation is generally aligned with the Convention's requirements, but there is also an ongoing review.

This approach is intended to have a dual benefit. First, it assists with the alignment of cybercrime legislation internationally. Second, the police can progress cyber-based investigations across borders with other participating countries, extending the reach and speed of investigations.[56]

The Police Electronic Crime Laboratory in Wellington has integrated with the National Cyber Crime Centre (NC3). The NC3 is a specialist e-crime response and investigation group that:

- provides a single reporting point for cybercrime enabling the collection and investigation of complaints;
- coordinates police response to cybercrime reported in New Zealand;
- coordinates police response to transnational cybercrime in which there can be any combination of New Zealand or overseas victims, offenders and technologies involved in the commission of an offence;
- proactively targets and electronically patrols places where crime occurs, focusing on high priority areas such as organised crime, violence, and online child exploitation.[57]

The *Crimes Act 1961* (NZ) contains limited cybercrime provisions.[58] Accessing a computer system for a dishonest purpose attracts a potential penalty of seven years' imprisonment where the person obtains any property, privilege, service, pecuniary advantage, benefit, or valuable consideration or causes loss to any other person. An attempt to do any of these things attracts a potential penalty of five years' imprisonment.[59] A person who 'intentionally accesses, directly or indirectly, any computer system without authorisation, knowing that he or she

54 *Crimes Act 1900* (NSW) ss308–308H; *Crimes Act 1958* (Vic) ss247–247H; *Criminal Code 1899* (Qld) ss228, 408E; *Summary Offences Act 1953* (SA) ss44–44A; *Criminal Code 1924* (Tas) ss257A–257F; *Criminal Code Act 1913* (WA) s440A; *Criminal Code 2002* (ACT) ss412–421; *Criminal Code Act* 1983 (NT) ss125A, 125B, 222, 276–276F.

55 *Criminal Code Act 1983* (NT) s276E.

56 See Howard Broad, New Zealand Police Commissioner, (2007) *Electronic crime strategy to 2010: Policing with confidence*: www.police.govt.nz/resources/2007/e-crime-strategy/e-crime-strategy.html.

57 Ibid.

58 Updated in 2003 by the *Crimes Amendment Act 2003* (NZ).

59 *Crimes Act 1961* (NZ) s249.

is not authorised to access that computer system or being reckless as to whether or not he or she is authorised to access that system' is liable to a maximum of two years' imprisonment.[60]

'Access', in relation to any computer system, means 'instruct, communicate with, store data in, receive data from, or otherwise make use of any of the resources of' the computer system:

> computer system —
> (a) means—
> (i) a computer; or
> (ii) 2 or more interconnected computers; or
> (iii) any communication links between computers or to remote terminals or another device; or
> (iv) 2 or more interconnected computers combined with any communication links between computers or to remote terminals or any other device;
> (b) includes any part of the items described in paragraph (a) and all related input, output, processing, storage, software, or communication facilities, and stored data.[61]

A person who 'intentionally or recklessly damages or alters any computer system if he or she knows or ought to know that danger to life is likely to result' attracts a potential penalty of 10 years' imprisonment.[62] A person who, 'intentionally or recklessly, and without authorisation', knowingly or recklessly 'damages, deletes, modifies, or otherwise interferes with any data or software in any computer system' is liable to penalty of seven years' imprisonment.[63]

Making, selling, distributing or possessing software for committing crime is punishable by up to two years' imprisonment.[64]

Child pornography – international

The International Child Pornography Conference held in Austria in 1999 sought to combat child pornography and exploitation on the internet. Initially the discussion revolved around the existing international obligations and commitments related to the protection of children, including the UN Convention on the Rights of the Child. The conference built and acted upon commitments undertaken at the Stockholm World Congress against the Commercial Sexual Exploitation of Children (1996) and ongoing initiatives in many countries and regions.

The UNICEF-sponsored Optional Protocol on the Sale of Children, Child Prostitution and Child Pornography became effective in January 2002. UNICEF estimates that one million children, mainly girls, are forced into the multi-billion

60 *Crimes Act 1961* (NZ) s252. Exemptions are made for the NZ Security Intelligence Service and the Government Communications Security Bureau: see *Crimes Act 1961* (NZ) ss253, 254.
61 *Crimes Act 1961* (NZ) s248 Definitions.
62 *Crimes Act 1961* (NZ) s250(1).
63 *Crimes Act 1961* (NZ) s250(2).
64 *Crimes Act 1961* (NZ) s251.

dollar commercial sex trade (this includes child pornography) every year. To date there are 96 signatories and 22 parties to the Optional Protocol. It requires signatories to criminalise violations of children's rights and calls for increased public awareness and international co-operation.

The UN Convention on the Rights of the Child[65] has been ratified by almost all nation states. The Convention recognises child pornography as a violation of children and requires parties to take legislative and practical measures to prevent the exploitative use of children in pornographic materials.

Internet gambling

The *Interactive Gambling Act 2001* (Cth) commenced in full in January 2002. The regulation of gambling is typically a state and territory matter, but the federal Parliament has power with regard to internet activities and other communications technologies. The Act is the government's response to the community's concern about the increase in gambling resulting from recent technologies.

The problem

Seventy per cent of Australians believe that gambling does more harm than good. Australia has the highest number of poker machines per head worldwide. In 2004–05 Australians lost $15.5 billion in gambling. Some 7000 businesses – including 2888 hotels, 2408 clubs and 13 casinos – provide gambling services. The Productivity Commission has found that around 290 000 Australians, or 2.1 per cent of the total population, are problem gamblers. It found that problem gamblers comprise 15 per cent of regular gamblers and account for one-third of all gambling expenditure annually. The commission's final report estimated that 1.5 million people are affected through consequential bankruptcy, divorce, suicide and lost time at work.[66]

The public's concern is that increased access to gambling through the internet and similar technologies will exacerbate problem gambling. The Productivity Commission described the new interactive technologies as a quantum leap in accessibility to gambling. A Department of Family and Community Services survey found that more than two-thirds of Australians support a ban on internet gambling. The Productivity Commission found that 92 per cent did not want any further expansion of poker machines.

65 See the UNICEF website: www.unicef.org/crc/.

66 The Productivity Commission is the Australian Government's independent research and advisory body on a range of economic, social and environmental issues affecting the welfare of Australians. Its role, expressed simply, is to help governments make better policies that are in the long-term interest of the Australian community: www.pc.gov.au.

Interactive Gambling Act 2001 (Cth)

The *Interactive Gambling Act 2001* (Cth) prohibits the provision of interactive gambling to people located in Australia. The Act defines 'interactive gambling service' as including internet casinos and internet poker machines, online ball-by-ball wagering on sporting events, and online scratch lotteries. The prohibition applies to casino-type gaming, betting on a sporting event after it has commenced and scratch lotteries online. Offences apply to both Australian and overseas interactive gambling service providers. Fines of up to $1.1 million per day apply.

The responsible minister has the power to designate foreign countries that Australian interactive gambling operators will be banned from providing their services to. Countries with similar laws have reciprocal powers, which means they can stop operators based in Australia from providing interactive gambling services to customers located in those countries.

The Act regulates interactive gambling services by:

- prohibiting interactive gambling services from being provided to customers in Australia;
- prohibiting Australian-based interactive gambling services from being provided to customers in designated countries; and
- establishing a complaints-based system to deal with internet gambling services where the relevant content (prohibited internet gambling content) is available for access by customers in Australia.[67]

A person may complain to the Australian Communications Media Authority (ACMA) about prohibited internet gambling content.[68] If the site on which the content appeared was hosted in Australia and ACMA considers the complaint warranted, it must refer the complaint to the Australian Federal Police (AFP). For content hosted outside Australia, ACMA must also notify internet service providers (ISPs) so that the ISPs can apply the industry standard: this might mean updating internet content-filtering software.[69]

The Act encourages the development of an industry code by ISPs. ACMA can establish an industry standard if there is no industry code or if an industry code is deficient. The industry codes and standards are available at the website of the Internet Industry Association.[70] Regulations may provide that civil proceedings do not lie against a person to recover money alleged to have been won or been paid in connection with an illegal interactive gambling service. The Act also prohibits the advertising of interactive gambling services.

The minister is required to undertake regular reviews of the impact of the Act. The reviews must take into account the growth of interactive gambling services, the social and commercial impact of interactive gambling services, and

67 *Interactive Gambling Act 2001* (Cth) ss15, 15A.
68 *Interactive Gambling Act 2001* (Cth) Part 3.
69 See Chapter 18.
70 See the website of the Internet Industry Association: www.iia.net.au.

the effect of the exemptions (there are excluded wagering services, excluded gaming services, services that have a designated broadcasting or datacasting link, and excluded lottery services).

Television broadcasters raised concerns that the legislation may unintentionally apply to certain game shows – such as *Video Hits* or *Classic Catches* – where viewers pay a fee via a 1900 phone call. Late amendments to the Act addressed these concerns (through exclusions). The minister has the power to impose additional conditions.

Comment

In Australia, betting on horse races over the internet remains legal, as do several other forms of online gambling, such as lotteries. However, the Act prohibits poker machine or roulette-style gambling. Senator Richard Alston (Liberal, Victoria) described the Act as 'strong action to combat the tragic economic and social consequences of gambling in this country'.[71] The chief executive of the Internet Industry Association, however, has stated that the legislation 'is technically inept and has no real prospects of protecting those whom it claims to protect'.[72] The latter comment seems to be closer to the mark. Most internet gambling sites are based offshore. The impact of the Act on access to such sites will be minimal. When searching for gambling sites, Australians do not look for Australian sites, as there are many 'reputable' sites in other countries, such as the United States. In 2000, before the Act was passed, Australia's leading internet casino – Lasseters, in Alice Springs – turned over $100 million. Lasseters have since closed their internet gambling services in Australia, stating they would relocate to Vanuatu, resulting in the loss of 45 jobs from Alice Springs and expertise in intellectual property and electronic commerce. Nevertheless the Act is a step, albeit a small step, towards management of the problems caused by gambling.

Cyberstalking

Various pieces of state legislation define stalking as 'continued and intentional conduct directed at another person that would cause a reasonable apprehension of violence or detriment to the stalked person or another person'. 'Cyberstalking', or stalking online, is not dealt with directly by legislation in Australia or New Zealand. However, stalking can involve use of the internet, email or other electronic communications to harass or threaten another person. Stalking behaviour includes posting improper messages on bulletin boards, forwarding viruses, threatening or offensive email, and electronic theft.

71 Senator Alston, press release, 27 March 2001.
72 See www.iia.net.au.

The Crimes Acts in most jurisdictions describe stalking in general terms, leaving it to the courts to consider specific instances, such as harassment by electronic means.[73] In Victoria and the Northern Territory the offence of stalking specifically includes 'telephoning, sending electronic messages to, or otherwise contacting, the victim or any other person'.[74] In 1999 the Queensland Criminal Code was amended to extend stalking to conduct utilising the 'telephone, fax, mail, email or other technology'.[75] In the ACT, stalking includes where the offender 'telephones, sends electronic messages to or otherwise contacts the stalked person; sends electronic messages about the stalked person to anybody else or makes electronic messages about the stalked person available to anybody else'.[76] However, even if the stalker can be identified, the enforcement of such laws can be problematic, as the offender may not be located within the jurisdiction.

In cyberspace terms, cyberstalking behaviour includes spamming, flaming, posting improper messages on bulletin boards, forwarding viruses, threatening or offensive email, and electronic theft. CyberAngels, an organisation assisting victims of cybercrime, estimates that there are approximately 63 000 internet stalkers and 474 000 victims worldwide. In Los Angeles, 20 per cent of the 600 cases dealt with by the Stalking and Threat Assessment Unit involve email and other electronic communications; in New York, 40 per cent of the work of the City Police Computer Investigations Unit in recent years has involved electronic harassment and threats. ISPs are reporting a marked increase in complaints regarding cyberstalking.

According to CyberAngels, cyberbullies employ a number of methods to threaten and disparage their targets. They include:

- Email messages: While this is the most common form of electronic communication, the use of this method for cyberbullying is less pervasive since most email programs allow the use of filters to block offending emails.
- Instant Messaging: Cyberbullies can and do use IM on computers and cell phones to send harassing and threatening messages to their targets.
- Chat rooms: Chat rooms allow cyberbullies to anonymously enter and write anything they want, mocking and insulting their victims in a forum that potentially has a large audience.
- Websites: Cyberbullies create websites or use social networking sites or blogs to mock, torment and harass the intended victims.
- Voting and Polling booths: Some websites offer users the opportunity to create online polling or voting booths. Cyberbullies use these to vote online for some insulting topics (for example: 'The ugliest, fattest, dumbest, etc, boy or girl at [stated] school').[77]

73 *Crimes Act 1900* (NSW) s562AB; *Criminal Law Consolidation Act 1935* (SA) s19AA; *Criminal Code Act 1913* (WA) s338E; *Criminal Code 1924* (Tas) s192; *Crimes Act 1991* (NZ) Part 11.

74 *Crimes Act 1958* (Vic) s21A; *Criminal Code Act 1983* (NT) s189.

75 See *Criminal Code 1899* (Qld) ss359A–359F.

76 *Crimes Act 1900* (ACT) s35(2)(f), (g), (h).

77 See www.cyberangels.org. CyberAngels is an online safety protection and educational non-profit body.

Cyberspace provides stalkers with new methods to contact victims. Threats can be sent electronically from anywhere in the world. Messages can be sent at random or set intervals automatically. The identity and location of the stalker can be concealed by using anonymous remailers that strip off the message header, making tracing difficult. The ease of the technology may encourage some to make threats. By using chat rooms and bulletin boards, and posting controversial material, stalkers can now enlist others to harass their victim – by impersonating them, or by publicising the victim's name, address and telephone number.

The nature of the internet and electronic communications makes it difficult for people to protect their personal information and privacy. Password-protected mailing lists and personal emails cannot really be regarded as private: communications can be intercepted, and many websites provide personal information such as silent phone numbers, identifying photographs of the victim and of his or her home, home and work addresses, financial data and other sensitive material.[78]

Even without reliable statistics, it would be relatively safe to predict that as users become more cyber-aware, abuses will increase. Legal and policy approaches to the problem have tended to lag behind the technology. Also, the jurisdictional problems of the internet present a real barrier to effective enforcement. Legal recourse will only be available when the law deals with cyberstalking in a meaningful way; until and unless that happens, it will be largely up to the consumer to use social and technical tools to prevent and control cyberstalking.

Technical responses available for the consumer include the following:

- using blocking and filtering software: this can delete email or chat room messages. The criteria used can include the name of the author, certain offensive words and so on;
- using sophisticated encryption programs: these can prevent messages being read by unauthorised people;
- using digital signatures and certificates: these will authenticate the author;
- using gender-neutral names; and
- changing passwords regularly.

International approach to cybercrime

Attacks against commercial websites have drawn international attention to the dangers presented by the internet and other computer networks. Cybercriminals and cyberterrorists can threaten business and government interests and cause vast damage.

78 On stalking generally see *Grosse v Purvis* [2003] QDC 151 and *DPP v Sutcliffe* [2001] VSC 43.

The Council of Europe's Convention on Cybercrime is the first international instrument to address various types of offending behaviour involving computer systems, networks or data.[79] It aims to harmonise national legislation in this field, facilitate investigations and allow co-operation between the authorities of different nation states.

The Convention includes provision for the co-ordinated criminalisation of computer hacking and hacking devices, illegal interception of data and interference with computer systems, computer-related fraud and forgery. It prohibits online child pornography, including the possession of such material after downloading, as well as the reproduction and distribution of copyright-protected material. The Convention defines offences, addresses questions related to the liability of individual and corporate offenders and determines minimum standards for the applicable penalties.

It deals with law enforcement issues including the power to carry out computer searches and seize computer data, to require data-subjects to produce data under their control and to preserve or obtain the expeditious preservation of vulnerable data by data-subjects'. These computer-specific investigative measures also imply co-operation by telecom operators and ISPs – their assistance is vital in terms of identifying computer criminals and securing evidence of their misdeeds.

Spam

Who could have predicted that Monty Python's use of the word 'Spam' continuously and ludicrously in a comedy skit would result in a new term, meaning 'unsolicited multiple postings of electronic mail', and in due course to federal legislation, the *Spam Act 2003* (Cth)? It is curious that the word 'spam' does not appear in the body of the Act, except for the title section. Prior to the legislation the possibility of liability for spamming was considered under the *Privacy Act 1988* (Cth), the *Criminal Code 1995* (Cth), the relevant state criminal legislation and at common law. The New Zealand equivalent is the *Unsolicited Electronic Messages Act 2007*.[80]

The problem

All email users have experienced spam. In Australia the reviewing Senate Committee described spam as a cancer. Spam can be offensive, intrusive, misleading and an invasion of privacy. It clogs email boxes and obscures legitimate email. Spam slows email systems and is often illegal or offensive. Minister Peter McGauran states that the legislation was a direct response to the groundswell

79 See Council of Europe, 'Convention on Cybercrime', ETS no. 185, Open for signature – Budapest, 23 November 2001: www.conventions.coe.int.

80 See Department of Internal Affairs, 'Anti-spam': www.dia.govt.nz/diawebsite.nsf/wpg_URL/Services-Anti-Spam-Index.

of business and community anger about the costly and disruptive occurrence of spam.

According to Star Internet, a large ISP in the United Kingdom, the cost in lost productivity is AUD$915 per employee per year. Other studies believe this figure to be conservative. Costs arise through increased download times and lost productivity. An EU study in 2001 estimates that the worldwide cost of spam is approximately AUD$18.4 billion; Ferris Research estimates that US companies lost US$8.9 billion in 2002.

The cost to spammers can be as little as 0.003 cent to send a single email,[81] and only 0.00032 cent to obtain one email address using appropriate harvesting software. Receiving positive responses of less than 1 per cent can be profitable for the spammer.

Spam Act 2003 (Cth)

This Act deals with unsolicited commercial electronic messages. In brief, a person must not send, or cause to be sent, a commercial electronic message unless the recipient has consented to the message being sent.[82] The word 'spam' is not used or defined in the Act other that in the title, though it is the colloquial term; the Act uses 'unsolicited commercial electronic message'. The word 'message' is used to ensure that the legislation extends to other sorts of electronic communications, such as SMS messages.[83]

Sender information

All commercial electronic messages must:

- clearly and accurately identify the person who authorised the sending of the message;
- include accurate information about how the recipient can readily contact the sender;
- comply with the regulations; and
- reasonably be likely to be valid for at least 30 days after the message has been sent.[84]

Unsubscribe function

Section 18 of the Act deals with this area. It states that all commercial electronic messages (solicited and unsolicited) must include the statement that the recipient may unsubscribe by replying using an electronic address set out in the message. The statement 'must be presented in a clear and conspicuous manner' and the electronic address must be 'reasonably likely to be capable of receiving' such a message for 'a period of at least 30 days after the message is sent'. In addition

81 www.noie.gov.au/publications/NOIE/spam/final_report/what.htm.
82 *Spam Act 2003* (Cth) s16.
83 The *Unsolicited Electronic Messages Act 2007* (NZ) became operational in 2007; see below.
84 *Spam Act 2003* (Cth) s17.

the section requires the electronic address to have been 'legitimately obtained'.[85] An unsubscribe message is defined as 'an electronic message to the effect that the relevant electronic account-holder does not want to receive any further commercial electronic messages from the sender'.[86]

Defences include the fact that the recipient consented, that the message was sent by mistake, or that the message only contained factual information.[87]

Factual information

The greatest weakness in the legislation is the exemption for sending 'factual information'. The Act does not regard email as spam if it contains factual information (with or without directly related comment) and additional peripheral information such as the sender's name, logo and contact details.[88]

'Factual information' is not defined in the Act. However, the explanatory memorandum states that the provision is designed 'to ensure that messages which may be seen to have some form of commercial element, but which are primarily aimed at providing factual information are not covered by the rules'. It gives as examples:

- an electronic message from a private law firm which includes an information sheet outlining the effects of a particular court decision;
- an electronic version of a neighbourhood watch newsletter which is sponsored by the local newsagent;
- an electronic newsletter from the local chamber of commerce which is sponsored by one of its members;
- factual information relating to bird-watching that is sponsored by a commercial entity.

Exemptions

Government bodies, political parties, religious organisations, charities and educational institutions are granted a special exemption from the application of the Act in certain circumstances.[89] Clauses 3 and 4 of Schedule 1 provide:

> **3 Government bodies, political parties, religious organisations and charities**
> For the purposes of this Act, an electronic message is a ***designated commercial electronic message*** if:
> (a) the sending of the message is authorised by any of the following bodies:
> (i) a government body;
> (ii) a registered political party;
> (iii) a religious organisation;
> (iv) a charity or charitable institution; and
> (b) the message relates to goods or services; and

85 *Spam Act 2003* (Cth) s18(1).
86 *Spam Act 2003* (Cth) s18(9).
87 *Spam Act 2003* (Cth) s18(2), (3), (4).
88 *Spam Act 2003* (Cth) Schedule 1, cl 2.
89 *Spam Act 2003* (Cth) Schedule 1, cll 3, 4.

(c) the body is the supplier, or prospective supplier, of the goods or services concerned.

4 Educational institutions

For the purposes of this Act, an electronic message is a ***designated commercial electronic message*** if:

(a) the sending of the message is authorised by an educational institution; and
(b) either or both of the following subparagraphs applies:
  (i) the relevant electronic account-holder is, or has been, enrolled as a student in that institution;
  (ii) a member or former member of the household of the relevant electronic account-holder is, or has been, enrolled as a student in that institution; and
(c) the message relates to goods or services; and
(d) the institution is the supplier, or prospective supplier, of the goods or services concerned.

Remedies

The main remedies for breaches of this Act are civil penalties and injunctions. The Act provides a tiered enforcement regime available to ACMA, which includes:

- a formal warning;[90]
- acceptance of an enforceable undertaking;[91]
- the issuing of an infringement notice;[92]
- application to the Federal Court for an injunction;[93] and
- the commencement of proceedings in the Federal Court for breach of a civil penalty provision.[94]

Search and seizure

The *Spam (Consequential Amendments) Act 2003* (Cth) makes amendments to the *Telecommunications Act 1997* (Cth) and the *Australian Communications and Media Authority Act 2005* (Cth) to enable investigation and enforcement of breaches of the Spam Act. The amendments would make it legal, in certain circumstances, for inspectors and police to enter homes and search and seize computers and other possessions without a search warrant and without the consent of the person whose home was entered. The wording is such that the same could be done to a recipient of spam.

Nicholson J examined the first application of the legislation in an Australian superior court in *ACMA v Clarity1 Pty Ltd*.[95] ACMA alleged that Clarity1 contravened section 16(1) of the Spam Act: that 'unsolicited commercial electronic messages' must not be sent. Clarity1 had sent or caused to be sent:

90 *Spam Act 2003* (Cth) s41.
91 *Spam Act 2003* (Cth) Part 6.
92 *Spam Act 2003* (Cth) Schedule 3.
93 *Spam Act 2003* (Cth) Part 5.
94 *Spam Act 2003* (Cth) Part 4.
95 [2006] 410 FCA.

- at least 213 443 382 Commercial Electronic Messages (CEM), of which 41 796 754 were successfully sent, to 5 664 939 unique electronic addresses; and
- at least 56 862 092 CEM, of which 33 199 806 were successfully sent, to 2 291 518 unique electronic addresses.[96]

On an analysis of 1 469 820 electronic addresses, 182 actually placed an order with Clarity1 for its materials.

Section 16(5) of the *Spam Act 2003* (Cth) relates to the burden of proof. The sender must prove that the relevant electronic account-holder (the recipient) consented to the sending of the message, or that the sender did not know or had not ascertained with reasonable diligence that the message had an Australian link, or that it was sent by mistake. In other words, it is the sender who is obliged to show (to the civil standard of proof) that there is sufficient evidence to raise an issue as to the existence or non-existence of a fact in issue.[97] An unexplained failure to call witnesses in relation to this burden of proof provision will allow the application of the rule in *Jones v Dunkel*:[98] an inference that the uncalled evidence or missing material would not have assisted the respondent's case. In such a case, 'considerable significance may attach if the absent witness is either the party or a senior executive of a corporate party closely engaged in the transactions in question and present in court during the hearing'.[99]

The term 'send' in section 16(1) includes an attempt to send, so it is immaterial whether or not the respondents had successfully sent the CEMs.

In Clarity1's favour Nicholson J identified three mitigating facts: each of the CEMs included a functional unsubscribe facility; from March 2001 to the hearing date, some 166 000 requests were made for removal from the defendant's lists, all of which were acted upon; and only 80 complaints were made to Clarity1, none of which related to a failure to remove an electronic address from the database.

Clarity1 raised many defences under the Act, many of which were highly questionable. For instance, Nicholson J dismissed Clarity1's claims that it acted as a charity or an educational institution. Clarity1 also claimed consent, arguing that:

- as the CEMs contained an unsubscribe facility and the recipients did not use the unsubscribe facility, it is reasonable to infer consent;
- consent may be inferred from the business relationship between Clarity1 and the recipients; and
- the recipients published their electronic addresses on the internet.

96 Para 58.
97 See JD Heydon, *Cross on evidence*, 6th edn, Butterworths, Sydney, 2000, at [7015].
98 (1959) 101 CLR 298.
99 Heydon, *Cross on evidence*, at [1215], citing *Dilosa v Latec Finance Pty Ltd* (1966) 84 WN (Pt 1) (NSW) 557 at 582.

Other than evidence by eight affidavits by deponents who indicated that they were 'happy' to receive the spam, there was no evidence supporting these arguments, other than by inference. However, neither that evidence nor Clarity1's claims indicate that Clarity1 was aware of such consent prior to dispatch.[100]

Clarity1 placed reliance on the Office of the Federal Privacy Commissioner's publication *Guidelines to the National Privacy Principles*, issued in September 2001, which stated that 'it may be possible to infer consent from the individual's failure to opt out'. However, in another passage the guidelines add:

> It is unlikely that consent to receive marketing material on-line could be implied from a failure to object to it. This is because it is usually difficult to conclude that the message has been read and . . . it is commonly considered that there are adverse consequences to an individual from opening or replying to email marketing – such as confirming [that] the individual's address exists.

Nicholson J concluded: 'The mere fact that Clarity1 sent a CEM to an electronic address and did not receive a response from the recipient does not provide a proper foundation for an inference of consent.'[101]

Clause 2 of Schedule 2 of the Act provides that consent may be inferred from a business or other relationship. Nicholson J referred to the government-issued publication *Spam Act 2003: A practical guide for business*, in particular the following passage:

> You may be able to reasonably infer consent after considering both the conduct of the addressee and their relationship with you. For example, if the addressee has an existing relationship with you and has previously provided their address then it would be reasonable to infer that consent has been provided . . . as long as it is consistent with the reasonable expectations of the addressee, and their conduct . . . initiated by a commercial activity (including provision, for a fee or free of charge, of information, goods, or of services) or other communication between you and potential addressee.[102]

Nicholson J reasoned that 'the receipt of electronic messages without more cannot give rise to the inference of consent because the receipt could be accounted for on many bases other than consent . . . There is no relationship when the communication is one sided'.[103] Even of the known 182 recipients who placed orders for materials, ACMA submitted that it does not follow even in these instances that the 182 purchasers gave consent to the receipt of the CEMs. Indeed the Explanatory Memorandum explains that even a pre-existing business transaction may be insufficient to imply consent. Consider a person who anonymously purchases groceries from Coles, Big W, Target and so forth. There is no consent to spam. Nevertheless, Nicholson J inferred conferred consent in the 'limited instances' of the 182 purchasers.[104]

100 [2006] 410 FCA para 72.
101 [2006] 410 FCA paras 75–79.
102 [2006] 410 FCA para 89.
103 [2006] 410 FCA para 92.
104 [2006] 410 FCA para 97.

Spam may be sent to persons who provide their electronic details by way of a 'conspicuous publication': where an 'electronic address has been conspicuously published; and it would be reasonable to assume that the publication occurred with the agreement' of the publisher of the electronic details.[105] However, such publication only constitutes consent if it is not accompanied by a statement to the effect that the electronic account-holder does not want to receive spam. Also, the spam must relate to work-related business, functions or duties. The court referred to the *Macquarie Dictionary* meaning of 'conspicuous': 'easy to be seen . . . readily attracting the attention'. There were several examples given in evidence of individuals whose email addresses were available from online yellow pages, their personal website and trade directories. However, ACMA countered with the example of an academic whose email address was published on his academic website, and pointed out that the CEMs at issue did not relate to academic activities. Nicholson J held that:

> it may be inferred from the sheer volume of the electronic addresses contained in the Databases and Lists that . . . the respondents did not consider whether the publication of the electronic address on the internet was done in circumstances that met each of the criteria set out in cl 4 of Sch 2 to the Spam Act.[106]

Clarity1 used address-harvesting software. However, they maintained that the harvesting occurred prior to 10 April 2004, when the Spam Act came into operation. Sections 20, 21 and 22 respectively provide that address-harvesting software and harvested address lists must not be 'supplied', 'acquired' or 'used'. The court naturally held that section 22 applied to 'use' after the commencement of the Act of lists acquired at any time, and thus there had been a contravention.

New Zealand response

The *Unsolicited Electronic Messages Act 2007* (NZ) came into effect in September 2007. It is based on the *Spam Act 2003* (Cth). The Act applies to email, instant messaging, SMS and MMS (text and image-based mobile phone messaging) of a commercial nature. It does not cover facsimiles, internet pop-ups or voice telemarketing. The Act prohibits unsolicited commercial electronic messages with a New Zealand link, and provides that commercial electronic messages must include accurate information about the person who authorised the sending of the message and a functional unsubscribe facility to enable the recipient to instruct the sender that no further messages are to be sent to the recipient. Address-harvesting software is prohibited.

The Department of Internal Affairs enforces the Act by investigating complaints and enforcing the provisions of the Act. The department is also charged with undertaking research into technologies used to send spam and advising the New Zealand government.

105 *Spam Act 2003* (Cth) Schedule 2, cl4.
106 [2006] 410 FCA para 108.

The first prosecution under the Act demonstrates the inadequacies of its penalty provisions. In *Chief Executive Department of Internal Affairs v Atkinson*[107] the defendant admitted sending more than 2 million contravening emails to NZ computers. Justice French imposed a NZ$100,000 penalty (the maximum possible is NZ$200,000). However, Atkinson grossed more than NZ$1.6 million. Although acknowledging that 'the deterrent effect of a penalty will be lost if it only marginally increases the cost of the illegal activity',[108] her Honour commented that the penalty should be 'substantially discounted' due to the 'co-operation and candour in an early stage with the authorities'.[109] The penalty amounted to a small percentage of Atkinson's overheads. The New Zealand legislature may need to review the penalty provisions of the Act.[110]

US response

The US federal anti-spam legislation, known as the CAN-SPAM Act, became law on 1 January 2004. The Act requires senders of unsolicited email advertisements to include an 'opt out' facility. The opt out provision may be a reply email address or an 'internet-based mechanism'. These emails must be identified as advertisements. Recipients cannot bring civil action; instead law enforcement authorities and ISPs can initiate civil actions on behalf of recipients to recover up to $2 million in damages. The sending of multiple commercial emails with misleading headers or which conceal the identity of the sender is prohibited, and convictions can result in significant fines or imprisonment.

EU response

In the European Union, in contrast, 'marketing' emails may be sent to recipients who have not given specific consent. However, recipients may opt out after a 10-day grace period.

The EU Privacy and Electronic Communications Directive was implemented in EU Member States on 31 October 2003. The directive prohibits unsolicited direct marketing to individuals by electronic communications including email unless consent has been given in advance. There is an exemption for an existing customer relationship: companies may continue to market products by email on an opt out basis. The UK version of the directive became operative on 11 December 2003.

107 High Court of New Zealand, unreported CIV-2008-409-002391 (December 2008), available at: www.courtsofnz.govt.nz/from/decisions/judgments.html.

108 Ibid., para 20.

109 Ibid., para 21. See *Unsolicited Electronic Messages Act 2007* (NZ) ss32, 45. Atkinson and another had previously been fined US$2.2 million in 2005 by the US Federal Trade Commission. Atkinson and his partner had controlled 35 000 computers, sending out more than 10 billion emails a day.

110 Atkinson and another had previously been fined US$2.2 million in 2005 by the US Federal Trade Commission. Atkinson and his partner had controlled 35 000 computers, sending out a potential 10 billion emails a day.

All organisations sending promotional or commercial emails should determine whether the email complies with legislation in the recipient's country and the country of dispatch, and ascertain if any recipient has opted out of receiving advertising by email.

Criticisms

The *Spam Act 2003* (Cth) bans some email which many would not regard as spam. For example, if you write an article in a magazine, a single unsolicited email to you asking that the article be reprinted elsewhere for a fee would be in contravention of the legislation. No single email should be regarded as spam. The Act also legitimises some email which almost all would agree should be regarded as spam. Using the 'factual information' loophole, car dealers could send bulk emails stating their name and address and that the latest model is now available. This loophole has been described as 'large enough to drive a truck through'.[111] Using the 'conspicuous publication' rule, one may be regarded as consenting to all spam if one's email address appears on a letterhead or business card.

Critics also argue that the search and seizure requirements should be amended to require the authorities to obtain a warrant. Further, the legislation will be ineffective against overseas spam. It is estimated that some 80–90 per cent of all spam comes from overseas.

National Do Not Call Register

The *Do Not Call Register Act 2006* (Cth) (and the *Do Not Call Register (Consequential Amendments) Act 2006* (Cth)) makes it an offence for telemarketers to call any number so registered. This is an attempt to end, or reduce, the more than one billion unwanted invasive telemarketer calls made annually in Australia. The legislation was enacted in response to a significant increase in the number of unsolicited calls in Australia, which led to 'rising community concerns about the inconvenience and intrusiveness of telemarketing'.[112]

The legislation requires ACMA to establish and oversee the Do Not Call Register. The federal government chose an opt out system – the growing direct marketing industry employs some 700 000 people. The approach taken is similar to the laws relating to spam: unsolicited telemarketing calls must not be made to a number on the Do Not Call Register unless express or inferred consent has been given by the telephone account holder or nominee. The scheme covers both calls made in Australia and calls made by overseas telemarketers on behalf of companies operating in Australia.

111 See comments by Electronic Frontiers Australia: www.efa.org.au.
112 Second Reading Speech, Do Not Call Register Bill 2006 (Cth).

'Designated telemarketing calls' are exempt from the Do Not Call Register. A telemarketing call is a designated telemarketing call if:

- the making of the call is authorised by a government body, a religious organisation, or a charity or charitable institution;
- the call relates to goods or services – the body is the supplier, or prospective supplier, of the goods and services; and
- is not restricted by the regulations.

There are also limited exemptions: for political parties, individual politicians and candidates, educational institutions and employers to their employees.

The following are not regarded as telemarketing calls:

- product recall calls;
- fault rectification calls;
- appointment rescheduling calls;
- appointment reminder calls;
- calls relating to payments;
- solicited calls; and
- calls not answered by the person to whom the call is made.

The penalty for a single person, organisation or business with no prior breach range from $1100 to $11 000. The maximum fine for repeated breach is $1 100 000. Breaches of legislation attract civil penalties. ACMA may also seek injunctive relief. Complaints of breaches may be made directly to ACMA.

Identity fraud

Rogues were impersonating others to commit fraud, using fake documents, lies and cunning, long before the advent of technology. In the electronic age, these rogues are using sophisticated means, sometimes not even leaving their computers. The resultant crimes include obtaining goods or services by deception, voter registration, terrorism, drug trafficking and identity fraud (the terms 'identity fraud' and 'identity theft' tend to be used interchangeably). Rogues may use such stolen or fictitious identities to avoid tax obligations, obtain government benefits or gain access to services, medical and otherwise.

Recent technology has resulted in a sharp increase in identity fraud: fraudsters are quick to utilise scanners, imaging equipment and colour printers. In 2003 the Australian Transaction Reports and Analysis Centre (AUSTRAC) estimated that identity fraud costs more than $1.1 billion per year in Australia. This figure is likely to be quite conservative as most studies and commentators agree that many instances of identity fraud are not reported. Many entities are reluctant to admit a failure in their computers and security systems.

Identity theft can range from finding identification cards in garbage bins to using card readers, hacking into computers, stealing and forging documents such as driver's licences and 'phishing'. Phishing is sending an email to a user falsely claiming to be an established legitimate enterprise in an attempt to scam the user into surrendering private information (such as bank account usernames and passwords) that will be used for identity theft.[113]

Shoulder surfing is where rogues spy on people at ATMs. Skimming is where credit card data is copied from the magnetic strip electronically. Examples include restaurants where the customer hands over the credit card for payment or where a rogue's reader is attached surreptitiously to an ATM and the PIN that is entered is observed via a minute hidden camera.

Australians use the internet for banking and financial transactions that require the provision of credit card or other account details over the internet. Many government agencies have websites that allow for online transactions and the exchange of personal information in the provision of government benefits or authorisations such as licences. For example, each year the Australian Taxation Office issues several hundred thousand tax file numbers, Centrelink processes 3–4 million new claims or re-grants, and the Department of Foreign Affairs issues more than one million passports. Each is a potential target.

Identity fraud and terrorism

A horrendous case of identity fraud involves the terrorists acts of September 11, 2001. Two of the terrorists bribed a legal secretary to complete and notarise false affidavits and residency certifications which they then used to obtain official identification papers from the US government, allowing them to board the planes which they then hijacked. In the days that followed, lists of the missing were published, allowing rogues to contact government departments claiming replacement identification documents. On obtaining these replacement driver's licences, they then obtained credit cards and purchased hundreds of millions of dollars' worth of goods and services.

Technological response

Possible technological solutions include biometrics such as fingerprint and iris recognition technology, digital signature authentication and encryption.

Governmental response

The Proof of Identity Steering Committee (POISC) is a cooperative effort by government and the financial sector to address these issues. POISC commissioned the Securities Industry Research Centre of Asia-Pacific (SIRCA) to conduct the first comprehensive study of the cost of identity fraud to the Australian community.

There is no Commonwealth legislation making it a criminal offence to merely steal or assume another person's identity, and the only state that has such a

113 See below.

law is South Australia.[114] Most relevant laws relate to the resultant effect of the action, not the initial preparatory action. South Australia has made identity theft a specific offence:

> **144B—False identity etc**
>
> (1) A person who –
>
> (a) assumes a false identity; or
>
> (b) falsely pretends –
>
> (i) to have particular qualifications; or
>
> (ii) to have, or to be entitled to act in, a particular capacity,
>
> makes a false pretence to which this section applies.
>
> (2) A person who assumes a false identity makes a false pretence to which this section applies even though the person acts with the consent of the person whose identity is falsely assumed.
>
> (3) A person who makes a false pretence to which this section applies intending, by doing so, to commit, or facilitate the commission of, a serious criminal offence is guilty of an offence and liable to the penalty appropriate to an attempt to commit the serious criminal offence.
>
> **144C—Misuse of personal identification information**
>
> (1) A person who makes use of another person's personal identification information intending, by doing so, to commit, or facilitate the commission of, a serious criminal offence, is guilty of an offence and liable to the penalty appropriate to an attempt to commit the serious criminal offence.
>
> (2) This section applies irrespective of whether the person whose personal identification information is used –
>
> (a) is living or dead; or
>
> (b) consents to the use of the personal identification information.

Other states rely on offences relating to forgery and dishonesty. The *Privacy Act 1988* (Cth) deals with the collection, storage and use of personal information but not the theft of that information. The *Crimes Legislation Amendment (Telecommunications Offences and Other Measures) Act (No. 2) 2004* (Cth) prohibits credit card skimming and internet banking fraud, including phishing. The *Financial Transaction Reports Act 1988* (Cth) and the Financial Transaction Reports Regulations 1990 provide that it is an offence to open an account in a false name by tendering a false identification document.

State and territory legislation deals with general offences such as fraud and false pretences. In Queensland, for example, relevant provisions of the *Criminal Code 1899* (Qld) include:

- section 408C(1) fraud;
- section 427(1) false pretences;
- section 441 falsifying records and producing false records;
- section 398 stealing; and
- section 408C misappropriation.[115]

114 *Criminal Law Consolidation Act 1935* (SA) Part 5A.

115 See www.police.qld.gov.au/pr/news/media/ and www.parliament.qld.gov.au/view/publications/documents/research/ResearchBriefs/2005/200503.pdf.

Phishing

Phishing is a form of criminal activity involving attempts to fraudulently acquire sensitive information such as passwords and credit card details by masquerading as a trustworthy person or business. The term 'phishing' was coined because fraudsters used sophisticated lures as bait to catch unsuspecting prey. Phishing dates back to 1996, when hackers phished for credit card numbers. Today phishing is more usually carried out by people masquerading as financial institutions and online payment services known to the target.

Initially fraudsters randomly selected a target in the hope that they would already have an account with the financial institution imitated. However, recent research has indicated that fraudsters gain access to records that list customers of a particular body, and then target those customers. This practice is known as 'spear phishing'.

The damage from phishing includes financial loss, loss of access to email facilities, theft of credit card numbers and social services numbers. The victim's credit rating may be negatively affected. The process is relatively easy for anyone with rudimentary internet skills.

In the United States in 2005, 1.2 million computer users suffered losses caused by phishing, of an estimated value of US$929 million; US businesses lose an estimated US$2 billion every year. The United Kingdom reports similar incidences. In 2007 the Australian Attorney General, Philip Ruddock, reported that identity fraud costs Australians at least $1 billion a year.

Phishing and other forms of identity fraud are typically dealt with by the relative recent amendments to the Commonwealth *Criminal Code 1995* (Cth) Part 10.6 – Telecommunications Services.

Further reading

O Akindemowo, *Information Technology Law in Australia*, Law Book Company, Sydney, 1999.

Australian Communications and Media Authority, 'Spam and e-Security', www.spam.acma.gov.au.

Australian Institute of Criminology, www.aic.gov.au/topics/cybercrime/.

Howard Broad, NZ Police Commissioner, (2007) *Electronic Crime Strategy to 2010: Policing with Confidence*, www.police.govt.nz/resources/2007/e-crime-strategy/e-crime-strategy.html.

Geraldine Chin, 'Technological change and the Australian Constitution', (2000) *MULR* 25.

Alan Davidson, 'Cybercrime – update and review', (2005) 25 *Proctor* 1–2.

JD Heydon, *Cross on evidence*, 6th edn, Butterworths, Sydney, 2000.

Internet Industry Association, www.iis.net.au.

Ministry for Communications and Information Technology, New Zealand Government, (2008) *The Digital Strategy 2.0*, www.digitalstrategy.govt.nz/Digital-Strategy-2/.

H Pontell, '"Pleased to meet you . . . won't you guess my name?" Identity fraud, cyber-crime and white collar delinquency', (2002) 23 *Adelaide Law Review*.

17

Evidence of electronic records

It is said that a verbal contract is worth the paper it is written on.[1] Although the common law recognises oral contracts, this oxymoron reflects the problems associated with proving not only the terms of a verbal contract, but also its very existence. The law of evidence is central to legal systems. Its restraints, rules and procedures aim to provide certainty and reliability within the criminal and civil justice systems. New and unanticipated problems emerge when these historic and sometimes antiquated rules of evidence are applied to electronic documents and records.

This chapter identifies the problems of applying paper-based rules of evidence in the electronic era. It first addresses how electronic information should be retained for the purposes of its potential use as evidence in legal proceedings and examines the relevant case law and statutory provisions. It then tackles the thorny problem of submitting to a court, as evidence, hard 'copies', typically printouts of electronic records such as electronic mail, particularly in circumstances where the electronic version has been deleted.

Evidence of electronic records

Courts typically establish facts before determining rights and imposing orders. The rules of evidence were developed to ensure a just basis for the making of such determinations. Evidence should be relevant, reliable and the best that is available. Notwithstanding modern statutes on evidence, case law from the 17th and 18th centuries still dictates the form in which evidence must be

1 This statement is traditionally attributed to movie mogul Samuel Goldwyn.

presented.[2] The rules relating to documentary evidence emerged when the quill and parchment were in use – copies of documents were originally made meticulously by hand. The assumption, when these rules were formulated, could only have been that any requirement for 'writing' and 'signature' would be paper based.

Background

The *Statute of Frauds* in 1677[3] required 'writing' and a 'signature' in specific circumstances in an attempt to prevent or reduce certain frauds which could arise through the use of oral evidence. The stated reason in the statute was for the 'prevention of many fraudulent Practices which are commonly endeavoured to be upheld by Perjury and Subornation of Perjury'.[4] The statute's central provision required:

> Noe Action shall be brought . . . unlesse the Agreement upon which such Action shall be brought or some Memorandum or Note thereof shall be in Writeing and signed by the partie to be charged therewith or some other person thereunto by him lawfully authorized.[5]

These provisions are effectively reproduced in modern statutes dealing with such matters as the disposition of land, intellectual property, insurance contracts and consumer protection. Naturally, the 'writing' and 'signature' elements, in 1677, referred to a tangible medium such as paper or parchment.

Where copies of 'originals' were hand made, the law treated such copies circumspectly, appreciating the real possibility of transcription error, or worse, deliberate and undetectable alteration. Modern photocopy machines make copies which are often indistinguishable from the original. Nevertheless remnants of the pre-electronic era remain in the law of evidence and copies continue to be treated with suspicion: an explanation as to why the original was not tendered as evidence can be required.

The efficiencies of electronics, computerisation and communications have led to streamlining and rationalisation in the preparation of business documents, often with little regard to legal consequences. Chapter 3 discussed the international legislative approach taken to ensure consumer and business confidence in electronic transactions. Electronic documents coming within the purvey of the Electronic Transactions Acts[6] are regarded as documents, subject to the same rules of evidence as paper documents. The legislature's approach here,

2 See JD Heydon, *Cross on evidence*, 7th edn, LexisNexis Butterworths, Sydney, 2004, Chapter 1 and John Forbes, *Evidence law in Queensland*, 7th edn, Lawbook Co., Sydney, 2008, Chapter 1.
3 *Statute of Frauds* 29 Car 2. c. 5.3.4.
4 Ibid., Preamble.
5 Ibid., s4.
6 *Electronic Transactions Act 1999* (Cth); *Electronic Transactions Act 2000* (NSW); *Electronic Transactions (Queensland) Act 2001* (Qld); *Electronic Transactions Act 2000* (SA); *Electronic Transactions Act 2000* (Tas); *Electronic Transactions (Victoria) Act 2000* (Vic); *Electronic Transactions Act 2003* (WA); *Electronic Transactions Act 2001* (ACT); *Electronic Transactions (Northern Territory) Act 2000* (NT).

as in other areas, has been reactive rather than proactive: to provide a means of accepting all forms of evidence, including emails, digital photographs, electronic banking records and logs, word processing documents, instant message histories, electronic accounting files, records of internet use and databases. Subject to appropriate levels of verification and integrity, any electronic record and document can be admitted. This is not new. The courts and the legislature have developed principles under which analogue records such as photographs, sound and filming recordings and facsimiles can be accepted.[7] However, in many circumstances the courts may reject the admission of or credibility of electronic records.

All commercial parties should retain documents in the form in which the documents are issued or presented, electronic or otherwise; not doing this risks admissibility problems should such documents be required as evidence in court. When a commercial dispute arises, the immediate concern is the ability to prove the issues before a court of law. We will now consider the common law principle known as the secondary evidence rule and evidentiary legislation.

Secondary evidence rule

Courts are not free to use all information as evidence.[8] In applying the common law secondary evidence rule, courts are often obligated to disregard otherwise relevant material. The two main exclusionary rules are the rule against hearsay and the rule as to secondary evidence of the contents of a document. The former excludes certain relevant information as untested and unreliable. Both rules include several exceptions. It is the latter rule which is less than satisfactory when applied to electronic records. The secondary evidence rule is often referred to as the 'best evidence' rule. In 1745 Lord Harwicke, in *Omychund v Barker*,[9] stated that no evidence was admissible unless it was 'the best that the nature of the case will allow'. The general rule is that a secondary evidence document will be inadmissible if the primary document is available.

There are a number of established exceptions. For example, where a party requires a document in the possession of another to be admitted into evidence, that party may issue a Notice to Produce, that is, a subpoena for documents. A copy of the original may be admitted as evidence if the party served with a Notice to Produce fails to produce the original. In the absence of the Notice to Produce, the original document is admissible but the copy is not. Other exceptions include consent, lost documents, where the production of the original is impossible and certain public documents.

7 *Butera v Director of Public Prosecutions for the State of Victoria* (1987) 164 CLR 180 at 186.
8 Heydon, *Cross on evidence*, p. 101 and Forbes, *Evidence Law in Queensland*, A.1.
9 (1745) 1 Atk, 21, 49; 26 ER 15 at 33.

The application of this rule to electronic records is fraught with danger. The courts need to determine whether electronic records on such media as videotapes, audiotapes, CDs, DVDs, a computer's hard disk, a floppy disk and electronic messages are documents for the purpose of the rule.

In relation to electronic messages debate rages over which version is the original and which is the copy: a minimum of eight 'copies' of an electronic message come into existence from creation to receipt. The question of which are admissible and which are not has not been resolved. Many argue that the original electronic message is the version created in the sender's computer and that the recipient only acquires a copy. Others would argue that the 'electronic record' as received forms the original, especially where it is changed along the path by, for example, data fluctuations altering a few characters or by the addition of underlying metadata. With paper documents, the recipient is typically in possession of the original.

In 2007 the High Court of Australia in *Golden Eagle International Trading Pty Ltd v Zhang*[10] reaffirmed the 'vitality' of the best evidence rule. In considering the applicability of historical versus prospective actuarial tables, the majority judgment stated:

> Despite criticism of it, the 'best evidence rule' has not fallen completely into desuetude. Subject to the exigencies of litigation, the circumstances of the parties, and the other settled and statutory rules of evidence, it has vitality. An aspect of the rule is that courts should act upon the least speculative and most current admissible evidence available. To prefer the prospective rather than the historical life expectancy tables is to do no more than that.[11]

In 1987 the High Court of Australia in *Butera v Director of Public Prosecutions for the State of Victoria*[12] considered the best evidence rule in the context of copied audiotapes. The majority held that:

> A copy of the tape is admissible provided the provenance of the original tape, the accuracy of the copying process and the provenance of the copy tape are satisfactorily proved.[13]

Such a formulation lends itself well to the electronic age. The High Court wrestled with the problematic consequences of a rule formulated as early as the 16th century, and decided to sidestep the best evidence rule and instead permit consideration of all the circumstances. In other words, the court gave itself discretion to consider the weight to be attributed to the evidence, something which the original rule was at pains to remove.

Ideally, authenticating an electronic document should be possible within the document itself, as with the signature on a traditional paper document. The

10 [2007] HCA 15.
11 [2007] HCA 15, para 4 per Gummow, Callinan and Crennan JJ. Also approved impliedly by Kirby and Hayne JJ at para 70.
12 (1987) 164 CLR 180.
13 (1987) 164 CLR 180 at 186.

header and other additional information in emails and other electronic documents corroborate and authenticate the nature of the document as well as features such as the sender, author, date and time.[14] The use of a digital signature or a secure or sophisticated electronic signature can help prove the origin of the document as well as authenticate it and verify its contents.

In *R v Maqsud Ali*[15] the British court discussed the application of the rule to photographs. The court considered that photographs are admissible 'on proof that they are relevant to the issues involved in the case and that the prints are taken from negatives that are untouched'.[16] The prints are in fact copies reproduced by means of mechanical and chemical devices. The court noted that evidence of things seen through telescopes or binoculars which could not be picked up by the naked eye have been admitted, and now there are devices for picking up, transmitting, and recording conversations. There should be 'no difference in principle between a tape recording and a photograph'.[17] This statement ought not be interpreted to mean that such recordings are admissible whatever the circumstances, but it did appear wrong to that court to deny advantages to the law of evidence available through new techniques and new devices, 'provided the accuracy of the recording can be proved', the voices recorded could be properly identified, the evidence is relevant and otherwise admissible and the court is satisfied that a tape recording is admissible in evidence. The court added, 'Such evidence should always be regarded with some caution and assessed in light of all the circumstances of each case.'[18]

In *R v Frolchenko*[19] in the Queensland Court of Appeal the issue of the signature on a document was briefly canvassed by Williams J. His honour noted that given modern methods of communication, such as email, many communications in writing will not bear a personal signature, but can still be authenticated by looking at such things as whether the name appears in typescript at the end of the document. The court stated that 'absence of an immediate challenge to the admissibility of the document on the ground that the party to the litigation was not responsible for its contents is material'.[20]

One possible factor that the court could take into account is the implementation and use of an audit trail by the organisation or individual who receives or sends the email message. This would go part of the way to demonstrating that the output from the system that is audited is what it purports to be.

14 C Reed, 'Authenticating electronic mail messages – some evidential problems', (1989) 52 *Modern Law Review* 649.
15 [1966] 1 QB 688.
16 [1966] 1 QB 688 at 701.
17 [1966] 1 QB 688 at 701.
18 [1966] 1 QB 688 at 703.
19 [1998] QCA 43.
20 [1998] QCA 43, per Williams J.

Evidence legislation

In Australia most states and territories apply the secondary evidence rule by force of legislation. Evidence legislation often provides an extremely complicated mechanism for adducing evidence of electronic records. For example, section 95 of the *Evidence Act 1977* (Qld) provides:

> S95(1) In any proceeding where direct oral evidence of a fact would be admissible, any statement contained in a document produced by a computer and tending to establish that fact shall, subject to this part, be admissible as evidence of that fact, if it is shown that the conditions mentioned in subsection (2) are satisfied in relation to the statement and computer in question.
>
> (2) The said conditions are—
>
> (a) that the document containing the statement was produced by the computer during a period over which the computer was used regularly to store or process information for the purposes of any activities regularly carried on over that period, whether for profit or not, by any person; and
>
> (b) that over that period there was regularly supplied to the computer in the ordinary course of those activities information of the kind contained in the statement or of the kind from which the information so contained is derived; and
>
> (c) that throughout the material part of that period the computer was operating properly or, if not, that any respect in which it was not operating properly or was out of operation during that part of that period was not such as to affect the production of the document or the accuracy of its contents; and
>
> (d) that the information contained in the statement reproduces or is derived from information supplied to the computer in the ordinary course of those activities.
>
> (3) Where over a period the function of storing or processing information for the purposes of any activities regularly carried on over that period as mentioned in subsection (2)(a) was regularly performed by computers, whether—
>
> (a) by a combination of computers operating over that period; or
>
> (b) by different computers operating in succession over that period; or
>
> (c) by different combinations of computers operating in succession over that period; or
>
> (d) in any other manner involving the successive operation over that period, in whatever order, of 1 or more computers and 1 or more combinations of computers;
>
> all the computers used for that purpose during that period shall be treated for the purposes of this part as constituting a single computer and references in this part to a computer shall be construed accordingly.
>
> (4) In any proceeding where it is desired to give a statement in evidence by virtue of this section, a certificate doing all or any of the following things, that is to say—
>
> (a) identifying the document containing the statement and describing the manner in which it was produced;

> (b) giving such particulars of any device involved in the production of that document as may be appropriate for the purpose of showing that the document was produced by a computer;
> (c) dealing with any of the matters to which the conditions mentioned in subsection (2) relate;
>
> and purporting to be signed by a person occupying a responsible position in relation to the operation of the relevant device or the management of the relevant activities (whichever is appropriate) shall be evidence of the matters stated in the certificate and for the purposes of this subsection it shall be sufficient for a matter to be stated to the best of the knowledge and belief of the person stating it.

Similar provisions appear in the statutes of South Australia[21] and the Northern Territory.[22]

The section has been criticised by many commentators.[23] In *R v Shephard*[24] the House of Lords considered the then UK equivalent of section 95, stating that oral evidence could be accepted in lieu of a certificate signed by a person with responsibility for the operation of the computer:

> Proof that the computer is reliable can be provided in two ways. Either by calling oral evidence or by tendering a written certificate in accordance with the terms of paragraph 8 of Schedule 3 (s 95(4) Queensland), subject to the power of the judge to require oral evidence.[25]

The House of Lords held that oral evidence as to the requirements of the section concerning the workings of the computer could be satisfied by 'the oral evidence of a person familiar with the operation of the computer who can give evidence of its reliability and such a person need not be a computer expert'.[26] As a consequence, it is most important for all organisations to consider engaging staff who become sufficiently familiar with the organisation's computer system to be in a position to give evidence of its reliability and of the process of record keeping:

> (a) information shall be taken to be supplied to a computer if it is supplied thereto in any appropriate form and whether it is so supplied directly or (with or without human intervention) by means of any appropriate equipment.
> (b) where, in the course of activities carried on by any person, information is supplied with a view to its being stored or processed for the purposes of those activities by a computer operated otherwise than in the course of those activities, that information, if duly supplied to that computer, shall be taken to be supplied to it in the course of those activities.

21 *Evidence Act 1929* (SA) ss34C, 55B.
22 *Evidence Act* (NT) s26D. For the NZ approach see *Evidence Act 2006* (NZ) subpart 8 – Documentary evidence and evidence produced by machine, device of technical process.
23 For example the Queensland Law Reform Commission's Issues Paper, 'The receipt of evidence by Queensland courts: Electronic records', WP No. 52, 58; www.qlrc.qld.gov.au/wp52.html.
24 [1993] AC 380.
25 [1993] AC 380, Lord Griffiths at 386.
26 [1993] AC 380 at 387.

(c) a document is to be taken to have been produced by a computer whether it was produced by it directly or (with or without human intervention) by means of appropriate equipment.[27]

Section 55 of the *Evidence Act 1958* (Vic) similarly provides:

S55 Admissibility of documentary evidence as to facts in issue

(1) In any legal proceeding (not being a criminal proceeding) where direct oral evidence of a fact would be admissible, any statement contained in a document and tending to establish that fact shall be admissible as evidence of that fact if—

(a) the maker of the statement had at the time of the making of the statement personal knowledge of the matters dealt with by the statement, and is called as a witness in the proceeding; or

(b) the document is, or forms part of, a record relating to any business and made in the course of that business from information supplied (whether directly or indirectly) by persons who had, or may reasonably be supposed to have had, personal knowledge of the matters dealt with in the information they supplied, and the person who supplied the information recorded in the statement in question is called as a witness in the proceeding.

The *Evidence Act 1906* (WA) contains a similar provision,[28] but an amendment was recently inserted to effectively remove the best evidence rule: section 73A provides that reproductions are admissible and the 'best evidence rule [is] modified'. Where a document 'accurately reproduces the contents of another document' it is admissible in evidence before a court 'in the same circumstances, and for the same purposes, as that other document, whether or not that other document still exists'.[29]

In determining whether a particular document accurately reproduces the contents of another, a court:

is not bound by the rules of evidence and—

(a) may rely on its own knowledge of the nature and reliability of the processes by which the reproduction was made;

(b) may make findings based on a certificate in the prescribed form signed by a person with knowledge and experience of the processes by which the reproduction was made;

(c) may make findings based on a certificate in the prescribed form signed by a person who has compared the contents of both documents and found them to be identical; or

(d) may act on any other basis it considers appropriate in the circumstances.[30]

27 *Evidence Act 1977* (Qld) s95(6).
28 *Evidence Act 1906* (WA) s79C.
29 *Evidence Act 1906* (WA) s73A(1).
30 *Evidence Act 1906* (WA) s73A(2).

The provision expressly applies to a reproduction made:

(a) by an instantaneous process;
(b) by a process in which the contents of a document are recorded by photographic, electronic or other means, and the reproduction is subsequently produced from that record;
(c) by a process prescribed for the purposes of this section; or
(d) in any other way.[31]

Legislation abolishing the 'original document' rule

In the early 1990s an attempt was made to unify all Australian federal, state and territory Evidence Acts. Today, the Evidence Acts of the Commonwealth, New South Wales, Tasmania and the ACT are identical.[32] In each of these jurisdictions the best evidence rule has been abolished. Section 51 of each of the Acts, entitled 'Original document rule abolished', provides:

> The principles and rules of the common law that relate to the means of proving the contents of documents are abolished.

Additionally the Acts make no reference to 'original'. The Acts define 'document' by reference to the contents rather than the medium. Reference to a document adduced into evidence is a reference to the contents of that document. Reference to a copy of a document adduced into evidence 'includes a reference to a document that is not an exact copy of the document in question but that is otherwise identical to the document in question in all *relevant* respects'.[33] This definition explains the rationale of the legislation. This approach, on its own, would be sufficient to impliedly abolish the best evidence rule.

However, following provisions elaborate. In addition to adducing evidence of the contents of a document by tendering the original, and by adducing evidence of an admission made by another party to the proceeding with respect to the contents of the documents, specific provisions permit a variety of copies to be adduced.

A party may tender a document that is, or purports to be, a copy of the document in question which has been produced, or purports to have been produced, by a device that reproduces the contents of documents. Second, if the document in question is an article or thing by which words are recorded in such a way as to be capable of being reproduced as sound, or in which words are recorded in a code, including shorthand writing, the evidence may be adduced by tendering a document that is, or purports to be, a transcript of the words. Third, if the document in question is an article or thing on or in which information is stored in such a way that it cannot be used by the court unless a device is used to retrieve,

31 *Evidence Act 1906* (WA) s73A(3).
32 *Evidence Act 1995* (Cth); *Evidence Act 1995* (NSW); *Evidence Act 2001* (Tas). The ACT applies the Commonwealth Act.
33 Evidence Acts of the Commonwealth, New South Wales, Tasmania and the ACT, s47. Emphasis added.

produce or collate it, evidence may be adduced by tendering a document that has been produced by use of the device. Fourth, evidence may be adduced by tendering a document that forms part of business records which purports to be a copy of or an extract from or a summary of, the document.[34]

As a result, all documents, including electronic documents and copies, are now admissible in those jurisdictions. However, most importantly it is left to the courts to determine the weight given to them.

In the High Court of Australia in *Butera v Director of Public Prosecutions for the State of Victoria*,[35] Dawson J stated that 'some modes of proof are better than others, but that . . . goes to weight rather than admissibility'.[36] As with oral evidence given by witnesses, the fact that certain evidence is admissible does not mean that it will be accepted by the court, or even be given the same weight throughout.

In *Eastman v R*[37] the Federal Court of Australia considered the application of section 48 in relation to the admission into evidence of an audiotape:

> The definition of '*document*' includes a record of information from which sounds can be reproduced. As the transcripts were admissible under s.48(1)(c) as evidence of the contents of the tapes, the procedure followed by the trial judge could not affect the admissibility of the tape recordings. As it was, the transcripts were received into evidence as an aide-memoire, and the jury was instructed to treat the transcripts only as an aid. Where, on the evidence adduced in a particular case, there is doubt or disagreement whether the transcript, or part of it, accurately deciphers the sounds captured on the tape, it seems to us that this should be the role of the transcript, notwithstanding the provisions of s.48(1)(c) of the Evidence Act. In the present case, the transcript[s] . . . were admissible for the purpose of assisting the jury.[38]

Section 48 applies equally to electronic records. This section permits copies or printouts to be tendered as evidence in these four jurisdictions. The expression 'device' is used to include a computer system, and correspondingly, 'tendering a document that was or purports to have been produced by use of the device' is satisfactory.

International perspective

The UN Commission on International Trade Law (UNCITRAL) Model Law of Electronic Commerce 1996 (Model Law)[39] includes a provision dealing with 'Admissibility and evidential weight of data messages'. The expression 'data messages' is defined to include 'information generated, sent, received or stored by electronic, optical or similar means, including, but not limited to, electronic

34 Evidence Acts of the Commonwealth, New South Wales, Tasmania and the ACT, s48(1) .
35 (1987) 164 CLR 180.
36 (1987) 164 CLR 180 at 195.
37 [1997] FCA 48.
38 [1997] FCA 48, Ground 11.
39 UNCITRAL website: www.uncitral.org/pdf/english/texts/electcom/05-89450_Ebook.pdf.

data interchange (EDI), electronic mail, telegram, telex or telecopy'.[40] Article 9 provides that the rules of evidence must not deny the admissibility of a data message in evidence on the sole ground that it is a data message, nor where the data message is the best evidence reasonably available, on the grounds that it is not in its original form. Specially, sub-article 9(2) states, 'Information in the form of a data message shall be given due evidential weight.'

In a manner reminiscent of Butera's case, the Model Law provides that in assessing the evidential weight of a data message, 'regard shall be had to the reliability of the manner in which the data message was generated, stored or communicated, to the reliability of the manner in which the integrity of the information was maintained, to the manner in which its originator was identified, and to any other relevant factor'.[41]

Although the Model Law has been used as a template in more than 50 jurisdictions internationally,[42] this particular provision was omitted from the Australian and New Zealand Electronic Transactions Acts, because it was thought that such a provision is best placed in a jurisdiction's evidence legislation and not in its electronic commerce legislation. For example, in Australia, the Model Law was the basis for the Electronic Transactions Acts in all nine jurisdictions;[43] however, this provision was omitted from the initial *Electronic Transactions Act 1999* (Cth) because the *Evidence Act 1995* (Cth) had a few years earlier abolished the best evidence rule, making an additional provision obsolete. The remaining eight Australian jurisdictions followed the lead of the Commonwealth, even though there was no similar redundancy in most jurisdictions' legislation at the time.

The *Electronic Transactions Act 2002* (NZ) also did not include this provision. The *Evidence Act 2006* (NZ) provides that where a party offers evidence that was produced 'wholly or partly by a machine, device, or technical process' (for example, scanning) and the machine, device, or technical process is 'of a kind that ordinarily does what a party asserts it to have done', it is presumed, in the absence of evidence to the contrary, 'that on a particular occasion the machine, device, or technical process did what that party asserts it to have done'.[44] Where the information is stored so that it cannot be accessed by the court unless the relevant 'machine, device, or technical process' is used, then 'a party may offer a document that was or purports to have been displayed, retrieved, or collated by use of the machine, device, or technical process'.[45]

The *Civil Evidence Act 1995* (UK) also effectively abolishes the best evidence rule. The substantive provision permits the admission of copies of any degree of remoteness from the original. A statement contained in a document may be proved either '(a) by the production of that document, or (b) whether or not that

40 Article 2.
41 Sub-article 9(2).
42 See Chapter 3.
43 The Commonwealth, six states and two territories; and New Zealand.
44 *Evidence Act 2006* (NZ) s137(1).
45 *Evidence Act 2006* (NZ) s137(2).

document is still in existence, by the production of a copy of that document or of the material part of it, authenticated in such manner as the court may approve'.[46] It is immaterial 'how many removes there are between a copy and the original'.[47] Section 13 encompasses electronic documents by providing that a 'document' means anything in which information of any description is recorded, and 'copy', in relation to a document, means anything onto which information recorded in the document has been copied, by whatever means and whether directly or indirectly.

In the United States, federal courts follow the Federal Uniform Rules of Evidence, while state courts generally follow state legislature rules. At the federal level the best evidence rule has been retained and is largely codified in the Federal Uniform Rules of Evidence.[48] To prove 'the content of a writing, recording, or photograph, the original writing, recording, or photograph is required, except as otherwise provided in these rules or by Act of Congress'.[49] Duplicates are admissible 'to the same extent as an original' unless a genuine question is raised as to the authenticity of the original or in circumstances in which it would be unfair to admit the duplicate in lieu of the original.[50] The official notes to this provision explain that when the only concern is 'with getting the words or other contents before the court with accuracy and precision, then a counterpart serves equally as well as the original'.[51] A 'duplicate' is defined as 'a counterpart produced by the same impression as the original, or from the same matrix, or by means of photography, including enlargements and miniatures, or by mechanical or electronic re-recording, or by chemical reproduction, or by other equivalent techniques which accurately reproduce the original'.[52]

Writings and recordings include not only 'handwriting, typewriting, printing, photostating, photographing', but also electronic forms such as 'magnetic impulse, mechanical or electronic recording, or other form of data compilation'.[53] Photographs include still photographs, x-rays, videotapes and motion pictures.[54]

The *Indian Evidence Act 1872* (India) codifies the secondary evidence rule providing that the contents of documents 'may be proved either by primary or by secondary evidence';[55] '[d]ocuments must be proved by primary evidence'[56] subject to the customary exceptions.[57] Section 65 provides:

46 *Civil Evidence Act 1995* (UK) s8.
47 *Civil Evidence Act 1995* (UK) s8.
48 Several US states (such as California) follow this federal approach.
49 Federal Uniform Rules of Evidence, article X, rule 1002.
50 Federal Uniform Rules of Evidence, article X, rule 1003.
51 Federal Uniform Rules of Evidence, article X, Official Note to rule 1003.
52 Federal Uniform Rules of Evidence, article X, rule 1003.
53 Federal Uniform Rules of Evidence, article X, rule 1001.
54 Federal Uniform Rules of Evidence, article X.
55 *Indian Evidence Act 1872* (India) s61.
56 *Indian Evidence Act 1872* (India) s64; see also s22.
57 *Indian Evidence Act 1872* (India) s65.

65 Cases in which secondary evidence relating to documents may be given—

Secondary evidence may be given of the existence, condition or contents of a document in the following cases:

(a) When the original is shown or appears to be in the possession or power of the person against whom the document is sought to be proved, or of any person out of reach of, or not subject to, the process of the Court, or of any person legally bound to produce it, and when, after the notice mentioned in Section 66, such person does not produce it;

(b) When the existence, condition or contents of the original have been proved to be admitted in writing by the person against whom it is proved or by his representative in interest;

(c) When the original has been destroyed or lost, or when the party offering evidence of its contents cannot, for any other reason not arising from his own default or neglect, produce it in reasonable time;

(d) When the original is of such a nature as not to be easily movable;

(e) When the original is a public document within the meaning of Section 74;

(f) When the original is a document of which a certified copy is permitted by this Act, or by any other law in force in India to be given in evidence;

(g) When the originals consist of numerous accounts or other documents which cannot conveniently be examined in Court, and the fact to be proved is the general result of the whole collection.

In cases (a), (c) and (d), any secondary evidence of the contents of the documents is admissible. In case (b), the written admission is admissible. In case (e) or (f), a certified copy of the document, but no other kind of secondary evidence, is admissible. In case (g), evidence may be given as to the general result of the documents by any person who has examined them, and who is skilled in the examination of such documents.

Primary evidence is defined as 'the document itself produced for the inspection of the Court'.[58] Secondary evidence is defined as: (1) certified copies as defined by the Act; (2) copies made from the original by mechanical processes which in themselves insure the accuracy of the copy and copies compared with such copies; (3) copies made from or compared with the original; (4) counterparts of documents as against the parties who did not execute them; and (5) oral accounts of the contents of a document given by some person who has himself seen it.[59]

Hard copies of electronic records as evidence

Where the original document is electronic, a question sometimes arises as to whether a hard copy, such as a printout of the electronic document, is admissible as evidence in a court of law if strictly applying the secondary evidence rule. For example, many organisations print out a hard copy of important emails as a 'permanent' record, and then delete the electronic version. These organisations

58 *Indian Evidence Act 1872* (India) s62.

59 *Indian Evidence Act 1872* (India) s63. The Act includes specific examples referred to as 'illustrations'.

risk a ruling that the printout is inadmissible, thus jeopardising their opportunity to provide relevant evidence before the court. Law firms that adopt this practice would be professionally negligent. The printouts are inferior copies of the electronic documents.

This practice is an unnecessary hangover from the early days of computers being used for commercial purposes, the 1970s and 1980s, when computer space was expensive and computer specialists urged and advised users to 'clean out' the computer space for premium efficiency. In the 21st century computer space is cheap and plentiful, and there are many electronic methods of archiving and storing such records, documents and communications.

The US case of *Armstrong v Executive of the President*[60] involved the status of a printout of an email. The court concluded that the printed version of the email contained less information than the electronic version. The missing information included the date of the transmission, the date of receipt, the detailed list of recipients and linkages between messages sent and replies received.

Should such a document be called into question, it is clear that, in jurisdictions where the secondary evidence rule has been abolished, greater weight would and should be given to the electronic document. Also, it may be reasonable for the court to question why the 'original' document was destroyed, particularly if it is thought that the electronic version contained information which would have assisted the court in substantiating the document's originality and accuracy. The requirement for documents to be kept for set periods of time under statutes of limitation applies equally to electronic documents.[61]

This issue has connections to the role of President Reagan in the Iran-Contra affair in the 1980s. In 1986 Oliver North and national security adviser John Poindexter electronically erased thousands of their email messages on their way out of the National Security Council (NSC). However, the system's backup tapes allowed investigators to recover these messages and use them as evidence in court proceedings. Journalist Scott Armstrong and others sought the NSC records. The Executive Office of the President argued that the entire NSC was exempt from otherwise applicable US Freedom of Information legislation. The subsequent lawsuit *Armstrong v Executive Office of the President*[62] was brought to prevent the backup records from being erased. Researchers were using the records to piece together the controversial arms sales to Iran and the funding of Nicaraguan rebels.

The Reagan Administration planned to delete the backups. Court rulings established that the archives and records laws for the retention of documents apply to email. Subsequently the Bush Administration[63] staged a midnight raid on Inauguration Eve in January 1993 to put the tapes beyond the law.

60 810 F. Supp.335 (1993).

61 See Catherine E Pasterczyk, 'E-Federal e-mail management: A records manager's view of *Armstrong v Executive Office of the President* and its aftermath', (1998) 32-2 *ARMA Records Management* 10-22.

62 1 F. 3d 1274 (DC Cir 1993).

63 The first Bush Administration.

Judge Richey pointed out that a paper copy of the electronic material does not contain all of the information included in the electronic version. He stated:

> paper and the computer versions of these electronic records are different . . . A note distributed over these computer system[s] includes information that is not reproduced on the paper copy regarding who has received the information and when the information was received, neither of which is reproduced on the paper copy . . . Material must be saved in a way that includes all the pertinent information contained therein . . . paper copies of these materials do not include all of the relevant information.[64]

There are no cases in Australia or New Zealand decided on this evidentiary issue, but the potential remains. As long as legal systems require standards for the quality of evidence, secondary evidence will be rejected or in appropriate cases given less weight for judicial determinations.

Originals and copies – envelopes and attachments

The secondary evidence rule is less than satisfactory when applied to electronic records. Who has the original when an email is sent? This is a complicated question and one which has not been adequately dealt with by the courts. There are a number of complicating factors.

First, there are multiple copies of the email created and copied. The sender has a copy in the space for Random Access Memory (RAM), temporary space and permanent space. On sending, the email is copied to the internet service provider (ISP), several routers on the internet, the recipient's ISP and finally the recipient's RAM space, temporary space and permanent space. The recipient may copy the email electronically to storage folders or archives.

Second, the recipient actually receives a different document from that sent. Hidden from the reader is additional information. This was identified in Armstrong's case (see above). Some of this information is not part of the email at the time that it is sent – the final email that is copied differs from the one that was sent.

Third, there is a requirement to consider the intention of the sender. Did the sender intend that the entire electronic record form the communication, or only the 'text' of the main message? This is not an entirely new problem. When a traditional letter is sent by post, to what extent does the sender intend including the envelope, postmark and stamps as part of the communication? The envelope and postmark can be evidence of sending, receipt and timing. The dilemma is the difference between the intended use of the sender and the actual use made by the recipient: it is not clear if the recipient's understanding and use should be seen as dependent on the intention of the sender. Should a dispute arise, both parties would be in a much stronger position if they retained the full

64 1 F. 3d 1274 (DC Cir 1993). See also www.eff.org/legal/cases/PROFS_case/richey.decision.

electronic version of, for example, the email, as recognised in Armstrong's case. Intention and use may both be gleaned from the format and structure used by the sender.

Many organisations, businesses and individuals prefer their communications to be more formal than emails. Some prepare a formal electronic letter, with the organisation's letterhead and even a digitised traditional signature, and attach it to the electronic communication, which may simply state that the 'document' is attached. It has been argued that the email would now operate in the same manner as an envelope, perhaps merely corroborating the time and place of the communication, but not forming part of the message. This practice raises several questions. Should the attachment be kept in its original electronic form or can it be printed out? It may appear that the sender simply prefers the appearance of a formal hardcopy letter. However, assuming that the attachment was a simple Microsoft Word file, there are several complicating factors.[65] Microsoft Word files contain a significant amount of statistical information in addition to the text: the date the file was created, the number of revisions, the date and times of revisions and in some cases prior drafts. This additional information could corroborate claims made by one party regarding a particular statement of affairs. The nature and extent of other formats can exacerbate such considerations. A printout of the attachment and deletion of the electronic file removes this potential corroborative evidence.

Fourth, the courts – and much of the public discussion – have failed to recognise the distinction between analogue and digital data. Data recorded onto audio and video tapes is typically analogue. Every time analogue data is copied the data is degraded. The picture may be less clear and the sound may have decreased quality and increased background noise. Photocopies typically scan the original and attempt to recreate it. Some modern photocopiers are excellent, but colours may be marginally copied or turned to black, white or grey; and a light pencil notation may become invisible on the copy. The reproduction is clearly inferior and less reliable. Digital copying, for example of DVDs and computer data, involves recording by the duplication of bits and bytes precisely. When the process operates correctly the copy is identical to the original. Copies of analogue data ought to be regarded with greater suspicion than digital copies, putting aside the question of tampering. Butera's case involved an analogue copy. The court ultimately required information on the provenance of the original tape, the accuracy of the copying process and the provenance of the copy tape. Despite the difference in the process and the result, the same approach is appropriate for digital copies. It is vital that organisations set up processes which record all steps, are secure and are capable of presentation as evidence when required.

65 These factors arise with many similar word processing programs.

Conclusion

As a general principle, in jurisdictions applying the secondary evidence rule, courts must dictate that the printout is an inadmissible copy. There are well-established exceptions to this rule, such as lost or destroyed documents, public documents and the application of a Notice to Produce. However, courts ought not admit evidence where the loss or destruction was a deliberate act of the party relying on the copy.[66] The probity of the inferior copy must in appropriate cases be questioned.

In jurisdictions where the secondary evidence rule has been abolished, greater weight should be given to the electronic document. Destroying the electronic record in preference to a printed hard copy risks the 'document' being inadmissible or being given less weight. It is reasonable for the court to question why 'original' electronic documents have been destroyed, as they contain information which would assist the court in substantiating originality and accuracy.

In legal proceedings, most parties, as a matter of practice and convenience, admit undisputed documents into evidence. Nevertheless, where the legitimate interests of the client would be furthered by challenging a hard copy of an electronic record, it is incumbent on the lawyer to do so and for the court to consider, if it remains appropriate, the principle of the secondary evidence rule.

All parties should retain documents in the form in which they were issued or presented. Attempting to change the medium may have undesirable if not unpredictable consequences.

Further reading

Alan Davidson, 'Armstrong's case 10 years on', (2003) 23 *Proctor* 34.

Catherine E Pasterczyk, 'E-Federal e-mail management: A records manager's view of *Armstrong v Executive Office of the President* and its aftermath', (1998) 32-2 *ARMA Records Management* 10-22.

Queensland Law Reform Commission, 'The receipt of evidence by Queensland courts: Electronic records', WP No. 52, 58; www.qlrc.qld.gov.au/wp52.html.

C Reed, 'Authenticating electronic mail messages – some evidential problems', (1989) 52 *Modern Law Review* 649.

66 See RA Brown, *Documentary evidence in Australia*, 2nd edn, LBC Information Services, Sydney, 1996, 128–29.

18

Censorship – Broadcast and online content regulation

The Australian Communications and Media Authority (ACMA) is responsible for administering legislation regarding the content and regulation of most forms of electronic communications. This chapter deals with the role of ACMA and the regulation of internet, television and radio communications.

The Australian Communications and Media Authority

On 1 July 2005 ACMA was formed by the merger of the Australian Broadcasting Authority (ABA) and the Australian Communications Authority (ACA). On 1 July 1997 the ACA was formed by the merger of the Australian Telecommunications Authority (AUSTEL) and the Spectrum Management Agency (SMA).

ACMA is responsible for the regulation of broadcasting, radiocommunications, telecommunications and online content. Its responsibilities include:

- promoting self-regulation and competition in the telecommunications industry, while protecting consumers and other users;
- fostering an environment in which electronic media respect community standards and responds to audience and user needs;
- managing access to the radiofrequency spectrum, including the broadcasting services bands;
- representing Australia's communications and broadcasting interests internationally.[1]

1 See the ACMA website: www.acma.gov.au/WEB/STANDARD/1001/pc = PC_100675.

The internet

ACMA administers the scheme for dealing with content on the internet, enforces Australia's anti-spam law and can make rules about accessing the internet via premium mobile phone services. (The term 'premium services' relates to a range of voice and data services that are charged at a premium or high rate. They can include voice services such as chat and dating lines, and internet websites containing adult content. The Minister for Broadband, Communications and the Digital Economy can direct ACMA to make a service provider determination relating to accessing premium services on mobile phones.)

Online content regulation is established under Schedules 5 and 7 of the *Broadcasting Services Act 1992* (Cth), which deal with offensive and illegal material on the internet and the protection children from exposure to material that is unsuitable. The Act gives ACMA the following functions:

- investigation of complaints about Internet content and Internet gambling services;
- encouraging development of codes of practice for the Internet industry, registering, and monitoring compliance with such codes;
- providing advice and information to the community about Internet safety issues, especially those relating to children's use of the Internet;
- undertaking research into Internet usage issues and informing itself and the Minister of relevant trends;
- liaising with relevant overseas bodies.[2]

The *Spam Act 2003* (Cth) bans the sending of unsolicited electronic messages, including email and SMS and the harvesting of email addresses.

Radio and television

ACMA plans the channels that radio and television services use, issues and renews licences, regulates the content of radio and television services and administers the ownership and control rules for broadcasting services. Up to date lists of currently licensed television and radio broadcasters are available on the ACMA website. A search by postcode facility allows access to a list of services that broadcast in that postcode area.

Most aspects of program content are governed by codes of practice developed by industry groups representing the various broadcasting sectors. Details of the codes, of ACMA program standards, of channels assigned for digital TV and other information about digital TV and radio can be found on the ACMA website.

Complaints about the content of programs on radio and television, including ABC and SBS services, must first be made to the broadcaster concerned.

2 See www.acma.gov.au/WEB/STANDARD/1001/pc = PC_100012.

Telephones

ACMA administers and allocates all phone numbers within Australia, enforces the anti-spam laws, and monitors and manages performance by carriage service providers against the Customer Service Guarantee and the universal service obligation.

Its role also includes:

- encouraging the development of codes of practice for the telecommunications industry, and registering and monitoring compliance with such codes;
- providing advice and information to consumers about their rights and safeguards;
- undertaking research into phone usage; and
- liaising with relevant overseas bodies and actively participating in international standardisation activities.[3]

ACMA licenses Australia's telecommunications carriers and is responsible for infrastructure regulation, including for codes covering the planning and installation of telecommunications facilities (these include those for mobile phones).

ACMA helps ensure that carriers and carriage service providers comply with their obligations to assist law enforcement and national security agencies. It also manages the effectiveness and efficiency of certain aspects of the 000 emergency call service. In addition, ACMA manages the contract for the service which allows people with a communication impairment to use the telephone in a way that is equivalent to the standard telephone service available to all other Australians.

Licences

Licences issued by ACMA include apparatus licences, class licences (these authorise users of specified parts of the radiofrequency spectrum to operate on a shared, uncoordinated basis), radiofrequency spectrum licences, licences to provide broadcasting services and carrier licences for telecommunications services.

ACMA also licenses amateur radio operators and marine radio operators and manages low-power open narrowcasting services.

ACMA issues the radiocommunications licences that underpin services that support everyday life, such as mobile phones, broadcasting services, wireless local area networks, GPS systems and two-way radios, and is responsible for compliance with those licensing requirements and for investigating complaints of interference to radiocommunications services.

3 See www.acma.gov.au/WEB/STANDARD/1001/pc = PC_2793 and the *Telecommunications Act 1997* (Cth).

Consumers

Consumer codes are registered by ACMA, making them enforceable. ACMA monitors compliance and the strategies used to raise consumer awareness of the codes.

ACMA monitors and reports on the performance of telecommunications service providers against the timeframes for the provision and repair of the standard telephone service specified in the Customer Service Guarantee. It also administers the Universal Service Obligation (USO), which ensures that all people in Australia, no matter where they live or carry on business, have reasonable access to a standard telephone service, payphones, prescribed carriage services and digital data services.

ACMA raises community awareness about communications and consumer safeguards. There is a range of ACMA information products and campaigns, including widely used toolkits about mobile phone, fixed phone and internet services.

ACMA administers complaint schemes for radio, television and internet content and is responsible for ensuring that carriers and eligible carriage service providers, including internet service providers (ISPs), join the Telecommunications Industry Ombudsman (TIO) scheme. One of the TIO's main roles is to investigate complaints made against its members.

Industry

ACMA works with industry, through the regulatory framework, to foster self-regulation, provide benefits to end-users, readily accommodate technological change and contribute to an efficient and internationally competitive communications sector.

Internet content

There are various degrees of censorship and restrictions applying to television, radio, books, newspapers and films. However, any user of the internet knows that material that breaches those standards is available online.

The internet contains the most splendid educational and cultural material. It also contains offensive and disgraceful material. Internationally there is divergence of opinion regarding appropriate standards in cultural areas such as dress, literature and films. There is, however, consensus regarding the repugnance of the portrayal of actual extreme violence and child pornography. But there remains disagreement in relation to the age of a 'child', the upper limit of which varies from 12 to 21. According to Article 1 of the UN Convention on the Rights of the Child, 'child' refers to a person under 18 years of age.

There are also disagreements as to the meaning of 'violence', 'offensive' and 'indecent'.[4]

No matter what one's definition is, though, there is consensus on the fact that offensive material is available on the internet. But keeping track of what is available is almost impossible for any government or organisation – Google, for example, has indexed more than one trillion web pages.

Some countries have made attempts to control specific kinds of offensive internet sites. German courts and legislators have sought to extend anti-Nazi law to other jurisdictions to restrict web-based Holocaust deniers. Similarly, France has pressured eBay and Yahoo to limit the sale of Nazi memorabilia. In Turkey the office of an ISP was demolished when it was found to have allowed images of women with uncovered arms and faces.

ABC (US) journalist Michael Malone wrote about child pornography:

> This is the very heart of darkness. These are images that are more than shocking and repulsive. They kill your soul, in part because you know that every poor child you see on these sites is dead, if not now at the hands of a sadist, then decades from now from drugs, alcoholism or suicide. The pictures first make you sick, then angry, and finally homicidal . . . There were already certain unspeakable images so burned into my brain that, even now, I wish I could take a scalpel and cut them out.[5]

Studies suggest that a majority of internet users have at some stage visited websites that may be regarded as pornographic: some suggest that 5–24 per cent of overall internet usage is spent accessing pornographic sites.[6] The figures vary according to the nature and type of survey, but with more than a trillion web pages, even the conservative figure yields several billion pages.

US cases

Reno v American Civil Liberties Union

The Clinton Administration recognised the potential danger of the booming internet pornography business and enacted the *Communications Decency Act 1996* (US).[7] However, in *Reno v American Civil Liberties Union*,[8] the US Supreme Court regarded the definitions of 'indecent' and 'patently offensive' as vague and broad and struck down most of the Act. This case is the most significant event in terms of pornography on the internet because it removed incentives for policing or enforcement of pornography online. However, it is also true that such material could, both before and after the case, be hosted off shore.

4 For example, see www.ageofconsent.com/ageofconsent.htm.
5 See www.caslon.com.au/censorshipguide2.htm.
6 See www.eff.org.
7 *Communications Decency Act 1996* (US) (CDA).
8 521 US 844 (1997). Janet Reno was the Attorney General for the Clinton Administration.

US v American Library Association

In June 2003 the US Supreme Court, in *US v American Library Association*,[9] held that anti-pornography filtering does not violate free speech rights. Congress can require public libraries that receive federal funds to provide public internet access to install filters. Chief Justice Rehnquist stated:

> Internet terminals are not acquired by a library in order to create a public forum for Web publishers to express themselves. Rather, a library provides such access for the same reasons it offers other library resources; to facilitate research, learning and recreational pursuits by furnishing materials of requisite and appropriate quality . . . The decisions by most libraries to exclude pornography from their print collections are not subject to heightened scrutiny; it would make little sense to treat libraries' judgments to block online pornography differently.[10]

The decision gives US legislators the power to control the billion-dollar industry which internet pornography has become since the decision in *Reno v American Civil Liberties Union*. The negative result may be the limits on the extent to which members of the public may be able to access legitimate internet content. Justice Breyer stated that web surfers can ask librarians to disable filters to access a particular site, but counsel argued that even people undertaking legitimate research will not ask a librarian to turn off the anti-pornography filter. One witness, Dr Bertman, who maintained a medical site, expressed concern that young persons with sexually transmitted diseases may have their access to legitimate sites curtailed. The Bush Administration has argued that federally funded libraries do not include X-rated movies and magazines, so they should be permitted to filter out such online material: this sets them up as an exception to the Reno case, with some control over such material returned to them.

Australia

Internet content has been the jurisdiction of ACMA[11] since January 2000. The *Communications Legislation Amendment (Content Services) Act 2007* (Cth) inserted Schedule 7 into the *Broadcasting Services Act 1992* (Cth) for the purpose of regulating all content services delivered via carriage services, irrespective of the platform, and whether they consist of user-generated content or otherwise. Schedule 7 commenced on 20 January 2008. In developing the new content rules, ACMA stated that it was 'guided by its disposition to allow adults to continue to read, hear and see what they want, while protecting children from exposure to inappropriate content, regardless of the delivery mechanism'.[12]

9 539 US 194 (2003).
10 539 US 194 (2003), 195.
11 At the time it was known as the Australian Broadcasting Authority (ABA).
12 ACMA Chair Chris Chapman, www.acma.org.au.

The provisions are similar to the previous obligations relating to stored content. The rules provide that after receiving a complaint and investigating internet or mobile content, ACMA may require the content service provider to either remove the content or place the content behind specified access restrictions.

Under Schedule 7 the following categories of online content are prohibited content:

- any online content that is classified RC or X18+ by the Classification Board (formerly the Office of Film and Literature Classification). This includes real depictions of actual sexual activity, child pornography, depictions of bestiality, material containing excessive violence or sexual violence, detailed instruction in crime, violence or drug use, and/or material that advocates the doing of a terrorist act;
- content which is classified R18+ and not subject to a restricted access system that prevents access by children. This includes depictions of simulated sexual activity, material containing strong, realistic violence and other material dealing with intense adult themes; and
- content which is classified MA15+, that is provided by a mobile premium service or a service that provides audio or video content upon payment of a fee, and that is not subject to a restricted access system. This includes material containing strong depictions of nudity, implied sexual activity, drug use or violence, very frequent or very strong coarse language, and other material that is strong in impact.[13]

If the content has not been classified by the Classification Board, but if it were to be classified, there is a substantial likelihood that the content would be prohibited content, it is defined as potential prohibited content.

In addition, Schedule 7 regulates for:

- providers of hosting services, live content services, link services and commercial content services to have in place access restrictions if providing R18+ and commercial MA15+ content;
- 'take-down', 'service cessation' and 'link deletion' notices to remove content or access to content that is the subject of a complaint; and
- a co-regulatory approach that provides for the development of industry codes to address issues including the classification of content, procedures for handling complaints about content and increasing awareness of potential safety issues associated with the use of content services.[14]

Mobile premium services including premium rate SMS and MMS[15] services and mobile content portals are regulated under the *Telecommunications Service Provider (Mobile Premium Services) Determination 2005 (No. 1)* (Cth), made

13 Classifications are based on criteria outlined in the *Classification (Publications, Films and Computer Games) Act 1995* (Cth), National Classification Code and the Guidelines for the Classification of Films and Computer Games 2005.

14 Internet content was previously regulated under the Restricted Access Systems Declaration 1999 under clause 4(1) of Schedule 5 to the BSA.

15 SMS is short for Short Message Service and MMS is short for Multimedia Message Service.

under subsection 99(1) of the *Telecommunications Act 1997* (Cth). A mobile carriage service provider must not supply content classified as MA15+ or R18+ to a customer unless the customer has requested access and has been verified as at least 18 years old. Content classified as MA15+ or R18+ must not be supplied by premium SMS or MMS services otherwise than on a number with a listed prefix.

Schedule 7 required ACMA to develop a new restricted access systems declaration to regulate access to MA15+ content and R18+ content with an Australian connection, amend the mobile premium services determination to remove the access restrictions and designated prefix requirements that are made redundant by Schedule 7 and vary the Telecommunications Numbering Plan 1997 to allow the transfer of the designated prefix requirements from the mobile premium services determination.

'Content service' is defined as '(a) a service that delivers content to persons having equipment appropriate for receiving that content, where the delivery of the service is by means of a carriage service; or (b) a service that allows end-users to access content using a carriage service'.[16]

Referral to law enforcement agencies

If, in the course of an investigation, ACMA is satisfied that content is prohibited content or potential prohibited content and is of a sufficiently serious nature to warrant referral to a law enforcement agency, ACMA must notify a member of an Australian police force. 'Sufficiently serious nature' is not defined, but would include evidence of child pornography or the commission of a crime such as sexual assault and common assault.

Take-down notices

If ACMA is satisfied that content hosted by a 'hosting service provider' is prohibited content and there exists an Australian connection, and if the content has been classified RC or X18+, or R18+ or MA15+ (without a restricted access system in place) by the Classification Board, ACMA must give the hosting service provider a written final take-down notice.[17] ACMA can declare that a specified access-control system is a restricted access system in relation to content for the purposes of Schedule 7.[18] Eligible electronic publications are exempt. An eligible electronic publication is defined as '(i) an electronic edition of a book, magazine or newspaper; or (ii) an audio recording of the text, or abridged text, of a book, magazine or newspaper; and (b) a print edition is was available to the public . . . in Australia then (c) the content is an eligible electronic publication'.[19]

16 *Broadcasting Services Act 1992* (Cth) Schedule 7, cl2.
17 *Broadcasting Services Act 1992* (Cth) Schedule 7, cl47(1).
18 *Broadcasting Services Act 1992* (Cth) Schedule 7, cl14.
19 *Broadcasting Services Act 1992* (Cth) Schedule 7, cl11.

If the content is not classified and ACMA is satisfied that, if the content were to be classified by the Classification Board, there is a substantial likelihood that the content would be classified RC or X18+, or R18+ or MA15+ (without a restricted access system in place), ACMA must issue an interim take-down notice. ACMA must then apply to the Classification Board to have the content classified, and depending upon the classification, withdraw the notice or issue a final take-down notice.[20]

Service-cessation notices

If ACMA is satisfied that content hosted by a live content service provider is prohibited content and the service provider has an Australian connection, and if the content has been classified RC or X18+, or R18+ or MA15+ (without a restricted access system in place) by the Classification Board, ACMA must give the live content service a written service-cessation notice.[21] Again, eligible electronic publications are exempt.

If the content is not classified and ACMA is satisfied that if the content were to be classified by the Classification Board there is a substantial likelihood that the content would be classified RC or X18+, or R18+ or MA15+ (without a restricted access system in place), ACMA must issue an interim service-cessation notice. The ACMA must then apply to the Classification Board to have the content classified, and depending upon the classification, withdraw the notice or issue a final service-cessation notice.[22]

Link-deletion notices

If ACMA is satisfied that 'end users in Australia can access content using a link provided by a links service', that 'the content is prohibited content and the links service has an Australian connection', and if the content has been classified RC or X18+, or R18+ or MA15+ (without a restricted access system in place), by the Classification Board, ACMA must give the hosting service provider a written final link-deletion notice.[23] Eligible electronic publications are exempt.

If the content is not classified and ACMA is satisfied that if the content were to be classified by the Classification Board there is a substantial likelihood that the content would be classified RC or X 18+, or R18+ or MA15+ (without a restricted access system in place), ACMA must issue an interim link-deletion notice. ACMA must then apply to the Classification Board to have the content classified, and depending upon the classification, withdraw the notice or issue a final link-deletion notice.[24]

20 *Broadcasting Services Act 1992* (Cth) Schedule 7, cl47(2).
21 *Broadcasting Services Act 1992* (Cth) Schedule 7, cl56 (1).
22 *Broadcasting Services Act 1992* (Cth) Schedule 7, cl56(2).
23 *Broadcasting Services Act 1992* (Cth) Schedule 7, cl62(1).
24 *Broadcasting Services Act 1992* (Cth) Schedule 7, cl62(2).

Industry codes

Schedule 7 encourages bodies and associations that represent sections of the internet industry to develop industry codes. A Code of Practice drafted by the Australian Internet Industry Association (IIA)[25] has been accepted as the legislative guideline for ISPs, content providers and internet users. While the Code is not mandatory, Schedule 7 makes provision for ACMA to direct an ISP or content host to comply with a registered code.

Complaints and investigations

Complaints regarding prohibited and potential prohibited content may be lodged with ACMA.[26] ACMA is required to act on all complaints unless satisfied that the complaint is frivolous, vexatious or not in good faith.[27] ACMA may also initiate investigations.

RC or X classifications include:

- material containing detailed instruction in crime, violence or drug use;
- child pornography;
- bestiality;
- excessively violent or sexually violent material; and
- real depictions of actual sexual activity.[28]

R classification includes:

- material containing excessive and/or strong violence or sexual violence;
- material containing implied or simulated sexual activity;
- material which deals with issues or contains depictions which require an adult perspective.[29]

Conclusion

Many content providers and hosts have little understanding of censorship law or of classification. The distinctions between classifications are intricate and difficult to ascertain, even for trained censors. This is one reason why the fee for classification is substantial and generally prohibitive. Content providers and hosts are not likely to be able to afford multiple applications. ACMA can request classification in response to a complaint, but not upon the request of a concerned content provider.

Some content providers and businesses use offshore service providers rather than risk dealing with Australian law and ACMA. The proposals in respect of offshore content regulation are unworkable.[30]

25 Available at www.iia.net.au. The current version of the Code of Practice is 10.4, but it is under review.
26 See generally *Broadcasting Services Act 1992* (Cth) Schedule 7, Part 3.
27 *Broadcasting Services Act 1992* (Cth) Schedule 7, cl43.
28 See the Classification Board's website: www.classification.gov.au.
29 Ibid.
30 See *Broadcasting Services Act 1992* (Cth) Schedule 5.

Given the global nature of the internet and online services, the scheme set up pursuant to Schedules 5 and 7 of the *Broadcasting Services Act 1992* (Cth) is largely a toothless tiger. Take-down notices, cessation-service notices and link-deletion notices will have an impact on Australian content, but most complaints made to ACMA are in relation to overseas content.[31] The result of decisions such as *Reno v ACLU* is that the lawmakers of nations such as the United States are powerless to enact controlling legislation. The international response is stalemated.

International comparison

UK

In *R v Westgarth Smith and Jayson*[32] the Court of Appeal held that merely browsing the internet or opening of email containing indecent photographs of children is an offence under the *Protection of Children Act 1978* (England and Wales).

EU

The EU's European Action Plan for Safe Use of the Internet is an attempt at a European policy. The January 1999 Paris conference on The Sexual Abuse of Children, Child Pornography and Paedophilia on the Internet was organised by UNESCO.[33] The site of the 1999 Vienna conference on Combating Child Pornography on the Internet points to a range of resources.[34]

Television

The regulation and enforcement of the telecommunications standards and online content provisions of the *Broadcasting Services Act 1992* (Cth) is the responsibility of ACMA.

Program content is governed largely by Codes of Practice developed by industry groups. ACMA registers codes once it is satisfied that broadcasters have undertaken appropriate public consultation and that the codes contain appropriate community safeguards. This minimises the need for the federal government to become entangled in the detail of content regulation. Instead, the television stations, under the watchful eye of ACMA, set the standard.

The Commercial Television Industry Code of Practice was developed by the commercial television industry and has been registered by ACMA. It came

31 See www.acma.gov.au.
32 [2002] EWCA Crim 683.
33 See unesdoc.unesco.org/images/0011/001147/114734eo.pdf.
34 See europa.eu/scadplus/leg/en/lvb/l33116.htm.

into effect in September 2007 and operates in conjunction with other ACMA standards regulating programs for children and the Australian content of programs and advertisements. Licensees are expected to comply fully with the code.

The code contains the Television Classification Guidelines, which set out the classification categories permitted on commercial free to air television and define the type of material that is suitable for each category, and covers:

- program, commercials, and program promotion classification;
- advertising time on television;
- accuracy, fairness and privacy in news and current affairs;
- disclosure of commercial arrangements in factual programming;
- loudness of advertisements; and
- complaints handling.[35]

ACMA's enforcement powers

The primary responsibility for ensuring that programs reflect community standards rests with the radio and television stations.[36] People who have complaints in this area need to first send them, in writing, to the station concerned. Complaints may be in relation to the terms of the Code of Practice or in relation to children's content, the proportion of Australian content, local content, sport (anti-siphoning) provisions, advertising and political content.[37] In stage two of what can be a five-stage process, if a station fails to answer such a complaint within 60 days, or the complainant considers the response unsatisfactory, the complainant may refer the complaint to ACMA.[38] ACMA acts as an independent adjudicator; it also accepts direct complaints.

Breaches of the codes are not necessarily breaches of the Act. In stage three, when a breach is found ACMA can request an informal undertaking from a licensee that certain steps will be taken to ensure no future breach of the relevant code provision. A wide range of possible undertakings can be requested. ACMA can impose an additional condition on a licence, such as corrective action. Compliance with the code could itself be made a licence condition.

Stage four arises when a station fails to comply with a licence condition. ACMA will then issue a notice directing compliance with the condition. Finally, at stage five, failure to comply with the direction notice can result in suspension or cancellation of the licence, or even referral to the Director of Public Prosecutions for prosecution and possible imposition of a fine by the Federal Court of Australia.

35 See www.acma.gov.au/WEB/STANDARD/1001/pc = PC_90096.
36 Commercial Television Industry Code of Practice, s1.1.1.
37 Commercial Television Industry Code of Practice, s7.2.
38 *Broadcasting Services Act 1992* (Cth) s148.

Television Classification Guidelines in practice

In March 2006, ACMA concluded a lengthy investigation of the 4 July 2005 episode of *Big Brother Uncut*. Big Brother is a reality television program broadcast by Network Ten. The program presents a group of housemates sharing a house for approximately three months, and competing for a cash prize. *Big Brother Uncut* is an MA version of the program shown late at night and contains content that would not be suitable for broadcast in other time periods.

ACMA investigated the episode after receiving two complaints from members of the public that had not been resolved by Network Ten. ACMA found that the episode breached the code, which requires that material be classified in accordance with the Television Classification Guidelines.

ACMA announced that the 4 July episode contained strong adult themes, combined with other classifiable elements such as sexual references, implied sexual behaviour and full frontal or partial nudity, with a cumulative intensity beyond that which could be justified by the storyline or program context.[39] ACMA found the episode as a whole was beyond the level of suitability for the MA15+ classification.

As a result, Network Ten provided ACMA with undertakings about the approach it would take in classifying the 2006 series of *Big Brother Uncut*. The undertakings were determined cooperatively between ACMA and Network Ten. ACMA warned that notwithstanding this 'refreshing course of action . . . it is imperative that broadcasters stay within the Code's limits'. The undertakings were intended to ensure that future programs did not contain material that exceeds the relevant classification criteria.

Radio broadcasting codes and breaches

Codes of practice

Codes of practice are developed under the *Broadcasting Services Act 1992* (Cth) by industry in consultation with ACMA. The Act states that codes may relate to, among other things: 'preventing the broadcasting of programs that, in accordance with community standards, are not suitable to be broadcast by that section of the industry'.[40]

In developing such codes, community attitudes to the portrayal in programs of matter that is likely to incite or perpetuate hatred against, or vilifies, any person or group on the basis of ethnicity, nationality, race, gender, sexual preference, age, religion or physical or mental disability are to be taken into account.[41]

39 See www.acma.gov.au/WEB/STANDARD/1001/pc = PC_100201.
40 *Broadcasting Services Act 1992* (Cth) s123(2)(a).
41 *Broadcasting Services Act 1992* (Cth) s123(2)(a).

Where a section of the industry develops its own code and ACMA is satisfied that it provides appropriate community safeguards, ACMA must include the code in the Register of Codes of Practice. The code must be endorsed by a majority of the broadcasting service providers in that section of the industry and the public must have been given an adequate opportunity to comment on the code.[42]

ACMA can investigate complaints about compliance with any code that is included in the Register. The Commercial Radio Code of Practice, for example, was reviewed late in 2008.[43]

Investigations

Under Part 11 of the *Broadcasting Services Act 1992* (Cth), ACMA investigates complaints relating to a possible breach by:

- a licensed broadcaster of the Act, the regulations, a licence condition, a class licence or a code of practice;[44] or
- the ABC or SBS of a code of practice.[45]

If a member of the public wishes to complain about something of concern they have heard on a program broadcast by a radio station, for instance, and the matter is covered by the Commercial Radio Code of Practice, the person must first make a written complaint to the station. However, if a complaint relates to a matter covered by a licence condition, it can be made directly to ACMA.

When making a complaint to ACMA, persons should provide a copy of their complaint to the station, a copy of the station's reply (if this has been received), and any other relevant correspondence with the station. ACMA states that it takes all complaints seriously, but it may rule that particular complaints are in fact frivolous, vexatious or not made in good faith. ACMA is required to acknowledge all complaints in writing.

For qualifying complaints (that is, those that are not outside jurisdiction, vexatious or frivolous), ACMA considers the information and gives the station an opportunity to respond. If it is a licence condition matter, no personal and private information provided in the complaint, such as name and address details, is disclosed to the licensee. This does not apply to code complaints, as these are made directly to the licensee in the first instance. When all relevant information is available, ACMA assesses the complaint. ACMA is required to notify the complainant of the results of its investigation.[46]

Under the Act, ACMA has discretion about whether or not to publish the report of an investigation. ACMA is not required to publish an investigation report if publication would disclose matter that is of a confidential character or that is

42 *Broadcasting Services Act 1992* (Cth) s123(2)(a).
43 The Commercial Radio Code of Practice is available at www.commercialradio.com.au.
44 *Broadcasting Services Act 1992* (Cth) s147.
45 *Broadcasting Services Act 1992* (Cth) s150.
46 *Broadcasting Services Act 1992* (Cth) Part 11.

likely to prejudice the fair trial of a person.[47] Where ACMA intends to publish an investigation report that may adversely affect the interests of a person, ACMA is required to give the person an opportunity to make representations in relation to the matter.[48] ACMA has usually not provided personal or private information in an investigation report.

ACMA has three program standards for commercial radio licensees. These commenced operation on 15 January 2001 and were extended indefinitely in March 2003. The standards are:

- Broadcasting Services (Commercial Radio Compliance Program) Standard 2000 – requires the establishment of compliance programs by licensees;
- Broadcasting Services (Commercial Radio Advertising) Standard 2000 – requires advertisements to be distinguished from other programs;
- Broadcasting Services (Commercial Radio Current Affairs Disclosure) Standard 2000 – requires the disclosure of commercial agreements by presenters of current affairs programs.[49]

Encouraging violence and brutality – the Alan Jones case

In 2007, ACMA found that prominent radio personality Alan Jones of 2GB Sydney had broadcast material likely to encourage violence or brutality and to vilify people of Lebanese and Middle-Eastern backgrounds on the basis of ethnicity. In doing so, both Alan Jones and the radio station had breached the Commercial Radio Code of Practice. This code applies to all Australian commercial radio stations.[50]

The complaint against Alan Jones and 2GB alleged breaches under subclause 1.3(a) and (e) of the code. Clause 1.3 provides:

> 1.3 A licensee must not broadcast a program which:
> (a) is likely to incite, encourage or present for its own sake violence or brutality;
> ...
> (b) is likely to incite or perpetuate hatred against or vilify any person or group on the basis of age, ethnicity, nationality, race, gender, sexual preference, religion or physical or mental disability.

Broadcasts may be made on these topics under 1.3(c)(ii), 1.3(d) or 1.3(e) if they are presented reasonably and in good faith for academic, artistic (including comedy or satire), religious instruction, scientific or research purposes or for other purposes in the public interest, including discussion or debate about any act or matter. (Clauses 1.5 to 1.8 deal with program content and language, including sex and sexual behaviour.)

47 *Broadcasting Services Act 1992* (Cth) s179(1).
48 *Broadcasting Services Act 1992* (Cth) s180.
49 See www.acma.gov.au/WEB/STANDARD//pc = PC_91766.
50 See internet.aca.gov.au/WEB/STANDARD/pc = PC_310133.

The ACMA report gives several quotes from the radio program over five days in December 2005 which contributed to its decision. The following is one, from 7 December 2005:

> AJ: My suggestion is to invite biker gangs to be present at Cronulla railway station when these Lebanese thugs arrive. The biker gangs have been much maligned but they do a lot of good things – it would be worth the price of admission to watch these cowards scurry back onto the train for the return trip to their lairs . . . and wouldn't it be brilliant if the whole event was captured on TV cameras and featured on the evening news so that we, their parents, family and friends can see who these bastards are . . . Australians old and new should not have to put up with this scum.

The licensee – John Singleton, the station's majority shareholder – argued that the material was presented in the public interest: discussion of factors contributing to unrest in the Cronulla area of southern Sydney in December 2005. However, ACMA was not persuaded that the relevant comments were presented reasonably and in good faith, and applied an 'ordinary listener test'.

During the investigation, Singleton submitted that ACMA's analysis of the code and findings revealed significant practical problems for commercial radio licensees, particularly talkback radio.

In its report ACMA stated that due to this and previous breaches, it would 'now move to pursue significantly heightened compliance measures in relation to the potential for future breaches of clause 1.3(e) by the licensee. ACMA will be writing to [the licensee] shortly about the proposed compliance action, details of which will be announced when finalised.'

At the conclusion of the investigation Singleton vehemently expressed dissent and frustration, focusing on what he described as the 'lack of recourse' available for a decision which would have enormous commercial impact. He said that if there was recourse, 'we will be taking [it], against either ACMA or their directors'.[51]

The licensee is permitted to appeal an ACMA decision in the Federal Court under the *Administrative Decisions Judicial Review Act 1977* (Cth), but only on matters of process. Singleton's last word on the issue was:

> [Y]ou cannot trash someone's individual reputation based on five letters from all around Australia, none of whom are from listeners, then ask Alan and 2GB to spend hundreds of thousands of dollars to defend these letters, which we did, and then make a finding, to which there is no charge, to which there is no answer, to which there is no appeal as far as I understand at this stage. Fair go, fair go.[52]

51 See cleaves.zapto.org/news/story-468.html?condense_comments = true&save_prefs = true.
52 For a list of disputes, see www.acma.gov.au/WEB/STANDARD//pc = PC_91716; for investigation report, see www.acma.gov.au/webwr/_assets/main/lib101068/2gb_%20report1485.pdf; for other materials and sources, see www.acma.gov.au; and see www.theaustralian.news.com.au/story/0,20867,21541268-7582,00.html.

Conclusion

ACMA plays a significant role in administering, investigating and reporting on many forms of electronic communications. In relation to mass forms of communications – television and radio – in Australia, its responsibility is substantial, but in many respects, the tools provided by the legislature are inadequate. The response to the *Big Brother* case above, for example, took almost 12 months. The subsequent season of the defaulting *Big Brother* series had ended by then. The 2GB radio example above similarly took so long that it permitted multiple breaches.

Perhaps more crucially, while the online content provisions of Australian legislation may reduce prohibited content that originates within Australia, they have no real impact on the amount of unclassified material reaching Australian users. This means that in the global context ACMA's effectiveness is somewhat limited. Cyberspace libertarians support this freedom of information flow, but the opposing view asks – at what cost?

19

An international perspective

International organisations, particularly trade organisations, were among the first to recognise the significance of the electronic commerce revolution and the corresponding need to provide rules and guidance. Several of these organisations' responses have become international practice or law.[1] The number and extent of organisations studying, reporting and implementing plans in relation to electronic commerce is astonishing. This chapter briefly examines some of the many electronic commerce policies implemented by selected international organisations.

As the methods and means of trading evolve in a changing technological and social environment, traders must meet the needs and demands of ever-developing communities. International trade has to deal with distance, varying modes of shipping, packaging, containerisation and time delay as well as different languages, customs, laws and infrastructure. Commercial parties must adjust to new modes of communication, and new industry practices. Banking practices, in particular, have become more sophisticated. In all fields, documents no longer have to be physically delivered and the requirement for written documentation to establish *bona fides* is diminishing.

Electronic commerce is by its nature 'international'. This means that the laws of one nation state will impact on those of others; that impact will vary according to each particular nation's place and importance in the world of international trade. Electronic commerce, as we have seen, raises questions regarding a number of issues: the security of transactions, standards, protection of intellectual property, taxation, trade law, privacy and many others.

1 For example the UNCITRAL Model Law on Electronic Commerce which provided the template for legislation in more than one hundred jurisdictions; see Chapter 3.

Trading, consumer and privacy issues have been addressed by a number of international organisations including APEC, the United Nations and the OECD. The UN Commission on International Trade Law (UNCITRAL) has developed one Recommendation,[2] one Treaty, and two Model Laws – the first for consumer protection in an electronic environment and the other for electronic signatures. These will both be discussed in detail below.

UN Commission on International Trade Law (UNCITRAL)

Established by the UN General Assembly in 1966, the commission is regarded as the core legal body of the UN system in the field of international trade law and the main vehicle by which the United Nations can perform an active role in the management of international trade. The general mandate of the commission is to further the progressive harmonisation and unification of international trade law and to remove unnecessary obstacles to international trade caused by inadequacies and divergence in national legislation affecting trade. The UNCITRAL Secretariat, located in Vienna, Austria, prepares legal texts in a number of key areas, such as international commercial dispute settlement, electronic commerce, insolvency, international payments, sale of goods, transport law, procurement and infrastructure development. UNCITRAL also provides technical assistance to law reform activities, including assisting member states to review and assess their law reform needs and to draft the legislation required to implement the UNCITRAL texts.

The commission has carried out work in eight different areas of trade law, including: international sale of goods and related transactions; international transport of goods; international payments (Legal Guide on Electronic Fund Transfers, Model Law on International Credit Transfers); international commercial arbitration and electronic commerce (Model Law). To develop the preparatory work on topics within the commission's program areas, UNCITRAL established three working groups, currently named the Working Group on International Contract Practices, the Working Group on Insolvency Law and the Working Group on Electronic Commerce.

UNCITRAL has released one Convention and two Model Laws that affect electronic commerce: the Model Law on Electronic Commerce and the Model Law on Electronic Signatures. The related Convention, the UN Convention on the Use of Electronic Communications in International Contracts, was not widely adopted and ultimately lapsed as a result of not being ratified by a sufficient number of member states.

2 UNCITRAL Recommendation on the Legal Value of Computer Records (1985).

The UNCITRAL Model Law on Electronic Commerce

The aim of the UNCITRAL Model Law on Electronic Commerce is to provide national legislatures with a template of internationally acceptable rules that will create a more secure legal environment for electronic commerce.[3] It is intended to facilitate the use of electronic communication and the storage of information, such as electronic data interchange and electronic mail. It provides standards to assess the legal value of electronic messages and legal rules for electronic commerce in specific areas such as carriage of goods. The decision by UNCITRAL to formulate model legislation on electronic commerce was a response to what appeared to be inadequate and outdated existing national legislation governing these areas. In some cases, such legislation imposed or implied restrictions on the use of modern means of communication by prescribing the use of 'written', 'signed' or 'original' documents.

It quickly became clear, however, that appropriate and relevant laws and practices are necessary in all countries where electronic data interchange and electronic mail are used, in order to facilitate access to international markets. The Model Law assists with this by providing a framework within which such national laws can be developed.

The UN General Assembly recommended:

> that all States give favourable consideration to the Model Law when they enact or revise their laws, in view of the need for uniformity of the law applicable to alternatives to paper-based methods of communication and storage of information; and that all efforts be made to ensure that the Model Law, together with the Guide, become generally known and available.[4]

The Model Law has gained significant international acceptance and has been the basis for electronic commerce legislation in more than 100 jurisdictions. It informs the discussion of electronic commerce laws in international *fora* and of domestic laws in many countries.

The UNCITRAL Model Law on Electronic Signatures[5]

In July 2001 UNCITRAL released its Model Law on Electronic Signatures. Like its predecessor, this Model Law provides national legislatures with a template. Its intention is to build on the principles underlying the signature provisions of the Model Law on Electronic Commerce.

The increased use of electronic authentication technologies as substitutes for handwritten signatures and other traditional authentication procedures has led to the need for a legal framework to deal with the issues that will almost certainly arise. In order to leverage the economic advantages of such technologies, business requires clarification on their use so as to reduce any uncertainty as to their

3 See Chapters 3 and 5.
4 UN General Assembly Resolution (A/RES/51/628), adopted 16 December 1996.
5 Available at the UNCITRAL website: www.uncitral.org/pdf/english/texts/electcom/ml-elecsig-e.pdf.

legal effect. In recognition of the risk that different legislative approaches would probably be taken in different countries, calls arose for a uniform set of basic rules. Legal harmony, as well as technical interoperability, was and remains the desired objective.

The Model Law on Electronic Signatures builds on the fundamental principles underlying Article 7 of the Model Law on Electronic Commerce – that there should be no discrimination on the basis of the media used in the transmission of a signature, and that an electronic signature should therefore be recognised as functionally equivalent to a handwritten signature.

It offers practical standards against which the technical reliability of electronic signatures may be measured, in order to add certainty and truly foster functional equivalence between electronic and 'original' signatures. In addition, it provides a link between such technical reliability and the legal effectiveness that may be expected from a given electronic signature.

The Model Law on Electronic Signatures adopts an approach under which the legal effectiveness of a given electronic signature technique may be pre-determined (or assessed prior to being actually used). It is thus intended to foster confidence in the fact that certain electronic signature techniques can be relied upon in legally significant transactions. By establishing a set of basic rules of conduct for the various parties that may become involved in the use of electronic signatures (that is, signatories, relying parties, and third-party certification service providers) that allows appropriate flexibility, the Model Law hopes to assist in shaping more harmonious commercial practices in cyberspace.

The objectives of the Model Law of Electronic Signatures, which include enabling or facilitating the use of electronic signatures and providing equal treatment to users of paper-based documentation and users of computer-based information, are essential for fostering economy and efficiency in international trade. A state that incorporates both the procedures prescribed in the Model Law of Electronic Signatures and the provisions of the Model Law on Electronic Commerce in national legislation will be creating a 'media-neutral' legal environment, in which the media involved in the transmission of a signature have no effect on the way the relevant law is interpreted.

This media-neutral approach is also used in the Model Law on Electronic Commerce; it is intended to provide legal guidelines for all situations where information is generated, stored or communicated, irrespective of the medium on which the information may be affixed.[6] The words 'a media-neutral environment', as used in that Model Law, reflect that principle of non-discrimination.

The Model Law of Electronic Signatures equally reflects the principle that there should be no discrimination among the various techniques used to communicate or store information electronically; this principle that is often referred to as 'technology neutrality'. One of its main features is that it adds certainty to the

6 See Guide to Enactment of the Model Law on Electronic Commerce, para 24.

operation of Article 7 of the Model Law on Electronic Commerce, mentioned above.

The UN Convention on the Use of Electronic Communications in International Contracts

Adopted by the General Assembly in 2005, the UN Convention on the Use of Electronic Communications in International Contracts aims to enhance legal certainty and commercial predictability where electronic communications are used in relation to international contracts. The Convention addresses the determination of: a party's location in an electronic environment; the time and place of dispatch and receipt of electronic communications; the use of automated message systems for contract formation; and the criteria to be used for establishing functional equivalence between electronic communications and paper documents, including 'original' paper documents, as well as between electronic authentication methods and handwritten signatures. The Convention complements and builds upon both the Model Law on Electronic Commerce and the Model Law on Electronic Signatures.

The Preamble to the Convention provides that signatories are:

> *Reaffirming* their belief that international trade on the basis of equality and mutual benefit is an important element in promoting friendly relations among States,
>
> *Noting* that the increased use of electronic communications improves the efficiency of commercial activities, enhances trade connections and allows new access opportunities for previously remote parties and markets, thus playing a fundamental role in promoting trade and economic development, both domestically and internationally,
>
> *Considering* that problems created by uncertainty as to the legal value of the use of electronic communications in international contracts constitute an obstacle to international trade,
>
> *Convinced* that the adoption of uniform rules to remove obstacles to the use of electronic communications in international contracts, including obstacles that might result from the operation of existing international trade law instruments, would enhance legal certainty and commercial predictability for international contracts and help States gain access to modern trade routes,
>
> *Being of the opinion* that uniform rules should respect the freedom of parties to choose appropriate media and technologies, taking account of the principles of technological neutrality and functional equivalence, to the extent that the means chosen by the parties comply with the purpose of the relevant rules of law,
>
> *Desiring* to provide a common solution to remove legal obstacles to the use of electronic communications in a manner acceptable to States with different legal, social and economic systems.

The Convention has many of the same shortcomings of the Electronic Transactions Acts and careful consideration of it should be undertaken before the Convention is acceded to. In April 2009 the Standing Committee of

Attorneys-General agreed to amend state and territory legislation to permit Australia to adopt the Convention.

UNCITRAL is responsible for continuously reviewing and explaining developments in internet technologies and for developing new law and legal frameworks for member states. As an organisation it works very hard to develop consensus on the most efficacious means by which to implement new technologies and on how to best develop procedures that will reduce risk for businesses seeking to leverage the huge competitive advantages inherent in the rapid transfer, collation and examination of enormous amounts of data. UNCITRAL will continue to provide advice and guidance on the use of hardware/software systems to transact business in cyberspace as they – and the internet itself – develop.

World Trade Organization (WTO)

The WTO has acted as a government of governments in trade-related matters. In July 2008 the WTO had 153 members. In March 1998 the WTO released a paper titled 'Electronic Commerce and the Role of the WTO'. The paper considers the benefits to international trade of the use of the internet and electronic forms of communication, and asserts that the internet, as an instrument for international trade, will fall under the WTO's General Agreement on Trade in Services (GATS). The stated goal of the WTO is not to create a set of new rules to govern the electronic marketplace but rather to use existing frameworks. The WTO agreements that deal with electronic commerce issues operate in conjunction with GATS, the Trade Related Aspects of Intellectual Property Rights (TRIPS) and the Agreement on Telecommunications Services. The electronic commerce provisions of the North American Free Trade Agreement (NAFTA) and other regional agreements have been based on the WTO principles.

The Interim Report on Electronic Commerce, produced by the WTO Council For Trade In Services, 1999, and the Work Program On Electronic Commerce, Interim Report to the General Council, Council for Trade in Services, 1999, reported:

> The electronic delivery of services falls within the scope of the GATS, since the Agreement applies to all services regardless of the means by which they are delivered . . . Measures affecting the electronic delivery of services are measures affecting trade in services and would therefore the covered by GATS obligations. All GATS provisions, whether relating to general obligations (e.g., MFN, transparency, domestic regulation, competition, payments and transfers, etc.) or specific commitments (market access, national treatment or additional commitments) are applicable to the supply of services through electronic means . . . There is still a need to clarify whether certain products delivered electronically might be classified as goods, and therefore subject to GATT disciplines, rather than as services.

The WTO is often stymied in its attempts to push a reformist agenda due to the *realpolitik* of international trade negotiations and its work in the internet milieu is no different. With strong regional and domestic considerations to take into account, member states struggle to find consensus on even small issues of trade so it is largely unsurprising that only limited progress has been made in the ongoing process of developing agreements on electronic commerce protocols. However the work continues and the future may yet see a comprehensive international trade agreement in the realm of electronic commerce.

General Agreement on Tariffs and Trade (GATT)

The GATT is a multilateral treaty that aims to promote trade among its members. Originally signed by 23 countries in 1947, the Uruguay round of the GATT, completed in 1994, now has 134 contracting parties, composed of both developing and developed states.

GATT entered into force on a temporary basis in 1948. It was expected to be superseded by the International Trade Organizations (see the ITO Havana Charter 1948). The ITO failed to find support when the US Senate did not ratify it in 1953. The principles of 'non-discrimination' and 'transparency', taken from the Havana Charter, are central to the GATT. One feature of the GATT that represents the non-discriminatory character of the treaty is the Most Favoured Nation clause, which provides that a party giving a trade advantage to one country is required to give that same advantage to all contracting parties. The trend in the international economy over time is for nation states to remove trade barriers, even though developing economies have largely tended to take a protectionist position in order to nurture their domestic producers. This is diminishing as general wealth increases globally.

In the electronic commerce domain, commentators have expressed the hope that nations will create regulations together to avoid the delays and problems associated with the implementation of the GATT principles. As mentioned above in relation to the WTO, the difficulties of developing sound policy are increased by the plethora of competing interests and contradictory objectives among the member states. Considerable difficulty has been experienced devising agreements that deal with tangibles; the intangible realm of the internet poses challenges that are unlikely to be resolved any more quickly.

The Organization for Economic Cooperation and Development (OECD)

The OECD comprises 30 member countries.[7] Its purpose is to provide a forum for its members to discuss matters of mutual interest and seek solutions to be applied within each member's own national context.

7 Australia, Austria, Belgium, Canada, the Czech Republic, Denmark, Finland, France, Germany, Greece, Hungary, Iceland, Ireland, Italy, Japan, Korea, Luxembourg, Mexico, The Netherlands, New Zealand, Norway, Poland, Portugal, the Slovak Republic, Spain, Sweden, Switzerland, Turkey, the United Kingdom and the United States.

For some 50 years, the OECD has been one of the world's largest and most reliable sources of comparable statistics, and economic and social data. The OECD monitors trends, analyses and forecasts economic developments and researches social changes and evolving patterns in trade, environment, agriculture, technology, taxation and privacy. Release of OECD statements often creates international headlines in the mainstream media and great respect is generally accorded its findings. Electronic commerce also features in the OECD's vision for economic growth and improved social conditions.

Electronic commerce is inherently trans-border, and OECD members met in 1998 to discuss a range of related issues, including how to build the trust of users and consumers in the use of technology, how to develop and implement policy on the collection of personal data and how to protect the privacy of individuals whose data is collected and stored.

The final resolution included a number of clauses to do with consumer protection issues, including: the use of secure encryption programs in the public and private sectors; full and fair disclosure of essential information; advertising online; complaints handling; and dispute resolution. As the leading international taxation research organisation, the OECD – along with other organisations, businesses and non-member countries – was also asked to develop a taxation framework applicable to electronic commerce. The OECD has implemented several Declarations resulting from these processes.

The Declaration on Protection of Privacy on Global Networks reaffirms privacy as a fundamental right and recognises the importance of the 1980 OECD Privacy Guidelines.[8] These Guidelines have provided many nation states with the foundation for protecting privacy and for the formation of appropriate laws. The Declaration encourages businesses to adopt policies and technical solutions in line with the Privacy Guidelines. The Declaration on Consumer Protection in the Context of Electronic Commerce underscores the importance, for the protection of consumers, of transparency in electronic commerce and its constituent transactions. The Declaration on Authentication for Electronic Commerce recognises the varying approaches taken by nation states to the formation of technology policy and media-specific requirements and espouses the principle that no discrimination should occur in relation to authentication approaches.

In 2008 the OECD issued the Seoul Declaration on the Future of the Internet Economy. The Declaration confirms the need for governments to work closely with business, civil society and technical experts on policies that promote competition, empower and protect consumers, and expand internet access and use worldwide. Under the Declaration the OECD also takes responsibility for assessing how its competition, consumer protection, privacy and security instruments are being applied by governments and business in relation to electronic commerce. The then OECD Secretary-General, Angel Gurría, stated:

> Given that this infrastructure has become critical to our economies and societies, we should all engage in developing better, more broad-based, governance arrangements

8 See Chapter 13.

> and policies . . . A more decentralised, networked approach to policy formulation for the Internet economy that includes the active participation of stakeholders.[9]

The Declaration sets out a roadmap to upgrade the communication policies that have helped the internet become the economic driver that it is today and seeks to ensure support for future development. It also commits the OECD to improving both the statistical indicators that measure access and use of the internet, and its networks, in order to provide reliable data and analysis.

The OECD's electronic commerce initiatives include:

- Implementing the OECD Privacy Guidelines in the electronic environment: focus on the internet (1997);
- Ministerial Declaration on Consumer Protection in the Context of Electronic (1998);
- Electronic Commerce: Existing Commitments for Online Supply of Services (1999);
- International and Regional Bodies: Activities and Initiatives in E-Commerce (2001);
- E-commerce: responding to challenges and opportunities (2001);
- OECD Cross-Border Fraud Guidelines (2003);
- Mobile Commerce (2007);
- Money Laundering & Terrorist Financing Vulnerabilities of Commercial Websites and Internet Payment Systems (2008); and
- Recommendation of the OECD Council Concerning Guidelines for Consumer Protection in the Context of Electronic Commerce (2008).[10]

The OECD is a global player in the ongoing development and use of the internet, and there is little doubt that it will continue to be so. Both governments and the marketplace have developed a high degree of confidence in its findings.

The Asia Pacific Economic Cooperation (APEC)

APEC was established in 1989 and has 21 members, including all the major economies of the Asia Pacific region. Begun as an informal dialogue group, APEC has become the primary regional vehicle for promoting open trade and practical economic cooperation. Its goal is to advance Asia-Pacific economic dynamism and sense of community. The 21 APEC members are referred to as 'Member Economies' and account for approximately 41 per cent of the world's population, approximately 55 per cent of world GDP and about 49 per cent of world trade.[11]

9 Closing Remarks by the OECD Secretary-General Angel Gurría at the meeting of the Future of the Internet Economy, available at www.oecd.org.

10 These and other OECD papers are available at www.oecd.org.

11 Notes from the website, www.apec.org. The members are: Australia, Brunei Darussalam, Canada, Chile, People's Republic of China, Hong Kong China, Indonesia, Japan, Republic of Korea, Malaysia, Mexico, New Zealand, Papua New Guinea, Peru, The Philippines, Russia, Singapore, Chinese Taipei, Thailand, the United States and Vietnam.

APEC has advanced a comprehensive set of initiatives to help its members economies leverage developments in electronic commerce. In 2000 it released the Action Agenda for the New Economy, which was intended to promote the 'right policy environment' and provide a framework to strengthen markets, electronic commerce, infrastructure, and knowledge and skills development for its members. It also aimed to work towards providing 'affordable quality access to telecommunications services and the internet' to communities in all APEC member countries.[12]

The e-APEC Strategy was endorsed in 2001. It is a long-term and action-oriented plan that has three pillars:

- to create an environment for strengthening market structures and institutions;
- to facilitate an environment for infrastructure investment and technology development;
- to enhance human capacity building and promote entrepreneurship.[13]

APEC's 2004 report 'Implementing the e-APEC Strategy – Progress and Recommendations for Further Action'[14] found that member economies had made significant improvements in e-infrastructure, broadband, e-government services and mobile services. It further found that increased competition and market-oriented policies had helped innovation and investment in internet-related infrastructure, research and development. The report stated that the e-APEC Strategy continues to grow in importance and exhorted member economies, especially those at earlier stages of implementation, to continue building skills and capacity.[15]

The following activities and projects implement the e-APEC Strategy.

The Electronic Commerce Steering Group developed a Data Privacy Pathfinder initiative to enable stakeholders to work together to protect private information in the APEC region and build confidence and trust in electronic commerce. Pathfinder supports business needs, reduces compliance costs, provides consumers with effective remedies, allows regulators to operate efficiently, and minimises regulatory burdens.

The Committee on Trade and Investment prepared APEC's Second Trade Facilitation Action Plan. The plan sets out a framework and timetable for achieving a 5 per cent reduction in trade transaction costs. It focuses on customs procedures, standards and conformance, electronic commerce and the mobility of business people.

The APEC Sub-Committee on Customs Procedures adopted the Single Window Strategic Plan and the Single Window Development Plan in 2007. The Strategic Plan provides a framework for the development of Single Window

12 See e-APEC Strategy: www.apec.org/apec/leaders__declarations/2001/appendix_2_eAPEC_strategy.html.
13 Ibid.
14 Available at: www.apec.org/apec/apec_groups/other_apec_groups/new_economy_-_e-apec.html.
15 See e-APEC Strategy: www.apec.org/apec/leaders__declarations/2001/appendix_2_eAPEC_strategy.html.

systems (uniform procedures among members aimed at achieving paperless trading) and enable seamless data sharing. It contains six recommendations and provides mechanisms for APEC members to work collaboratively.

The APEC Telecommunications and Information Working Group aims to expand collaboration in dealing with issues of electronic security, to build confidence in the use of electronic networks, and to support policy and regulatory reforms that facilitate competition and the expanded reach of networks.

The Electronic Commerce Steering Group continues relevant work to implement APEC's Strategies and Actions Toward a Cross-Border Paperless Trading Environment to enable the electronic transmission of trade-related information across the region by 2020.

Summary

APEC is well placed to assist with the development of cooperative activity among its member states. The range of powerful, wealthy economies it represents are highly motivated to find ways to trade at a constant or diminishing cost, and therefore support the development of shared policy on electronic commerce procedures. APEC deals with the highest levels of government, which is an advantage that many other organisations do not share. Almost certainly, as policy in the electronic commerce arena develops, APEC will be at the forefront.

International Chamber of Commerce (ICC)

Established in 1919, the ICC, which is based in Paris, was intended to act as a bulwark against the rising tide of nationalist fervour and protectionism that developed in the international political arena after the cessation of hostilities in Europe in 1918.[16] This aim has essentially remained unchanged over the last 90 years. The ICC's operational objectives are to serve world business by:

(1) promoting trade and investment;
(2) opening markets for goods and services, and
(3) supporting the free flow of capital.[17] (The ICC is also a powerful advocate of self-regulation in the marketplace.[18])

Among its many accomplishments has been the establishment of several agencies, including: the International Court of Arbitration (1923) for resolving cross-border trade disputes;[19] the International Maritime Bureau; and the ICC Commercial Crime Bureau.[20] The ICC also publishes widely on all manner of

16 www.iccwbo.org/id93/index.html.
17 www.iccwbo.org/id93/index.html.
18 www.iccwbo.org/policy/banking/iccjjdi/index.html.
19 www.iccwbo.org/court.
20 Charles del Busto, (1994) *ICC Guide to Documentary Credit Operations for the UCP 500*, 112.

business-related topics, makes representations to the United Nations and to national governments across the globe, and works to further the interests of trade generally.[21]

The ICC develops and publishes rules that govern the conduct of business across international borders and provides many essential services to the international business community. It has members from 63 national committees and 7000 member companies and associations from over 130 countries. It presents views to governments and works with its members to address a plethora of issues, including those related to electronic commerce. Two of the proposed guides and model laws proposed by the ICC in this area are the GUIDEC (General Usage for International Digitally Ensured Commerce) and the ICC Model Clause for Use in Contracts Involving Transborder Data Flows.

The GUIDEC, issued in 1997, is a general framework for 'ensuring' and 'certification' of digital messages, based upon existing civil and common law and pertinent international principles. It governs the use of public key cryptography for digital signatures and the role of a trusted third party, called a certifier, who issues validation certificates – which the receiving parties can rely upon – that establish that holders of public keys are who they purport to be. These certificates also verify that the ensuring party has the ability to contract and bind either himself or the party they are authorised to represent.

The ICC has indicated that these guidelines will be updated as digital technology develops; it has also issued revised guidelines for advertising and marketing on the internet.

The ICC, as an organisation of practitioners, professionals and corporations, continues to be one of the most important international *fora* in which the practicalities of implementing and using international electronic commerce law are discussed and debated. It also works to ensure that governments and intergovernmental organisations understand that regulations and rules impacting on its members must be both cost-effective and efficacious. As a lobby group, the ICC has significant worldwide influence on lawmakers and advisors to all levels of government.

International Organization for Standardization (ISO)

The ISO is a worldwide federation of national standards bodies. Its mission is to promote the development of standardisation with a view to facilitating the international exchange of goods and services, and to develop cooperation in the sphere of intellectual, scientific, technological and economic activity. The ISO covers all fields except electrical and electronic engineering, which is the

21 See Garth Wooler, *Legal and practice perspectives on letters of credit under UCP600*, Wooler, Brisbane, 2007.

responsibility of the International Electrotechnical Commission (IEC). A joint ISO/IEC committee (JTC 1) carries out work in the information technology field. The technical work of ISO is highly decentralised, in line with the very international nature of its membership – there are over 2700 technical committees, subcommittees and working groups under its auspices. In these committees, representatives of industry, research institutes, government authorities, consumer bodies and international organisations from around the world come together to resolve problems of global standardisation.

In the domain of electronic commerce the ISO has released standards for the management of electronic banking,[22] and Electronic Data Interchange for Administration, Commerce and Transport (EDIFACT).[23]

International Labour Organization (ILO)

The ILO is a Geneva-based specialised agency of the United Nations founded in 1919 and built on the principle that universal and lasting peace can be established only if it is based on social justice. It provides the international institutional framework for the formulation of policies and programs to promote basic human rights, improve working and living conditions and enhance employment opportunities. This involves the creation of international labour standards backed by a unique system to supervise their application, and an extensive program of international technical cooperation, training, education, research and publishing activities. The ILO is unique among world organisations in that employers' and workers' representatives – the 'social partners' of the economy – have an equal voice with government in shaping its policies and programs.

The annual International Labour Conference provides an international forum for discussion of world labour and social problems and sets minimum international labour standards and the broad policies of the organisation. Every two years it also adopts the ILO's biennial work program and budget (it is financed by member states). The day-to-day work of the ILO is guided by its governing body, which is made up of 28 government members, 14 worker members and 14 employer members. The ILO's secretariat, operational headquarters, research centre and publishing house are in Geneva; administration and management are decentralised in offices in more than 40 countries.

The ILO's general role in terms of electronic commerce is to analyse and monitor the impact of information and communication technologies (ICT) on employment (loss and creation of jobs), enterprise development, work organisation, working time arrangements, working conditions and industrial relations. In this context, key issues are telework, protection of workers' personal data and

22 ISO 9564-1: 2002.
23 ISO 9735: 2002.

protection of the rights of categories of workers particularly affected by ICT, and the role of such technologies in globalisation and the restructuring of national economies.[24]

International Telecommunication Union (ITU)

The ITU provides a focal point for public and private sectors to cooperate in the development and standardisation of telecommunications. With 191 member states and more than 700 affiliated organisations and individuals, it has emerged as the leading UN forum for consideration of matters associated with radiocommunication and other technologies. The ITU also organises the World Summit on the Information Society. In addition, it establishes the international regulations and treaties governing all terrestrial and space uses of the frequency spectrum within which countries set their national legislation.

In relation to electronic commerce it develops standards to facilitate the interconnection of telecommunication systems on a worldwide scale regardless of the type of technology used. The ITU also fosters the expansion of telecommunications services and infrastructure in developing countries by recommending medium-term policies and strategies to member nations' administrations.

UN Centre for Trade Facilitation and Electronic Business (UN/CEFACT)

The UN Economic Commission for Europe's (UN/ECE) activities in trade facilitation have global scope, and to meet the requirements of these ever widening responsibilities the agency was restructured in 1997. Seeking to 'secure the interoperability for the exchange of information between the public and private sector',[25] the ECE formed UN/CEFACT. This organisation enables participants from all over the world to work together, on an equal footing, to improve business processes and to ensure the effective transfer of trade information.

UN/CEFACT provides a forum for institutional cooperation in formulating and recommending international trade facilitation strategies and for reconciling official governmental and commercial requirements. It encourages global participation in its work and takes a thorough approach to the technical and policy areas of trade facilitation. The participation of many private sector associations in CEFACT's work at the policy level, and of hundreds of private sector technical experts in CEFACT working groups, is forging new cooperative relationships between private and public organisations.

24 See, for example, www.ilo.org/public/english/bureau/inst/papers/2000/dp123/index.htm.
25 www.unece.org/cefact/about.htm.

CEFACT 'supports activities dedicated to improving the ability of business, trade and administrative organizations, from developed, developing and transitional economies, to exchange products and relevant services effectively'.

UN Conference on Trade and Development (UNCTAD)

UNCTAD was established in 1964 and is the principal organ of the UN General Assembly in the field of trade and development. Its main goals are to maximise the trade, investment and development opportunities of developing countries, and to help them face the challenges arising from globalisation and integration into the world economy. In 1992, UNCTAD launched its Trade Efficiency Initiative. The main objective of the initiative is to help developing countries and their SMEs integrate participate in international trade. This means simplifying and harmonising trade procedures worldwide and providing traders or potential traders with access to information networks and better business practices.

UNCTAD's annual Information Economy Report focuses on national and international policy and strategy options for developing economies. The E-Commerce and Development Report, also issued annually, focuses on the trends in ICT, and the impact of these technologies on the global economy.[26] UNCTAD also monitors and records a significant amount of raw data on international trade and regularly releases statistical reports on a wide range of matters, including e-tourism[27] and the information economy.[28]

UN Educational, Scientific and Cultural Organisation (UNESCO)

UNESCO's constitution was adopted by the London Conference in November 1945, and entered into effect in November 1946. The organisation's 'main objective is to contribute to peace and security in the world by promoting collaboration among nations through education, science, culture and communication' in order to further universal respect for: justice; the rule of law, and human rights.[29]

UNESCO deals with a very wide range of matters and the realm of electronic commerce takes a high profile in many of them. Access to high-quality, value-added education, rapid transfer of scientific development, and the cultural impact of technology on indigenous peoples are only some of the issues where electronic commerce and technology have already had a significant impact.

26 www.unctad.org/Templates/StartPage.asp?intItemID = 2629&lang = 1.
27 www.unctadxi.org/templates/Startpage____1195.aspx.
28 www.unctad.org/Templates/webflyer.asp?docid = 9479&intItemID = 2096&lang = 1.
29 See www.unesco.org.

The UNESCO Observatory on the Information Society[30] was developed between 1997 and 1999 to monitor and report on the international developments – ethical, legal and societal – in the evolution of a variety of technologies, including electronic commerce systems. The observatory's original mandate included assessing access to information in the public domain, shifts in electronic commerce behaviours, privacy and confidentiality issues, and cyber-violence. It has since extended to include capacity building, infrastructure issues and more.

The UNESCO Observatory has become a gateway to many online resources and the originator of many policies on the management and implementation of internet technology worldwide.

Universal Postal Union (UPU)

The UPU was founded in 1874 and was brought into relationship with the United Nations in 1948. Based in Berne, Switzerland, the UPU now has 191 member countries. As a Specialised Agency of the United Nations, the UPU aims to organise and improve postal services throughout the world and to ensure international collaboration in this area. Among its aims, as set out in the Universal Postal Convention and the General Regulations, are the formation of a single territory by all signatory nations for the purposes of postal communication and uniformity of postal rates and units of weight.

Few areas of commerce have been as affected by the rapid development of electronic data management as the postal services industry. Powerful advances in barcode-reading technologies and satellite tracking systems have enabled postal organisations to implement management systems that increase efficiencies exponentially.

The UPU monitors developments in electronic commerce from within its Postal Technology Centre. This agency advises the UPU's members on developments 'in the field of information systems and electronic communications, particularly those involving products and services based on electronic data interchange'.[31]

World Bank

The International Bank for Reconstruction and Development, frequently called the World Bank, was established in 1944 at the UN Monetary and Financial Conference. The World Bank's goal is to reduce poverty and improve living standards by promoting sustainable growth and investment in people. It does

30 www.unesco-ci.org/cgi-bin/portals/information-society/page.cgi?d = 1.
31 www.upu.int/faq/en/upu_activities/how_is_the_postal_sector_responding_to_technological_developments.html.

this by providing loans, technical assistance and policy guidance to developing country members. The World Bank Group is made up of five organisations: the International Bank for Reconstruction and Development (IBRD); the International Development Association (IDA); the International Finance Corporation (IFC); the Multilateral Investment Guarantee Agency (MIGA); and the International Centre for Settlement of Investment Disputes (ICSID). These agencies raise most of their money on the world's financial markets through selling public or private bonds and other debt securities to pension funds, insurance companies, corporations and other banks, and individuals.

Many of the World Bank's programs relate to the expansion and implementation of electronic infrastructure that will help countries become part of the global internet commerce system. The Information for Development Program (InfoDev) is a global program managed by the World Bank which assists developing countries and economies in transition. It takes advantage of the opportunities the information revolution offers for accelerating social and economic development by providing funding for field-testing a range of new development ideas. To be selected for this funding, programs must: promote market-based solutions to development problems; improve education and health; reduce poverty and exclusion of low-income countries and social groups; promote the protection of the environment; and increase the efficiency, accountability and transparency of governments.

World Customs Organisation (WCO)

The WCO is an independent intergovernmental body with worldwide membership whose mission is to enhance the effectiveness and efficiency of customs administration. It establishes, maintains, supports and promotes international instruments for the harmonisation and uniform application of simplified and effective customs systems and procedures governing the movement of commodities, people and conveyances across customs frontiers.

The WCO reinforces members' efforts to secure (through control and enforcement) compliance with national legislation in order to maximise the effectiveness of members' cooperation with each other and with international agencies. It also tries to help members meet the challenges of adapting to changing circumstances, by promoting communication and cooperation among members and with other international organisations, and by fostering human resource development, improvement in the management and working methods of customs administration, and the sharing of best practices.

World Intellectual Property Organisation (WIPO)

WIPO is an intergovernmental organisation responsible for the protection of intellectual property throughout the world. WIPO administers a large number of multilateral treaties dealing with the legal and administrative aspects of intellectual property, develops new international treaties and has an extensive program of cooperation under which technical assistance is extended to developing countries.

WIPO also provides services to the private sector, under international agreements: simplified and cost-effective means of obtaining international protection for patents, trademarks and industrial designs, and dispute-resolution services for parties involved in international disputes concerning intellectual property. The fees generated by WIPO through these services account for about 88 per cent of its budget.

It can be argued that there is no greater issue in the world of electronic commerce and the internet generally than respect for intellectual property rights. Concerned organisations and individuals as diverse as Microsoft Corporation and the rock band U2 continue to express deep concern about IP theft, especially where the theft appears to be propagated, or at best tolerated, by the government agencies of nation states. WIPO engages with parties at all levels to mitigate the effect of such practices.

Appendix A

Electronic Transactions (Victoria) Act 2000

No. 20 of 2000
Version as at 19 October 2007

Part 1—Preliminary

1 Purposes

The purposes of this Act are—

(a) to recognise that transactions effected electronically are not by that reason alone invalid;

(b) to provide for the meeting of certain legal requirements as to writing and signatures by electronic communication;

(c) to permit documents to be produced to another person by electronic communication;

(d) to permit the recording and retention of information and documents in electronic form;

(e) to provide for the determination of time and place of dispatch and receipt of electronic communications;

(f) to stipulate when an electronic communication will bind its purported originator.

2 Commencement

This Act comes into operation on 1 September 2000.

3 Definitions

(1) In this Act—

consent includes consent that can reasonably be inferred from the conduct of the person concerned, but does not include consent given subject to conditions unless the conditions are complied with;

data includes the whole or part of a computer program within the meaning of the *Copyright Act 1968* of the Commonwealth;

data storage device means any article or material (for example, a disk) from which information is capable of being reproduced, with or without the aid of any other article or device;

electronic communication means—
(a) a communication of information in the form of data, text or images by means of guided or unguided electromagnetic energy, or both; or
(b) a communication of information in the form of sound by means of guided or unguided electromagnetic energy, or both, where the sound is processed at its destination by an automated voice recognition system;

information means information in the form of data, text, images or sound;

information system means a system for generating, sending, receiving, storing or otherwise processing electronic communications;

information technology requirements includes software requirements;

law of this jurisdiction means any law in force in this jurisdiction, whether written or unwritten, but does not include a law of the Commonwealth;

non-profit body means a body that is not carried on for the purposes of profit or gain to its individual members and is, by the terms of the body's constitution, prohibited from making any distribution, whether in money, property or otherwise, to its members;

place of business, in relation to a government, an authority of a government or a non-profit body, means a place where any operations or activities are carried out by that government, authority or body;

this jurisdiction means Victoria;

transaction includes any transaction in the nature of a contract, agreement or other arrangement, and also includes any transaction of a non-commercial nature.

(2) Notes do not form part of this Act.

4 Object

The object of this Act is to provide a regulatory framework that—
(a) recognises the importance of the information economy to the future economic and social prosperity of Australia; and
(b) facilitates the use of electronic transactions; and
(c) promotes business and community confidence in the use of electronic transactions; and
(d) enables business and the community to use electronic communications in their dealings with government.

5 Outline of Act

(1) The following is an outline of this Act—
(a) for the purposes of a law of this jurisdiction, a transaction is not invalid because it took place by means of one or more electronic communications;

(b) the following requirements imposed under a law of this jurisdiction can generally be met in electronic form—
(i) a requirement to give information in writing;
(ii) a requirement to provide a signature;
(iii) a requirement to produce a document;
(iv) a requirement to record information;
(v) a requirement to retain a document;

(c) for the purposes of a law of this jurisdiction, provision is made for determining the time and place of the dispatch and receipt of an electronic communication;

(d) the purported originator of an electronic communication is bound by it for the purposes of a law of this jurisdiction only if the communication was sent by the purported originator or with the authority of the purported originator.

(2) Subsection (1) is intended only as a guide to the general scheme and effect of this Act.

6 Crown to be bound

This Act binds the Crown in right of Victoria and, in so far as the legislative power of Parliament permits, the Crown in all its other capacities.

Part 2—Application of Legal Requirements to Electronic Communications

Division 1—General rule about validity of transactions for the purposes of laws of this jurisdiction

7 Validity of electronic transactions

(1) For the purposes of a law of this jurisdiction, a transaction is not invalid because it took place wholly or partly by means of one or more electronic communications.

(2) The general rule in subsection (1) does not apply in relation to the validity of a transaction to the extent to which another, more specific, provision of this Part deals with the validity of the transaction.

(3) The regulations may provide that subsection (1) does not apply to a specified transaction or specified class of transactions.

(4) The regulations may provide that subsection (1) does not apply to a specified law of this jurisdiction.

Division 2—Requirements under laws of this jurisdiction

8 Writing

(1) If, by or under a law of this jurisdiction, a person is required to give information in writing, that requirement is taken to have been met if the person gives the information by means of an electronic communication, where—

(a) at the time the information was given, it was reasonable to expect that the information would be readily accessible so as to be useable for subsequent reference; and

(b) the person to whom the information is required to be given consents to the information being given by means of an electronic communication.

(2) If, by or under a law of this jurisdiction, a person is permitted to give information in writing, the person may give the information by means of an electronic communication, where—

(a) at the time the information was given, it was reasonable to expect that the information would be readily accessible so as to be useable for subsequent reference; and

(b) the person to whom the information is permitted to be given consents to the information being given by means of an electronic communication.

(3) This section does not affect the operation of any other law of this jurisdiction that makes provision for or in relation to requiring or permitting information to be given, in accordance with particular information technology requirements—

(a) on a particular kind of data storage device; or

(b) by means of a particular kind of electronic communication.

(4) This section applies to a requirement or permission to give information, whether the expression 'give', 'send' or 'serve', or any other expression, is used.

(5) For the purposes of this section, *giving information* includes, but is not limited to, the following—

(a) making an application;

(b) making or lodging a claim;

(c) giving, sending or serving a notification;

(d) lodging a return;

(e) making a request;

(f) making a declaration;

(g) lodging or issuing a certificate;

(h) making, varying or cancelling an election;

(i) lodging an objection;

(j) giving a statement of reasons.

Note

Section 12 provides for exemptions from this section.

9 *Signatures*

(1) If, by or under a law of this jurisdiction, the signature of a person is required, that requirement is taken to have been met in relation to an electronic communication if—

(a) a method is used to identify the person and to indicate the person's approval of the information communicated; and

(b) having regard to all the relevant circumstances at the time the method was used, the method was as reliable as was appropriate for the purposes for which the information was communicated; and

(c) the person to whom the signature is required to be given consents to that requirement being met by way of the use of the method mentioned in paragraph (a).

(2) This section does not affect the operation of any other law of this jurisdiction that makes provision for or in relation to requiring—

(a) an electronic communication to contain an electronic signature (however described); or

(b) an electronic communication to contain a unique identification in an electronic form; or

(c) a particular method to be used in relation to an electronic communication to identify the originator of the communication and to indicate the originator's approval of the information communicated.

Note

Section 12 provides for exemptions from this section.

10 Production of document

(1) If, by or under a law of this jurisdiction, a person is required to produce a document that is in the form of paper, an article or other material, that requirement is taken to have been met if the person produces, by means of an electronic communication, an electronic form of the document, where—

(a) having regard to all the relevant circumstances at the time the communication was sent, the method of generating the electronic form of the document provided a reliable means of assuring the maintenance of the integrity of the information contained in the document; and

(b) at the time the communication was sent, it was reasonable to expect that the information contained in the electronic form of the document would be readily accessible so as to be useable for subsequent reference; and

(c) the person to whom the document is required to be produced consents to the production, by means of an electronic communication, of an electronic form of the document.

(2) If, by or under a law of this jurisdiction, a person is permitted to produce a document that is in the form of paper, an article or other material, then, instead of producing the document in that form, the person may produce, by means of an electronic communication, an electronic form of the document, where—

(a) having regard to all the relevant circumstances at the time the communication was sent, the method of generating the electronic form of the document provided a reliable means of assuring the maintenance of the integrity of the information contained in the document; and

(b) at the time the communication was sent, it was reasonable to expect that the information contained in the electronic form of the document would be readily accessible so as to be useable for subsequent reference; and

(c) the person to whom the document is permitted to be produced consents to the production, by means of an electronic communication, of an electronic form of the document.

(3) For the purposes of this section, the integrity of information contained in a document is maintained if, and only if, the information has remained complete and unaltered, apart from—

(a) the addition of any endorsement; or

(b) any immaterial change—

which arises in the normal course of communication, storage or display.

(4) This section does not affect the operation of any other law of this jurisdiction that makes provision for or in relation to requiring or permitting electronic forms of documents to be produced, in accordance with particular information technology requirements—

(a) on a particular kind of data storage device; or

(b) by means of a particular kind of electronic communication.

Note

Section 12 provides for exemption from this section.

11 Retention of information and documents

(1) If, by or under a law of this jurisdiction, a person is required to record information in writing, that requirement is taken to have been met if the person records the information in electronic form, where—

(a) at the time of the recording of the information, it was reasonable to expect that the information would be readily accessible so as to be useable for subsequent reference; and

(b) if the regulations require that the information be recorded on a particular kind of data storage device, that requirement has been met.

(2) If, by or under a law of this jurisdiction, a person is required to retain, for a particular period, a document that is in the form of paper, an article or other material, that requirement is taken to have been met if the person retains, or causes another person to retain, an electronic form of the document throughout that period, where—

(a) having regard to all the relevant circumstances at the time of the generation of the electronic form of the document, the method of generating the electronic form of the document provided a reliable means of assuring the maintenance of the integrity of the information contained in the document; and

(b) at the time of the generation of the electronic form of the document, it was reasonable to expect that the information contained in the electronic form of the document would be readily accessible so as to be useable for subsequent reference; and

(c) if the regulations require that the electronic form of the document be retained on a particular kind of data storage device, that requirement has been met throughout that period.

(3) For the purposes of subsection (2), the integrity of information contained in a document is maintained if, and only if, the information has remained complete and unaltered, apart from—

(a) the addition of any endorsement; or

(b) any immaterial change—

which arises in the normal course of communication, storage or display.

(4) If, by or under a law of this jurisdiction, a person (**the first person**) is required to retain, for a particular period, information that was the subject of an electronic communication, that requirement is taken to have been met if the first person retains, or causes another person to retain, in electronic form, the information throughout that period, where—

(a) at the time of commencement of the retention of the information, it was reasonable to expect that the information would be readily accessible so as to be useable for subsequent reference; and

(b) having regard to all the relevant circumstances at the time of commencement of the retention of the information, the method of retaining the information in electronic form provided a reliable means of assuring the maintenance of the integrity of the information contained in the electronic communication; and

(c) throughout that period, the first person also retains, or causes the other person to retain, in electronic form, such additional information obtained by the first person as is sufficient to enable the identification of the following—

(i) the origin of the electronic communication;

(ii) the destination of the electronic communication;

(iii) the time when the electronic communication was sent;

(iv) the time when the electronic communication was received; and

(d) at the time of commencement of the retention of the additional information covered by paragraph (c), it was reasonable to expect that the additional information would be readily accessible so as to be useable for subsequent reference; and

(e) if the regulations require that the information be retained on a particular kind of data storage device, that requirement has been met throughout that period.

(5) For the purposes of subsection (4), the integrity of information that was the subject of an electronic communication is maintained if, and only if, the information has remained complete and unaltered, apart from—

(a) the addition of any endorsement; or

(b) any immaterial change—

which arises in the normal course of communication, storage or display.

Note

Section 12 provides for exemption from this section.

12 Exemptions from this Division

(1) The regulations may provide that this Division, or a specified provision of this Division, does not apply to a specified requirement or specified class of requirements.

(2) The regulations may provide that this Division, or a specified provision of this Division, does not apply to a specified permission or specified class of permissions.

(3) The regulations may provide that this Division, or a specified provision of this Division, does not apply to a specified law of this jurisdiction.

Division 3—Other provisions relating to laws of this jurisdiction

13 Time and place of dispatch and receipt of electronic communications

(1) For the purposes of a law of this jurisdiction, if an electronic communication enters a single information system outside the control of the originator, then, unless otherwise agreed between the originator and the addressee of the electronic communication, the dispatch of the electronic communication occurs when it enters that information system.

(2) For the purposes of a law of this jurisdiction, if an electronic communication enters successively 2 or more information systems outside the control of the originator, then, unless otherwise agreed between the originator and the addressee of the electronic communication, the dispatch of the electronic communication occurs when it enters the first of those information systems.

(3) For the purposes of a law of this jurisdiction, if the addressee of an electronic communication has designated an information system for the purpose of receiving electronic communications, then, unless otherwise agreed between the originator and the addressee of the electronic communication, the time of receipt of the electronic communication is the time when the electronic communication enters that information system.

(4) For the purposes of a law of this jurisdiction, if the addressee of an electronic communication has not designated an information system for the purpose of receiving electronic communications, then, unless otherwise agreed between the originator and the addressee of the electronic communication, the time of receipt of the electronic communication is the time when the electronic communication comes to the attention of the addressee.

(5) For the purposes of a law of this jurisdiction, unless otherwise agreed between the originator and the addressee of an electronic communication—

(a) the electronic communication is taken to have been dispatched from the originator's place of business; and

(b) the electronic communication is taken to have been received at the addressee's place of business.

(6) For the purposes of the application of subsection (5) to an electronic communication—

(a) if the originator or addressee has more than one place of business, and one of those places has a closer relationship to the underlying transaction, it is to be assumed that that place of business is the originator's or addressee's only place of business; and

(b) if the originator or addressee has more than one place of business, but paragraph (a) does not apply, it is to be assumed that the originator's or addressee's principal place of business is the originator's or addressee's only place of business; and

(c) if the originator or addressee does not have a place of business, it is to be assumed that the originator's or addressee's place of business is the place where the originator or addressee ordinarily resides.

(7) The regulations may provide that this section does not apply to a specified electronic communication or specified class of electronic communications.

(8) The regulations may provide that this section does not apply to a specified law of this jurisdiction.

14 Attribution of electronic communications

(1) For the purposes of a law of this jurisdiction, unless otherwise agreed between the purported originator and the addressee of an electronic communication, the purported originator of the electronic communication is bound by that communication only if the communication was sent by the purported originator or with the authority of the purported originator.

(2) Subsection (1) does not affect the operation of a law of this jurisdiction that makes provision for—

(a) conduct engaged in by a person within the scope of the person's actual or apparent authority to be attributed to another person; or

(b) a person to be bound by conduct engaged in by another person within the scope of the other person's actual or apparent authority.

(3) The regulations may provide that this section does not apply to a specified electronic communication or specified class of electronic communications.

(4) The regulations may provide that this section does not apply to a specified law of this jurisdiction.

Part 3—Miscellaneous

15 Regulations

The Governor in Council may make regulations for or with respect to any matter or thing required or permitted by this Act to be prescribed or necessary to be prescribed to give effect to this Act.

Appendix B

UNCITRAL Model Law on Electronic Commerce

[Original: Arabic, Chinese, English, French, Russian, Spanish]

Part one. Electronic commerce in general

Chapter I. General provisions

*Article 1. Sphere of application**

This Law** applies to any kind of information in the form of a data message used in the context*** of commercial**** activities.

*The Commission suggests the following text for States that might wish to limit the applicability of this Law to international data messages:

'This Law applies to a data message as defined in paragraph (1) of article 2 where the data message relates to international commerce.'

**This Law does not override any rule of law intended for the protection of consumers.

***The Commission suggests the following text for States that might wish to extend the applicability of this Law:

'This Law applies to any kind of information in the form of a data message, except in the following situations: [. . .].'

****The term 'commercial' should be given a wide interpretation so as to cover matters arising from all relationships of a commercial nature, whether contractual or not.

Relationships of a commercial nature include, but are not limited to, the following transactions: any trade transaction for the supply or exchange of goods or services; distribution agreement; commercial representation or agency; factoring; leasing; construction of works; consulting; engineering; licensing; investment; financing; banking; insurance; exploitation agreement or concession; joint venture and other forms of industrial or business cooperation; carriage of goods or passengers by air, sea, rail or road.

Article 2. Definitions

For the purposes of this Law:

(a) 'Data message' means information generated, sent, received or stored by electronic, optical or similar means including, but not limited to, electronic data interchange (EDI), electronic mail, telegram, telex or telecopy;
(b) 'Electronic data interchange (EDI)' means the electronic transfer from computer to computer of information using an agreed standard to structure the information;
(c) 'Originator' of a data message means a person by whom, or on whose behalf, the data message purports to have been sent or generated prior to storage, if any, but it does not include a person acting as an intermediary with respect to that data message;
(d) 'Addressee' of a data message means a person who is intended by the originator to receive the data message, but does not include a person acting as an intermediary with respect to that data message;
(e) 'Intermediary', with respect to a particular data message, means a person who, on behalf of another person, sends, receives or stores that data message or provides other services with respect to that data message;
(f) 'Information system' means a system for generating, sending, receiving, storing or otherwise processing data messages.

Article 3. Interpretation

(1) In the interpretation of this Law, regard is to be had to its international origin and to the need to promote uniformity in its application and the observance of good faith.
(2) Questions concerning matters governed by this Law which are not expressly settled in it are to be settled in conformity with the general principles on which this Law is based.

Article 4. Variation by agreement

(1) As between parties involved in generating, sending, receiving, storing or otherwise processing data messages, and except as otherwise provided, the provisions of chapter III may be varied by agreement.
(2) Paragraph (1) does not affect any right that may exist to modify by agreement any rule of law referred to in chapter II.

Chapter II. Application of legal requirements to data messages

Article 5. Legal recognition of data messages

Information shall not be denied legal effect, validity or enforceability solely on the grounds that it is in the form of a data message.

Article 5 bis. Incorporation by reference

(as adopted by the Commission at its thirty-first session, in June 1998)

Information shall not be denied legal effect, validity or enforceability solely on the grounds that it is not contained in the data message purporting to give rise to such legal effect, but is merely referred to in that data message.

Article 6. Writing

(1) Where the law requires information to be in writing, that requirement is met by a data message if the information contained therein is accessible so as to be usable for subsequent reference.

(2) Paragraph (1) applies whether the requirement therein is in the form of an obligation or whether the law simply provides consequences for the information not being in writing.

(3) The provisions of this article do not apply to the following: [....].

Article 7. Signature

(1) Where the law requires a signature of a person, that requirement is met in relation to a data message if:

 (a) a method is used to identify that person and to indicate that person's approval of the information contained in the data message; and

 (b) that method is as reliable as was appropriate for the purpose for which the data message was generated or communicated, in the light of all the circumstances, including any relevant agreement.

(2) Paragraph (1) applies whether the requirement therein is in the form of an obligation or whether the law simply provides consequences for the absence of a signature.

(3) The provisions of this article do not apply to the following: [...].

Article 8. Original

(1) Where the law requires information to be presented or retained in its original form, that requirement is met by a data message if:

 (a) there exists a reliable assurance as to the integrity of the information from the time when it was first generated in its final form, as a data message or otherwise; and

 (b) where it is required that information be presented, that information is capable of being displayed to the person to whom it is to be presented.

(2) Paragraph (1) applies whether the requirement therein is in the form of an obligation or whether the law simply provides consequences for the information not being presented or retained in its original form.

(3) For the purposes of subparagraph *(a)* of paragraph (1):

 (a) the criteria for assessing integrity shall be whether the information has remained complete and unaltered, apart from the addition of any endorsement and any change which arises in the normal course of communication, storage and display; and

 (b) the standard of reliability required shall be assessed in the light of the purpose for which the information was generated and in the light of all the relevant circumstances.

(4) The provisions of this article do not apply to the following: [...].

Article 9. Admissibility and evidential weight of data messages

(1) In any legal proceedings, nothing in the application of the rules of evidence shall apply so as to deny the admissibility of a data message in evidence:
(a) on the sole ground that it is a data message; or,
(b) if it is the best evidence that the person adducing it could reasonably be expected to obtain, on the grounds that it is not in its original form.

(2) Information in the form of a data message shall be given due evidential weight. In assessing the evidential weight of a data message, regard shall be had to the reliability of the manner in which the data message was generated, stored or communicated, to the reliability of the manner in which the integrity of the information was maintained, to the manner in which its originator was identified, and to any other relevant factor.

Article 10. Retention of data messages

(1) Where the law requires that certain documents, records or information be retained, that requirement is met by retaining data messages, provided that the following conditions are satisfied:
(a) the information contained therein is accessible so as to be usable for subsequent reference; and
(b) the data message is retained in the format in which it was generated, sent or received, or in a format which can be demonstrated to represent accurately the information generated, sent or received; and
(c) such information, if any, is retained as enables the identification of the origin and destination of a data message and the date and time when it was sent or received.

(2) An obligation to retain documents, records or information in accordance with paragraph (1) does not extend to any information the sole purpose of which is to enable the message to be sent or received.

(3) A person may satisfy the requirement referred to in paragraph (1) by using the services of any other person, provided that the conditions set forth in subparagraphs *(a)*, *(b)* and *(c)* of paragraph (1) are met.

Chapter III. Communication of data messages

Article 11. Formation and validity of contracts

(1) In the context of contract formation, unless otherwise agreed by the parties, an offer and the acceptance of an offer may be expressed by means of data messages. Where a data message is used in the formation of a contract, that contract shall not be denied validity or enforceability on the sole ground that a data message was used for that purpose.

(2) The provisions of this article do not apply to the following: [. . .].

Article 12. Recognition by parties of data messages

(1) As between the originator and the addressee of a data message, a declaration of will or other statement shall not be denied legal effect, validity or enforceability solely on the grounds that it is in the form of a data message.

(2) The provisions of this article do not apply to the following: [. . .].

Article 13. Attribution of data messages

(1) A data message is that of the originator if it was sent by the originator itself.

(2) As between the originator and the addressee, a data message is deemed to be that of the originator if it was sent:

(a) by a person who had the authority to act on behalf of the originator in respect of that data message; or

(b) by an information system programmed by, or on behalf of, the originator to operate automatically.

(3) As between the originator and the addressee, an addressee is entitled to regard a data message as being that of the originator, and to act on that assumption, if:

(a) in order to ascertain whether the data message was that of the originator, the addressee properly applied a procedure previously agreed to by the originator for that purpose; or

(b) the data message as received by the addressee resulted from the actions of a person whose relationship with the originator or with any agent of the originator enabled that person to gain access to a method used by the originator to identify data messages as its own.

(4) Paragraph (3) does not apply:

(a) as of the time when the addressee has both received notice from the originator that the data message is not that of the originator, and had reasonable time to act accordingly; or

(b) in a case within paragraph (3)(b), at any time when the addressee knew or should have known, had it exercised reasonable care or used any agreed procedure, that the data message was not that of the originator.

(5) Where a data message is that of the originator or is deemed to be that of the originator, or the addressee is entitled to act on that assumption, then, as between the originator and the addressee, the addressee is entitled to regard the data message as received as being what the originator intended to send, and to act on that assumption. The addressee is not so entitled when it knew or should have known, had it exercised reasonable care or used any agreed procedure, that the transmission resulted in any error in the data message as received.

(6) The addressee is entitled to regard each data message received as a separate data message and to act on that assumption, except to the extent that it duplicates another data message and the addressee knew or should have known, had it exercised reasonable care or used any agreed procedure, that the data message was a duplicate.

Article 14. Acknowledgement of receipt

(1) Paragraphs (2) to (4) of this article apply where, on or before sending a data message, or by means of that data message, the originator has requested or has agreed with the addressee that receipt of the data message be acknowledged.

(2) Where the originator has not agreed with the addressee that the acknowledgement be given in a particular form or by a particular method, an acknowledgement may be given by

(a) any communication by the addressee, automated or otherwise, or
(b) any conduct of the addressee sufficient to indicate to the originator that the data message has been received.

(3) Where the originator has stated that the data message is conditional on receipt of the acknowledgement, the data message is treated as though it has never been sent, until the acknowledgement is received.

(4) Where the originator has not stated that the data message is conditional on receipt of the acknowledgement, and the acknowledgement has not been received by the originator within the time specified or agreed or, if no time has been specified or agreed, within a reasonable time, the originator:

(a) may give notice to the addressee stating that no acknowledgement has been received and specifying a reasonable time by which the acknowledgement must be received; and
(b) if the acknowledgement is not received within the time specified in subparagraph *(a)*, may, upon notice to the addressee, treat the data message as though it had never been sent, or exercise any other rights it may have.

(5) Where the originator receives the addressee's acknowledgement of receipt, it is presumed that the related data message was received by the addressee. That presumption does not imply that the data message corresponds to the message received.

(6) Where the received acknowledgement states that the related data message met technical requirements, either agreed upon or set forth in applicable standards, it is presumed that those requirements have been met.

(7) Except in so far as it relates to the sending or receipt of the data message, this article is not intended to deal with the legal consequences that may flow either from that data message or from the acknowledgement of its receipt.

Article 15. Time and place of dispatch and receipt of data messages

(1) Unless otherwise agreed between the originator and the addressee, the dispatch of a data message occurs when it enters an information system outside the control of the originator or of the person who sent the data message on behalf of the originator.

(2) Unless otherwise agreed between the originator and the addressee, the time of receipt of a data message is determined as follows:

(a) if the addressee has designated an information system for the purpose of receiving data messages, receipt occurs:
   (i) at the time when the data message enters the designated information system; or
   (ii) if the data message is sent to an information system of the addressee that is not the designated information system, at the time when the data message is retrieved by the addressee;
(b) if the addressee has not designated an information system, receipt occurs when the data message enters an information system of the addressee.

(3) Paragraph (2) applies notwithstanding that the place where the information system is located may be different from the place where the data message is deemed to be received under paragraph (4).

(4) Unless otherwise agreed between the originator and the addressee, a data message is deemed to be dispatched at the place where the originator has its place of business, and is deemed to be received at the place where the addressee has its place of business. For the purposes of this paragraph:

(a) if the originator or the addressee has more than one place of business, the place of business is that which has the closest relationship to the underlying transaction or, where there is no underlying transaction, the principal place of business;

(b) if the originator or the addressee does not have a place of business, reference is to be made to its habitual residence.

(5) The provisions of this article do not apply to the following: [. . .].

Part two. Electronic commerce in specific areas

Chapter I. Carriage of goods

Article 16. Actions related to contracts of carriage of goods

Without derogating from the provisions of part one of this Law, this chapter applies to any action in connection with, or in pursuance of, a contract of carriage of goods, including but not limited to:

(a)

(i) furnishing the marks, number, quantity or weight of goods;

(ii) stating or declaring the nature or value of goods;

(iii) issuing a receipt for goods;

(iv) confirming that goods have been loaded;

(b)

(i) notifying a person of terms and conditions of the contract;

(ii) giving instructions to a carrier;

(c)

(i) claiming delivery of goods;

(ii) authorizing release of goods;

(iii) giving notice of loss of, or damage to, goods;

(d) giving any other notice or statement in connection with the performance of the contract;

(e) undertaking to deliver goods to a named person or a person authorized to claim delivery;

(f) granting, acquiring, renouncing, surrendering, transferring or negotiating rights in goods;

(g) acquiring or transferring rights and obligations under the contract.

Article 17. Transport documents

(1) Subject to paragraph (3), where the law requires that any action referred to in article 16 be carried out in writing or by using a paper document, that

requirement is met if the action is carried out by using one or more data messages.

(2) Paragraph (1) applies whether the requirement therein is in the form of an obligation or whether the law simply provides consequences for failing either to carry out the action in writing or to use a paper document.

(3) If a right is to be granted to, or an obligation is to be acquired by, one person and no other person, and if the law requires that, in order to effect this, the right or obligation must be conveyed to that person by the transfer, or use of, a paper document, that requirement is met if the right or obligation is conveyed by using one or more data messages, provided that a reliable method is used to render such data message or messages unique.

(4) For the purposes of paragraph (3), the standard of reliability required shall be assessed in the light of the purpose for which the right or obligation was conveyed and in the light of all the circumstances, including any relevant agreement.

(5) Where one or more data messages are used to effect any action in subparagraphs *(f)* and *(g)* of article 16, no paper document used to effect any such action is valid unless the use of data messages has been terminated and replaced by the use of paper documents. A paper document issued in these circumstances shall contain a statement of such termination. The replacement of data messages by paper documents shall not affect the rights or obligations of the parties involved.

(6) If a rule of law is compulsorily applicable to a contract of carriage of goods which is in, or is evidenced by, a paper document, that rule shall not be inapplicable to such a contract of carriage of goods which is evidenced by one or more data messages by reason of the fact that the contract is evidenced by such data message or messages instead of by a paper document.

(7) The provisions of this article do not apply to the following: [. . .].

Appendix C

Selected provisions *Copyright Act 1968* (Cth)

Section 10 Interpretation

(1) In this Act, unless the contrary intention appears:

'access control technological protection measure' means a device, product, technology or component (including a computer program) that:

(a) is used in Australia or a qualifying country:

  (i) by, with the permission of, or on behalf of, the owner or the exclusive licensee of the copyright in a work or other subject-matter; and

  (ii) in connection with the exercise of the copyright; and

(b) in the normal course of its operation, controls access to the work or other subject-matter;

but does not include such a device, product, technology or component to the extent that it:

(c) if the work or other subject-matter is a cinematograph film or computer program (including a computer game)—controls geographic market segmentation by preventing the playback in Australia of a non-infringing copy of the work or other subject-matter acquired outside Australia; or

(d) if the work is a computer program that is embodied in a machine or device—restricts the use of goods (other than the work) or services in relation to the machine or device.

For the purposes of this definition, ***computer program*** has the same meaning as in section 47AB.

'adaptation' means:

(a) in relation to a literary work in a non-dramatic form a version of the work (whether in its original language or in a different language) in a dramatic form;

(b) in relation to a literary work in a dramatic form a version of the work (whether in its original language or in a different language) in a non-dramatic form;
(ba) in relation to a literary work being a computer program—a version of the work (whether or not in the language, code or notation in which the work was originally expressed) not being a reproduction of the work;
(c) in relation to a literary work (whether in a non-dramatic form or in a dramatic form):
  (i) a translation of the work; or
  (ii) a version of the work in which a story or action is conveyed solely or principally by means of pictures; and
(d) in relation to a musical work—an arrangement or transcription of the work.

'artistic work' means:
(a) a painting, sculpture, drawing, engraving or photograph, whether the work is of artistic quality or not;
(b) a building or a model of a building, whether the building or model is of artistic quality or not; or
(c) a work of artistic craftsmanship whether or not mentioned in paragraph (a) or (b);
but does not include a circuit layout within the meaning of the *Circuit Layouts Act 1989*.

'author', in relation to a photograph, means the person who took the photograph.

'broadcast' means a communication to the public delivered by a broadcasting service within the meaning of the *Broadcasting Services Act 1992*.

Note: A broadcasting service does not include the following:
(a) a service (including a teletext service) that provides only data or only text (with or without associated images); or
(b) a service that makes programs available on demand on a point-to-point basis, including a dial-up service.

'carriage service provider' has the same meaning as in the *Telecommunications Act 1997*.

'carrier' has the same meaning as in the *Telecommunications Act 1997*.

'circumvention device' for a technological protection measure means a device, component or product (including a computer program) that:
(a) is promoted, advertised or marketed as having the purpose or use of circumventing the technological protection measure; or
(b) has only a limited commercially significant purpose or use, or no such purpose or use, other than the circumvention of the technological protection measure; or
(c) is primarily or solely designed or produced to enable or facilitate the circumvention of the technological protection measure.
For the purposes of this definition, ***computer program*** has the same meaning as in section 47AB.

'circumvention service' for a technological protection measure means a service that:

(a) is promoted, advertised or marketed as having the purpose or use of circumventing the technological protection measure; or

(b) has only a limited commercially significant purpose or use, or no such purpose or use, other than the circumvention of the technological protection measure; or

(c) is primarily or solely designed or produced to enable or facilitate the circumvention of the technological protection measure.

'communicate' means make available online or electronically transmit (whether over a path, or a combination of paths, provided by a material substance or otherwise) a work or other subject-matter, including a performance or live performance within the meaning of this Act.

'computer program' means a set of statements or instructions to be used directly or indirectly in a computer in order to bring about a certain result.

'controls access': a device, product, technology or component (including a computer program) controls access to a work or other subject-matter if it requires the application of information or a process, with the permission of the owner or exclusive licensee of the copyright in the work or other subject-matter, to gain access to the work or other subject-matter.

'copy', in relation to a cinematograph film, means any article or thing in which the visual images or sounds comprising the film are embodied.

'dramatic work' includes:

(a) a choreographic show or other dumb show; and

(b) a scenario or script for a cinematograph film;

but does not include a cinematograph film as distinct from the scenario or script for a cinematograph film.

'electronic literary or music item' means:

(a) a book in electronic form; or

(b) a periodical publication in electronic form; or

(c) sheet music in electronic form;

regardless of whether there is a printed form.

'electronic rights management information', in relation to a work or other subject-matter, means information that:

(a) is electronic; and

(b) either:

 (i) is or was attached to, or is or was embodied in, a copy of the work or subject-matter; or

 (ii) appears or appeared in connection with a communication, or the making available, of the work or subject-matter; and

(c) either:

 (i) identifies the work or subject-matter, and its author or copyright owner (including such information represented as numbers or codes); or

(ii) identifies or indicates some or all of the terms and conditions on which the work or subject-matter may be used, or indicates that the use of the work or subject-matter is subject to terms or conditions (including such information represented as numbers or codes).

'exclusive licence' means a licence in writing, signed by or on behalf of the owner or prospective owner of copyright, authorizing the licensee, to the exclusion of all other persons, to do an act that, by virtue of this Act, the owner of the copyright would, but for the licence, have the exclusive right to do, and exclusive licensee has a corresponding meaning.

'infringing copy' means:

(a) in relation to a work—a reproduction of the work, or of an adaptation of the work, not being a copy of a cinematograph film of the work or adaptation;
(b) in relation to a sound recording—a copy of the sound recording not being a sound-track associated with visual images forming part of a cinematograph film;
(c) in relation to a cinematograph film—a copy of the film;
(d) in relation to a television broadcast or a sound broadcast—a copy of a cinematograph film of the broadcast or a record embodying a sound recording of the broadcast; and
(e) in relation to a published edition of a work—a facsimile copy of the edition; being an article (which may be an electronic reproduction or copy of the work, recording, film, broadcast or edition) the making of which constituted an infringement of the copyright in the work, recording, film, broadcast or edition or, in the case of an article imported without the licence of the owner of the copyright, would have constituted an infringement of that copyright if the article had been made in Australia by the importer, but does not include;
(f) a non-infringing book whose importation does not constitute an infringement of that copyright; or
(g) a non-infringing accessory whose importation does not constitute an infringement of that copyright; or
(h) a non-infringing copy of a sound recording whose importation does not infringe that copyright; or
(i) a non-infringing copy of a computer program whose importation does not infringe that copyright; or
(j) a non-infringing copy of an electronic literary or music item whose importation does not infringe that copyright.

'literary work' includes:

(a) a table, or compilation, expressed in words, figures or symbols; and
(b) a computer program or compilation of computer programs.

'material form', in relation to a work or an adaptation of a work, includes any form (whether visible or not) of storage of the work or adaptation, or a substantial part of the work or adaptation, (whether or not the work or adaptation, or a substantial part of the work or adaptation, can be reproduced).

'private and domestic use' means private and domestic use on or off domestic premises.

'record' includes a disc, tape, paper, electronic file or other device in which sounds are embodied.

'retransmission', in relation to a broadcast, means a retransmission of the broadcast, where:

(a) the content of the broadcast is unaltered (even if the technique used to achieve retransmission is different to the technique used to achieve the original transmission); and

(b) either:

  (i) in any case—the retransmission is simultaneous with the original transmission; or

  (ii) if the retransmission is in an area that has, wholly or partly, different local time to the area of the original transmission—the retransmission is delayed until no later than the equivalent local time.

'simulcasting' means simultaneously broadcasting a broadcasting service in both analog and digital form in accordance with the requirements of the *Broadcasting Services Act 1992* or of any prescribed legislative provisions relating to digital broadcasting.

'sound broadcast' means sounds broadcast otherwise than as part of a television broadcast.

'sound recording' means the aggregate of the sounds embodied in a record.

'sound-track', in relation to visual images forming part of a cinematograph film, means:

(a) the part of any article or thing, being an article or thing in which those visual images are embodied, in which sounds are embodied; or

(b) a disc, tape or other device in which sounds are embodied and which is made available by the maker of the film for use in conjunction with the article or thing in which those visual images are embodied.

'technological protection measure' means:

(a) an access control technological protection measure; or

(b) a device, product, technology or component (including a computer program) that:

  (i) is used in Australia or a qualifying country by, with the permission of, or on behalf of, the owner or the exclusive licensee of the copyright in a work or other subject-matter; and

  (ii) in the normal course of its operation, prevents, inhibits or restricts the doing of an act comprised in the copyright;

  but does not include such a device, product, technology or component to the extent that it:

  (iii) if the work or other subject-matter is a cinematograph film or computer program (including a computer game)–controls geographic market segmentation by preventing the playback in Australia of a non-infringing copy of the work or other subject-matter acquired outside Australia; or

(iv) if the work is a computer program that is embodied in a machine or device– restricts the use of goods (other than the work) or services in relation to the machine or device.

For the purposes of this definition, ***computer program*** has the same meaning as in section 47AB.

'television broadcast' means visual images broadcast by way of television, together with any sounds broadcast for reception along with those images.

'work' means a literary, dramatic, musical or artistic work.

'writing' means a mode of representing or reproducing words, figures or symbols in a visible form, and written has a corresponding meaning . . .

Section 31 Nature of copyright in original works

(1) For the purposes of this Act, unless the contrary intention appears, copyright, in relation to a work, is the exclusive right:
- (a) in the case of a literary, dramatic or musical work, to do all or any of the following acts:
  - (i) to reproduce the work in a material form;
  - (ii) to publish the work;
  - (iii) to perform the work in public;
  - (iv) to communicate the work to the public;
  - (v) to make an adaptation of the work;
  - (vi) to do, in relation to a work that is an adaptation of the first-mentioned work, any of the acts specified in relation to the first-mentioned work in subparagraphs (i) to (iv), inclusive; and
- (b) in the case of an artistic work, to do all or any of the following acts:
  - (i) to reproduce the work in a material form;
  - (ii) to publish the work;
  - (iii) to communicate the work to the public; and
- (c) in the case of a literary work (other than a computer program) or a musical or dramatic work, to enter into a commercial rental arrangement in respect of the work reproduced in a sound recording; and
- (d) in the case of a computer program, to enter into a commercial rental arrangement in respect of the program.

(2) The generality of subparagraph (1)(a)(i) is not affected by subparagraph (1)(a)(vi).

(3) Paragraph (1)(d) does not extend to entry into a commercial rental arrangement in respect of a machine or device in which a computer program is embodied if the program is not able to be copied in the course of the ordinary use of the machine or device.

(4) The reference in subsection (3) to a device does not include a device of a kind ordinarily used to store computer programs (for example, a floppy disc, a device of the kind commonly known as a CD ROM, or an integrated circuit).

(5) Paragraph (1)(d) does not extend to entry into a commercial rental arrangement if the computer program is not the essential object of the rental.

(6) Paragraph (1)(c) does not extend to entry into a commercial rental arrangement if:

(a) the copy of the sound recording concerned was purchased by a person (the record owner) before the commencement of Part 2 of the *Copyright (World Trade Organization Amendments) Act 1994*; and

(b) the commercial rental arrangement is entered into in the ordinary course of a business conducted by the record owner; and

(c) the record owner was conducting the same business, or another business that consisted of, or included, the making of commercial rental arrangements of the same kind, when the copy was purchased.

(7) Paragraph (1)(d) does not extend to entry into a commercial rental arrangement in respect of a computer program if:

(a) the copy of the computer program was purchased by a person (the program owner) before the commencement of Part 2 of the *Copyright (World Trade Organization Amendments) Act 1994*; and

(b) the commercial rental arrangement is entered into in the ordinary course of a business conducted by the program owner; and

(c) the program owner was conducting the same business, or another business that consisted of, or included, the making of commercial rental arrangements in respect of computer programs, when the copy was purchased.

Section 32 Original works in which copyright subsists

(1) Subject to this Act, copyright subsists in an original literary, dramatic, musical or artistic work that is unpublished and of which the author:

(a) was a qualified person at the time when the work was made; or

(b) if the making of the work extended over a period—was a qualified person for a substantial part of that period.

(2) Subject to this Act, where an original literary, dramatic, musical or artistic work has been published:

(a) copyright subsists in the work; or

(b) if copyright in the work subsisted immediately before its first publication—copyright continues to subsist in the work;

if, but only if:

(c) the first publication of the work took place in Australia;

(d) the author of the work was a qualified person at the time when the work was first published; or

(e) the author died before that time but was a qualified person immediately before his or her death.

(3) Notwithstanding the last preceding subsection but subject to the remaining provisions of this Act, copyright subsists in:

(a) an original artistic work that is a building situated in Australia; or

(b) an original artistic work that is attached to, or forms part of, such a building.

(4) In this section, ***qualified person*** means an Australian citizen or a person resident in Australia.

Section 33 Duration of copyright in original works

(1) This section has effect subject to subsection 32(2) and to section 34.

(2) Subject to this section, copyright that subsists in a literary, dramatic, musical or artistic work by virtue of this Part continues to subsist until the end of 70 years after the end of the calendar year in which the author of the work died.

(3) If, before the death of the author of a literary work (other than a computer program) or a dramatic or musical work:

(a) the work had not been published;

(b) the work had not been performed in public;

(c) the work had not been broadcast; and

(d) records of the work had not been offered or exposed for sale to the public;

the copyright in the work continues to subsist until the end of 70 years after the end of the calendar year in which the work is first published, performed in public, or broadcast, or records of the work are first offered or exposed for sale to the public, whichever is the earliest of those events to happen.

(4) A reference in the last preceding subsection to the doing of an act in relation to a work shall be read as including a reference to the doing of that act in relation to an adaptation of the work.

(5) If, before the death of the author of an engraving, the engraving had not been published, the copyright in the engraving continues to subsist until the end of 70 years after the end of the calendar year in which the engraving is first published.

Section 116B Removal or alteration of electronic rights management information

(1) This section applies if:

(a) either:

(i) a person removes, from a copy of a work or other subject-matter in which copyright subsists, any electronic rights management information that relates to the work or other subject-matter; or

(ii) a person alters any electronic rights management information that relates to a work or other subject-matter in which copyright subsists; and

(b) the person does so without the permission of the owner or exclusive licensee of the copyright; and

(c) the person knew, or ought reasonably to have known, that the removal or alteration would induce, enable, facilitate or conceal an infringement of the copyright in the work or other subject-matter.

(2) If this section applies, the owner or exclusive licensee of the copyright may bring an action against the person.

(3) In an action under subsection (2), it must be presumed that the defendant knew, or ought reasonably to have known, that the removal or alteration to which the action relates would have the effect referred to in paragraph (1)(c) unless the defendant proves otherwise.

Section 116C Distribution to the public etc. of works whose electronic rights management information has been removed or altered

(1) This section applies if:

(a) a person does any of the following acts in relation to a work or other subject-matter in which copyright subsists without the permission of the owner or exclusive licensee of the copyright:

(i) distributes a copy of the work or other subject-matter to the public;

(ii) imports into Australia a copy of the work or other subject-matter for distribution to the public;

(iii) communicates a copy of the work or other subject-matter to the public; and

(b) either:

(i) any electronic rights management information that relates to the work or other subject-matter has been removed from the copy of the work or subject-matter; or

(ii) any electronic rights management information that relates to the work or other subject-matter has been altered; and

(c) the person knew that the electronic rights management information had been so removed or altered without the permission of the owner or exclusive licensee of the copyright; and

(d) the person knew, or ought reasonably to have known, that the act referred to in paragraph (a) that was done by the person would induce, enable, facilitate or conceal an infringement of the copyright in the work or other subject-matter.

(2) If this section applies, the owner or exclusive licensee of the copyright may bring an action against the person.

(3) In an action under subsection (2), it must be presumed that the defendant:

(a) had the knowledge referred to in paragraph (1)(c); and

(b) knew, or ought reasonably to have known, that the doing of the act to which the action relates would have the effect referred to in paragraph (1)(d);

unless the defendant proves otherwise.

Section 116CA Distribution and importation of electronic rights management information that has been removed or altered

(1) This section applies if:

(a) a person does either of the following acts in relation to electronic rights management information that relates to a work or other subject-matter in which copyright subsists:

(i) distributes the electronic rights management information;

(ii) imports into Australia the electronic rights management information for distribution; and

(b) the person does so without the permission of the owner or exclusive licensee of the copyright; and

(c) either:
  (i) the information has been removed from a copy of the work or subject-matter without the permission of the owner or exclusive licensee of the copyright; or
  (ii) the information has been removed from a copy of the work or subject-matter with the permission of the owner or exclusive licensee of the copyright but the information has been altered without that permission; and
(d) the person knew that the information had been removed or altered without that permission; and
(e) the person knew, or ought reasonably to have known, that the act referred to in paragraph (a) that was done by the person would induce, enable, facilitate or conceal an infringement of the copyright.

(2) If this section applies, the owner or exclusive licensee of the copyright may bring an action against the person.

(3) In an action under subsection (2), it must be presumed that the defendant:
(a) had the knowledge referred to in paragraph (1)(d); and
(b) knew, or ought reasonably to have known, that the doing of the act to which the action relates would have the effect referred to in paragraph (1)(e);

unless the defendant proves otherwise.

Section 116D Remedies in actions under this Subdivision

(1) The relief that a court may grant in an action under this Subdivision includes an injunction (subject to such terms, if any, as the court thinks fit) and either damages or an account of profits.

(2) If, in an action under this Subdivision, the court is satisfied that it is proper to do so, having regard to:
(a) the flagrancy of the defendant's actions that are the subject of the action; and
(b) any benefit shown to have accrued to the defendant as a result of those acts; and
(c) any other relevant matters;

the court may, in assessing damages, award such additional damages as it considers appropriate in the circumstances.

Appendix D

ICANN Uniform Dispute Resolution Policy (UDRP)

1. **Purpose.** This Uniform Domain Name Dispute Resolution Policy (the 'Policy') has been adopted by the Internet Corporation for Assigned Names and Numbers ('ICANN'), is incorporated by reference into your Registration Agreement, and sets forth the terms and conditions in connection with a dispute between you and any party other than us (the registrar) over the registration and use of an Internet domain name registered by you. Proceedings under Paragraph 4 of this Policy will be conducted according to the Rules for Uniform Domain Name Dispute Resolution Policy (the 'Rules of Procedure'), which are available at www.icann.org/udrp/udrp-rules-24oct99.htm, and the selected administrative-dispute-resolution service provider's supplemental rules.
2. **Your Representations.** By applying to register a domain name, or by asking us to maintain or renew a domain name registration, you hereby represent and warrant to us that (a) the statements that you made in your Registration Agreement are complete and accurate; (b) to your knowledge, the registration of the domain name will not infringe upon or otherwise violate the rights of any third party; (c) you are not registering the domain name for an unlawful purpose; and (d) you will not knowingly use the domain name in violation of any applicable laws or regulations. It is your responsibility to determine whether your domain name registration infringes or violates someone else's rights.
3. **Cancellations, Transfers, and Changes.** We will cancel, transfer or otherwise make changes to domain name registrations under the following circumstances:
   a. subject to the provisions of Paragraph 8, our receipt of written or appropriate electronic instructions from you or your authorized agent to take such action;

b. our receipt of an order from a court or arbitral tribunal, in each case of competent jurisdiction, requiring such action; and/or
c. our receipt of a decision of an Administrative Panel requiring such action in any administrative proceeding to which you were a party and which was conducted under this Policy or a later version of this Policy adopted by ICANN. (See Paragraph 4(i) and (k) below.)

We may also cancel, transfer or otherwise make changes to a domain name registration in accordance with the terms of your Registration Agreement or other legal requirements.

4. **Mandatory Administrative Proceeding.** This Paragraph sets forth the type of disputes for which you are required to submit to a mandatory administrative proceeding. These proceedings will be conducted before one of the administrative-dispute-resolution service providers listed at www.icann.org/udrp/approved-providers.htm (each, a 'Provider').

a. **Applicable Disputes.** You are required to submit to a mandatory administrative proceeding in the event that a third party (a 'complainant') asserts to the applicable Provider, in compliance with the Rules of Procedure, that:
(i) your domain name is identical or confusingly similar to a trademark or service mark in which the complainant has rights; and
(ii) you have no rights or legitimate interests in respect of the domain name; and
(iii) your domain name has been registered and is being used in bad faith.

In the administrative proceeding, the complainant must prove that each of these three elements are present.

b. **Evidence of Registration and Use in Bad Faith**. For the purposes of Paragraph 4(a)(iii), the following circumstances, in particular but without limitation, if found by the Panel to be present, shall be evidence of the registration and use of a domain name in bad faith:
(i) circumstances indicating that you have registered or you have acquired the domain name primarily for the purpose of selling, renting, or otherwise transferring the domain name registration to the complainant who is the owner of the trademark or service mark or to a competitor of that complainant, for valuable consideration in excess of your documented out-of-pocket costs directly related to the domain name; or
(ii) you have registered the domain name in order to prevent the owner of the trademark or service mark from reflecting the mark in a corresponding domain name, provided that you have engaged in a pattern of such conduct; or
(iii) you have registered the domain name primarily for the purpose of disrupting the business of a competitor; or
(iv) by using the domain name, you have intentionally attempted to attract, for commercial gain, Internet users to your web site or other

on-line location, by creating a likelihood of confusion with the complainant's mark as to the source, sponsorship, affiliation, or endorsement of your web site or location or of a product or service on your web site or location.

c. **How to Demonstrate Your Rights to and Legitimate Interests in the Domain Name in Responding to a Complaint**. When you receive a complaint, you should refer to Paragraph 5 of the Rules of Procedure in determining how your response should be prepared. Any of the following circumstances, in particular but without limitation, if found by the Panel to be proved based on its evaluation of all evidence presented, shall demonstrate your rights or legitimate interests to the domain name for purposes of Paragraph 4(a)(ii):
   (i) before any notice to you of the dispute, your use of, or demonstrable preparations to use, the domain name or a name corresponding to the domain name in connection with a bona fide offering of goods or services; or
   (ii) you (as an individual, business, or other organization) have been commonly known by the domain name, even if you have acquired no trademark or service mark rights; or
   (iii) you are making a legitimate noncommercial or fair use of the domain name, without intent for commercial gain to misleadingly divert consumers or to tarnish the trademark or service mark at issue.
d. **Selection of Provider**. The complainant shall select the Provider from among those approved by ICANN by submitting the complaint to that Provider. The selected Provider will administer the proceeding, except in cases of consolidation as described in Paragraph 4(f).
e. **Initiation of Proceeding and Process and Appointment of Administrative Panel**. The Rules of Procedure state the process for initiating and conducting a proceeding and for appointing the panel that will decide the dispute (the 'Administrative Panel').
f. **Consolidation**. In the event of multiple disputes between you and a complainant, either you or the complainant may petition to consolidate the disputes before a single Administrative Panel. This petition shall be made to the first Administrative Panel appointed to hear a pending dispute between the parties. This Administrative Panel may consolidate before it any or all such disputes in its sole discretion, provided that the disputes being consolidated are governed by this Policy or a later version of this Policy adopted by ICANN.
g. **Fees**. All fees charged by a Provider in connection with any dispute before an Administrative Panel pursuant to this Policy shall be paid by the complainant, except in cases where you elect to expand the Administrative Panel from one to three panelists as provided in Paragraph 5(b)(iv) of the Rules of Procedure, in which case all fees will be split evenly by you and the complainant.

h. **Our Involvement in Administrative Proceedings**. We do not, and will not, participate in the administration or conduct of any proceeding before an Administrative Panel. In addition, we will not be liable as a result of any decisions rendered by the Administrative Panel.

i. **Remedies**. The remedies available to a complainant pursuant to any proceeding before an Administrative Panel shall be limited to requiring the cancellation of your domain name or the transfer of your domain name registration to the complainant.

j. **Notification and Publication**. The Provider shall notify us of any decision made by an Administrative Panel with respect to a domain name you have registered with us. All decisions under this Policy will be published in full over the Internet, except when an Administrative Panel determines in an exceptional case to redact portions of its decision.

k. **Availability of Court Proceedings**. The mandatory administrative proceeding requirements set forth in Paragraph 4 shall not prevent either you or the complainant from submitting the dispute to a court of competent jurisdiction for independent resolution before such mandatory administrative proceeding is commenced or after such proceeding is concluded. If an Administrative Panel decides that your domain name registration should be canceled or transferred, we will wait ten (10) business days (as observed in the location of our principal office) after we are informed by the applicable Provider of the Administrative Panel's decision before implementing that decision. We will then implement the decision unless we have received from you during that ten (10) business day period official documentation (such as a copy of a complaint, file-stamped by the clerk of the court) that you have commenced a lawsuit against the complainant in a jurisdiction to which the complainant has submitted under Paragraph 3(b)(xiii) of the Rules of Procedure. (In general, that jurisdiction is either the location of our principal office or of your address as shown in our Whois database. See Paragraphs 1 and 3(b)(xiii) of the Rules of Procedure for details.) If we receive such documentation within the ten (10) business day period, we will not implement the Administrative Panel's decision, and we will take no further action, until we receive (i) evidence satisfactory to us of a resolution between the parties; (ii) evidence satisfactory to us that your lawsuit has been dismissed or withdrawn; or (iii) a copy of an order from such court dismissing your lawsuit or ordering that you do not have the right to continue to use your domain name.

5. **All Other Disputes and Litigation.** All other disputes between you and any party other than us regarding your domain name registration that are not brought pursuant to the mandatory administrative proceeding provisions of Paragraph 4 shall be resolved between you and such other party through any court, arbitration or other proceeding that may be available.

6. **Our Involvement in Disputes.** We will not participate in any way in any dispute between you and any party other than us regarding the registration and use of your domain name. You shall not name us as a party or otherwise

include us in any such proceeding. In the event that we are named as a party in any such proceeding, we reserve the right to raise any and all defenses deemed appropriate, and to take any other action necessary to defend ourselves.

7. **Maintaining the Status Quo.** We will not cancel, transfer, activate, deactivate, or otherwise change the status of any domain name registration under this Policy except as provided in Paragraph 3 above.

8. **Transfers During a Dispute.**
   a. **Transfers of a Domain Name to a New Holder.** You may not transfer your domain name registration to another holder (i) during a pending administrative proceeding brought pursuant to Paragraph 4 or for a period of fifteen (15) business days (as observed in the location of our principal place of business) after such proceeding is concluded; or (ii) during a pending court proceeding or arbitration commenced regarding your domain name unless the party to whom the domain name registration is being transferred agrees, in writing, to be bound by the decision of the court or arbitrator. We reserve the right to cancel any transfer of a domain name registration to another holder that is made in violation of this subparagraph.
   b. **Changing Registrars.** You may not transfer your domain name registration to another registrar during a pending administrative proceeding brought pursuant to Paragraph 4 or for a period of fifteen (15) business days (as observed in the location of our principal place of business) after such proceeding is concluded. You may transfer administration of your domain name registration to another registrar during a pending court action or arbitration, provided that the domain name you have registered with us shall continue to be subject to the proceedings commenced against you in accordance with the terms of this Policy. In the event that you transfer a domain name registration to us during the pendency of a court action or arbitration, such dispute shall remain subject to the domain name dispute policy of the registrar from which the domain name registration was transferred.

9. **Policy Modifications.** We reserve the right to modify this Policy at any time with the permission of ICANN. We will post our revised Policy at least thirty (30) calendar days before it becomes effective. Unless this Policy has already been invoked by the submission of a complaint to a Provider, in which event the version of the Policy in effect at the time it was invoked will apply to you until the dispute is over, all such changes will be binding upon you with respect to any domain name registration dispute, whether the dispute arose before, on or after the effective date of our change. In the event that you object to a change in this Policy, your sole remedy is to cancel your domain name registration with us, provided that you will not be entitled to a refund of any fees you paid to us. The revised Policy will apply to you until you cancel your domain name registration.

Appendix E

.au DISPUTE RESOLUTION POLICY (auDRP)

Notes

1. This policy has been adapted from the Uniform Dispute Resolution Policy (UDRP) of the Internet Corporation for Assigned Names and Numbers (ICANN). Some sections of this policy are substantively different from the UDRP. For an explanation of the differences, see the report of the auDA Dispute Resolution Working Group at http://www.auda.org.au/policy/audrp.
2. This policy is intended to operate between the registrar and its licensee (the domain name holder or registrant). Thus, the policy uses 'we' and 'our' to refer to the registrar and it uses 'you' and 'your' to refer to the domain name holder.

* * *

1. **Purpose.** The.au Dispute Resolution Policy ('auDRP') is incorporated by reference into your Registrant Agreement, and sets forth the terms and conditions in connection with a dispute between you and any party other than us (the registrar) over the registration and use of an Internet domain name registered by you in one of the open *.au second level domains* (2LDs). Proceedings under Paragraph 4 of this Policy will be conducted according to the Rules for the auDRP (the 'auDRP Rules'), which are at Schedule B of this document, and the selected administrative dispute resolution service provider's supplemental rules.

2. **Your Representations.** By applying to register a domain name, or by asking us to maintain or renew a domain name registration, you hereby represent and warrant to us that:

a. the statements that you made in your domain name application are complete and accurate, including those as to your eligibility for a domain name in the open 2LD;

b. to your knowledge, the registration of the domain name will not infringe upon or otherwise violate the rights of any third party;
c. you are not registering the domain name for an unlawful purpose; and
d. you will not knowingly use the domain name in violation of any applicable laws or regulations.

It is your responsibility to determine whether your domain name registration infringes or violates someone else's rights.

3. **Cancellations, Transfers, and Changes.** We will cancel, transfer or otherwise make changes to domain name registrations under the following circumstances:

a. subject to the provisions of Paragraph 8, our receipt of written or appropriate electronic instructions from you or your authorised agent to take such action;
b. our receipt of an order from a court or arbitral tribunal, in each case of competent jurisdiction, requiring such action; and/or
c. our receipt of a decision of an Administrative Panel requiring such action in any administrative proceeding to which you were a party and which was conducted under this Policy or a later version of this Policy adopted by auDA, subject to Paragraph 4(i) and (k) below.

We may also cancel, transfer or otherwise make changes to a domain name registration in accordance with the terms of your Registrant Agreement or other legal requirements.

4. **Mandatory Administrative Proceeding.** This Paragraph sets forth the type of disputes for which you are required to submit to a mandatory administrative proceeding. These proceedings will be conducted before one of the administrative dispute resolution service providers listed on the auDA web site at http://www.auda.org.au/policy/audrp (each, a 'Provider').

a. Applicable Disputes. You are required to submit to a mandatory administrative proceeding in the event that a third party (a 'complainant') asserts to the applicable Provider, in compliance with the Rules of Procedure that:
   (i) your domain name is identical or confusingly similar to a name [Note 1], trademark or service mark in which the complainant has rights; and
   (ii) you have no rights or legitimate interests in respect of the domain name [Note 2]; and
   (iii) your domain name has been registered or subsequently used in bad faith.

   In an administrative proceeding, the complainant bears the onus of proof.
b. Evidence of Registration or Use in Bad Faith. For the purposes of Paragraph 4(a)(iii), the following circumstances, in particular but without limitation, if found by the Panel to be present, shall be evidence of the registration and use of a domain name in bad faith:
   (i) circumstances indicating that you have registered or you have acquired the domain name primarily for the purpose of selling, renting, or otherwise transferring the domain name registration

to another person for valuable consideration in excess of your documented out-of-pocket costs directly related to the domain name; or

(ii) you have registered the domain name in order to prevent the owner of a name, trademark or service mark from reflecting that name or mark in a corresponding domain name; or

(iii) you have registered the domain name primarily for the purpose of disrupting the business or activities of another person; or

(iv) by using the domain name, you have intentionally attempted to attract, for commercial gain, Internet users to a web site or other online location, by creating a likelihood of confusion with the complainant's name or mark as to the source, sponsorship, affiliation, or endorsement of that web site or location or of a product or service on that web site or location.

c. How to Demonstrate Your Rights to and Legitimate Interests in the Domain Name in Responding to a Complaint. When you receive a complaint, you should refer to Paragraph 5 of the auDRP Rules in determining how your response should be prepared. Any of the following circumstances, in particular but without limitation, if found by the Panel to be proved based on its evaluation of all evidence presented, is to be taken to demonstrate your rights or legitimate interests to the domain name for purposes of Paragraph 4(a)(ii):

(i) before any notice to you of the subject matter of the dispute, your bona fide use of, or demonstrable preparations to use, the domain name or a name corresponding to the domain name in connection with an offering of goods or services (not being the offering of domain names that you have acquired for the purpose of selling, renting or otherwise transferring); or

(ii) you (as an individual, business, or other organisation) have been commonly known by the domain name, even if you have acquired no trademark or service mark rights; or

(iii) you are making a legitimate non-commercial or fair use of the domain name, without intent for commercial gain to misleadingly divert consumers or to tarnish the name, trademark or service mark at issue.

d. Selection of Provider. The complainant must select the Provider from among those approved by auDA by submitting the complaint to that Provider. The selected Provider will administer the proceeding, except in cases of consolidation as described in Paragraph 4(f).

e. Initiation of Proceeding and Process and Appointment of Administrative Panel. The auDRP Rules state the process for initiating and conducting a proceeding and for appointing the panel that will decide the dispute (the 'Administrative Panel').

f. Consolidation. In the event of multiple disputes between you and a complainant, either you or the complainant may petition to consolidate the disputes before a single Administrative Panel. This petition shall be made to the first Administrative Panel appointed to hear a pending dispute between the parties. This Administrative Panel may consolidate before it any or all such disputes in its sole discretion, provided that the disputes being consolidated are governed by this Policy or a later version of this Policy adopted by auDA.

g. Fees. All fees charged by a Provider in connection with any dispute before an Administrative Panel pursuant to this Policy shall be paid by the complainant, except in cases where you elect to expand the Administrative Panel from one to three panelists as provided in Paragraph 5(b)(iv) of the Rules of Procedure, in which case all fees will be borne evenly by you and the complainant.
h. Our Involvement in Administrative Proceedings. We do not, and will not, participate in the administration or conduct of any proceeding before an Administrative Panel. In addition, we will not be liable as a result of any decisions rendered by the Administrative Panel.
i. Remedies. The remedies available to a complainant pursuant to any proceeding before an Administrative Panel shall be limited to requiring the cancellation of your domain name or the transfer of your domain name registration to the complainant (provided that the complainant is otherwise eligible to hold that domain name).
j. Notification and Publication. The Provider shall notify us of any decision made by an Administrative Panel with respect to a domain name you have registered with us. All decisions under this Policy will be published in full over the Internet, except when an Administrative Panel determines in an exceptional case to redact portions of its decision.
k. Availability of Court Proceedings. The mandatory administrative proceeding requirements set forth in Paragraph 4 shall not prevent either you or the complainant from submitting the dispute to a court of competent jurisdiction for independent resolution before such mandatory administrative proceeding is commenced or after such proceeding is concluded. If an Administrative Panel decides that your domain name registration should be cancelled or transferred, we will wait ten (10) business days (as observed in the location of our principal office) after we are informed by the applicable Provider of the Administrative Panel's decision before implementing that decision. We will then implement the decision unless we have received from you during that ten (10) business day period official documentation (such as a copy of a complaint, sealed by the registrar of the court) that you have commenced a lawsuit against the complainant. If we receive such documentation within the ten (10) business day period, we will not implement the Administrative Panel's decision, and we will take no further action, until we receive:
   (i) evidence satisfactory to us of a resolution between the parties;
   (ii) evidence satisfactory to us that your lawsuit has been dismissed, withdrawn or abandoned; or
   (iii) a copy of an order from such court dismissing your lawsuit or ordering that you do not have the right to continue to use your domain name.

5. **All Other Disputes and Litigation.** All other disputes between you and any party other than us regarding your domain name registration that are not brought pursuant to the mandatory administrative proceeding provisions of Paragraph 4 shall be resolved between you and such other party through any court, arbitration or other proceeding that may be available.
6. **Our Involvement in Disputes.** We will not participate in any way in any dispute between you and any party other than us regarding the registration and use of your domain name. You shall not name us as a party or otherwise include us in any such proceeding. In the event that we are named as a party in any such proceeding, we reserve the right to raise any and all defenses deemed appropriate, and to take any other action necessary to defend ourselves.
7. **Maintaining the Status Quo.** We will not cancel, transfer, activate, deactivate, or otherwise change the status of any domain name registration under this Policy except as provided in Paragraph 3 above.
8. **Transfers During a Dispute.**
   a. Transfers of a Domain Name to a New Holder. You may not transfer your domain name registration to another holder:
      (i) during a pending administrative proceeding brought pursuant to Paragraph 4 or for a period of fifteen (15) business days (as observed in the location of our principal place of business) after such proceeding is concluded; or
      (ii) during a pending court proceeding or arbitration commenced regarding your domain name unless the party to whom the domain name registration is being transferred agrees, in writing, to be bound by the decision of the court or arbitrator.

   We reserve the right to cancel any transfer of a domain name registration to another holder that is made in violation of this subparagraph.

   b. Changing Registrars. You may not transfer your domain name registration to another registrar during a pending administrative proceeding brought pursuant to Paragraph 4 or for a period of fifteen (15) business days (as observed in the location of our principal place of business) after such proceeding is concluded. You may transfer administration of your domain name registration to another registrar during a pending court action or arbitration, provided that the domain name you have registered with us shall continue to be subject to the proceedings commenced against you in accordance with the terms of this Policy. In the event that you transfer a domain name registration to us during the pendency of a court action or arbitration, such dispute shall remain subject to the domain name dispute policy of the registrar from which the domain name registration was transferred.
9. **Policy Modifications.** This Policy may only be modified by auDA.

Notes

[1] For the purposes of this policy, auDA has determined that a 'name ... in which the complainant has rights' refers to:

a) the complainant's company, business or other legal or trading name, as registered with the relevant Australian government authority; or
b) the complainant's personal name.

[2] For the purposes of this policy, auDA has determined that 'rights or legitimate interests in respect of the domain name' are not established merely by a registrar's determination that the respondent satisfied the relevant eligibility criteria for the domain name at the time of registration.

Appendix F

National Privacy Principles

1 Collection

1.1 An organisation must not collect personal information unless the information is necessary for one or more of its functions or activities.

1.2 An organisation must collect personal information only by lawful and fair means and not in an unreasonably intrusive way.

1.3 At or before the time (or, if that is not practicable, as soon as practicable after) an organisation collects personal information about an individual from the individual, the organisation must take reasonable steps to ensure that the individual is aware of:

(a) the identity of the organisation and how to contact it; and

(b) the fact that he or she is able to gain access to the information; and

(c) the purposes for which the information is collected; and

(d) the organisations (or the types of organisations) to which the organisation usually discloses information of that kind; and

(e) any law that requires the particular information to be collected; and

(f) the main consequences (if any) for the individual if all or part of the information is not provided.

1.4 If it is reasonable and practicable to do so, an organisation must collect personal information about an individual only from that individual.

1.5 If an organisation collects personal information about an individual from someone else, it must take reasonable steps to ensure that the individual is or has been made aware of the matters listed in subclause 1.3 except to the extent that making the individual aware of the matters would pose a serious threat to the life or health of any individual.

2 *Use and disclosure*

2.1 An organisation must not use or disclose personal information about an individual for a purpose (the ***secondary purpose***) other than the primary purpose of collection unless:

(a) both of the following apply:

(i) the secondary purpose is related to the primary purpose of collection and, if the personal information is sensitive information, directly related to the primary purpose of collection;

(ii) the individual would reasonably expect the organisation to use or disclose the information for the secondary purpose; or

(b) the individual has consented to the use or disclosure; or

(c) if the information is not sensitive information and the use of the information is for the secondary purpose of direct marketing:

(i) it is impracticable for the organisation to seek the individual's consent before that particular use; and

(ii) the organisation will not charge the individual for giving effect to a request by the individual to the organisation not to receive direct marketing communications; and

(iii) the individual has not made a request to the organisation not to receive direct marketing communications; and

(iv) in each direct marketing communication with the individual, the organisation draws to the individual's attention, or prominently displays a notice, that he or she may express a wish not to receive any further direct marketing communications; and

(v) each written direct marketing communication by the organisation with the individual (up to and including the communication that involves the use) sets out the organisation's business address and telephone number and, if the communication with the individual is made by fax, telex or other electronic means, a number or address at which the organisation can be directly contacted electronically; or

(d) if the information is health information and the use or disclosure is necessary for research, or the compilation or analysis of statistics, relevant to public health or public safety:

(i) it is impracticable for the organisation to seek the individual's consent before the use or disclosure; and

(ii) the use or disclosure is conducted in accordance with guidelines approved by the Commissioner under section 95A for the purposes of this subparagraph; and

(iii) in the case of disclosure—the organisation reasonably believes that the recipient of the health information will not disclose the health information, or personal information derived from the health information; or

(e) the organisation reasonably believes that the use or disclosure is necessary to lessen or prevent:

(i) a serious and imminent threat to an individual's life, health or safety; or

(ii) a serious threat to public health or public safety; or

(ea) if the information is genetic information and the organisation has obtained the genetic information in the course of providing a health service to the individual:

(i) the organisation reasonably believes that the use or disclosure is necessary to lessen or prevent a serious threat to the life, health or safety (whether or not the threat is imminent) of an individual who is a genetic relative of the individual to whom the genetic information relates; and

(ii) the use or disclosure is conducted in accordance with guidelines approved by the Commissioner under section 95AA for the purposes of this subparagraph; and

(iii) in the case of disclosure—the recipient of the genetic information is a genetic relative of the individual; or

(f) the organisation has reason to suspect that unlawful activity has been, is being or may be engaged in, and uses or discloses the personal information as a necessary part of its investigation of the matter or in reporting its concerns to relevant persons or authorities; or

(g) the use or disclosure is required or authorised by or under law; or

(h) the organisation reasonably believes that the use or disclosure is reasonably necessary for one or more of the following by or on behalf of an enforcement body:

(i) the prevention, detection, investigation, prosecution or punishment of criminal offences, breaches of a law imposing a penalty or sanction or breaches of a prescribed law;

(ii) the enforcement of laws relating to the confiscation of the proceeds of crime;

(iii) the protection of the public revenue;

(iv) the prevention, detection, investigation or remedying of seriously improper conduct or prescribed conduct;

(v) the preparation for, or conduct of, proceedings before any court or tribunal, or implementation of the orders of a court or tribunal.

Note 1: It is not intended to deter organisations from lawfully co-operating with agencies performing law enforcement functions in the performance of their functions.

Note 2: Subclause 2.1 does not override any existing legal obligations not to disclose personal information. Nothing in subclause 2.1 requires an organisation to disclose personal information; an organisation is always entitled not to disclose personal information in the absence of a legal obligation to disclose it.

Note 3: An organisation is also subject to the requirements of National Privacy Principle 9 if it transfers personal information to a person in a foreign country.

2.2 If an organisation uses or discloses personal information under paragraph 2.1(h), it must make a written note of the use or disclosure.

2.3 Subclause 2.1 operates in relation to personal information that an organisation that is a body corporate has collected from a related body corporate as if the organisation's primary purpose of collection of the information were the primary purpose for which the related body corporate collected the information.

2.4 Despite subclause 2.1, an organisation that provides a health service to an individual may disclose health information about the individual to a person who is responsible for the individual if:

(a) the individual:
  (i) is physically or legally incapable of giving consent to the disclosure; or
  (ii) physically cannot communicate consent to the disclosure; and
(b) a natural person (the ***carer***) providing the health service for the organisation is satisfied that either:
  (i) the disclosure is necessary to provide appropriate care or treatment of the individual; or
  (ii) the disclosure is made for compassionate reasons; and
(c) the disclosure is not contrary to any wish:
  (i) expressed by the individual before the individual became unable to give or communicate consent; and
  (ii) of which the carer is aware, or of which the carer could reasonably be expected to be aware; and
(d) the disclosure is limited to the extent reasonable and necessary for a purpose mentioned in paragraph (b).

2.5 For the purposes of subclause 2.4, a person is ***responsible*** for an individual if the person is:

(a) a parent of the individual; or
(b) a child or sibling of the individual and at least 18 years old; or
(c) a spouse or de facto spouse of the individual; or
(d) a relative of the individual, at least 18 years old and a member of the individual's household; or
(e) a guardian of the individual; or
(f) exercising an enduring power of attorney granted by the individual that is exercisable in relation to decisions about the individual's health; or
(g) a person who has an intimate personal relationship with the individual; or
(h) a person nominated by the individual to be contacted in case of emergency.

2.6 In subclause 2.5:

child of an individual includes an adopted child, a step-child and a foster-child, of the individual.

parent of an individual includes a step-parent, adoptive parent and a foster-parent, of the individual.

relative of an individual means a grandparent, grandchild, uncle, aunt, nephew or niece, of the individual.

sibling of an individual includes a half-brother, half-sister, adoptive brother, adoptive sister, step-brother, step-sister, foster-brother and foster-sister, of the individual.

3 Data quality

An organisation must take reasonable steps to make sure that the personal information it collects, uses or discloses is accurate, complete and up-to-date.

4 Data security

4.1 An organisation must take reasonable steps to protect the personal information it holds from misuse and loss and from unauthorised access, modification or disclosure.

4.2 An organisation must take reasonable steps to destroy or permanently de-identify personal information if it is no longer needed for any purpose for which the information may be used or disclosed under National Privacy Principle 2.

5 Openness

5.1 An organisation must set out in a document clearly expressed policies on its management of personal information. The organisation must make the document available to anyone who asks for it.

5.2 On request by a person, an organisation must take reasonable steps to let the person know, generally, what sort of personal information it holds, for what purposes, and how it collects, holds, uses and discloses that information.

6 Access and correction

6.1 If an organisation holds personal information about an individual, it must provide the individual with access to the information on request by the individual, except to the extent that:

(a) in the case of personal information other than health information—providing access would pose a serious and imminent threat to the life or health of any individual; or

(b) in the case of health information—providing access would pose a serious threat to the life or health of any individual; or

(c) providing access would have an unreasonable impact upon the privacy of other individuals; or

(d) the request for access is frivolous or vexatious; or

(e) the information relates to existing or anticipated legal proceedings between the organisation and the individual, and the information would not be accessible by the process of discovery in those proceedings; or

(f) providing access would reveal the intentions of the organisation in relation to negotiations with the individual in such a way as to prejudice those negotiations; or

(g) providing access would be unlawful; or
(h) denying access is required or authorised by or under law; or
(i) providing access would be likely to prejudice an investigation of possible unlawful activity; or
(j) providing access would be likely to prejudice:
  (i) the prevention, detection, investigation, prosecution or punishment of criminal offences, breaches of a law imposing a penalty or sanction or breaches of a prescribed law; or
  (ii) the enforcement of laws relating to the confiscation of the proceeds of crime; or
  (iii) the protection of the public revenue; or
  (iv) the prevention, detection, investigation or remedying of seriously improper conduct or prescribed conduct; or
  (v) the preparation for, or conduct of, proceedings before any court or tribunal, or implementation of its orders;

  by or on behalf of an enforcement body; or
(k) an enforcement body performing a lawful security function asks the organisation not to provide access to the information on the basis that providing access would be likely to cause damage to the security of Australia.

6.2 However, where providing access would reveal evaluative information generated within the organisation in connection with a commercially sensitive decision-making process, the organisation may give the individual an explanation for the commercially sensitive decision rather than direct access to the information.

Note: An organisation breaches subclause 6.1 if it relies on subclause 6.2 to give an individual an explanation for a commercially sensitive decision in circumstances where subclause 6.2 does not apply.

6.3 If the organisation is not required to provide the individual with access to the information because of one or more of paragraphs 6.1(a) to (k) (inclusive), the organisation must, if reasonable, consider whether the use of mutually agreed intermediaries would allow sufficient access to meet the needs of both parties.

6.4 If an organisation charges for providing access to personal information, those charges:
(a) must not be excessive; and
(b) must not apply to lodging a request for access.

6.5 If an organisation holds personal information about an individual and the individual is able to establish that the information is not accurate, complete and up-to-date, the organisation must take reasonable steps to correct the information so that it is accurate, complete and up-to-date.

6.6 If the individual and the organisation disagree about whether the information is accurate, complete and up-to-date, and the individual asks the organisation to associate with the information a statement claiming that the information is not accurate, complete or up-to-date, the organisation must take reasonable steps to do so.

6.7 An organisation must provide reasons for denial of access or a refusal to correct personal information.

7 Identifiers

7.1 An organisation must not adopt as its own identifier of an individual an identifier of the individual that has been assigned by:
(a) an agency; or
(b) an agent of an agency acting in its capacity as agent; or
(c) a contracted service provider for a Commonwealth contract acting in its capacity as contracted service provider for that contract.

7.1A However, subclause 7.1 does not apply to the adoption by a prescribed organisation of a prescribed identifier in prescribed circumstances.

Note: There are prerequisites that must be satisfied before those matters are prescribed: see subsection 100(2).

7.2 An organisation must not use or disclose an identifier assigned to an individual by an agency, or by an agent or contracted service provider mentioned in subclause 7.1, unless:
(a) the use or disclosure is necessary for the organisation to fulfil its obligations to the agency; or
(b) one or more of paragraphs 2.1(e) to 2.1(h) (inclusive) apply to the use or disclosure; or
(c) the use or disclosure is by a prescribed organisation of a prescribed identifier in prescribed circumstances.

Note: There are prerequisites that must be satisfied before the matters mentioned in paragraph (c) are prescribed: see subsections 100(2) and (3).

7.3 In this clause:

identifier includes a number assigned by an organisation to an individual to identify uniquely the individual for the purposes of the organisation's operations. However, an individual's name or ABN (as defined in the *A New Tax System (Australian Business Number) Act 1999*) is not an ***identifier***.

8 Anonymity

Wherever it is lawful and practicable, individuals must have the option of not identifying themselves when entering transactions with an organisation.

9 Transborder data flows

An organisation in Australia or an external Territory may transfer personal information about an individual to someone (other than the organisation or the individual) who is in a foreign country only if:

(a) the organisation reasonably believes that the recipient of the information is subject to a law, binding scheme or contract which effectively upholds principles for fair handling of the information that are substantially similar to the National Privacy Principles; or

(b) the individual consents to the transfer; or

(c) the transfer is necessary for the performance of a contract between the individual and the organisation, or for the implementation of pre-contractual measures taken in response to the individual's request; or
(d) the transfer is necessary for the conclusion or performance of a contract concluded in the interest of the individual between the organisation and a third party; or
(e) all of the following apply:
  (i) the transfer is for the benefit of the individual;
  (ii) it is impracticable to obtain the consent of the individual to that transfer;
  (iii) if it were practicable to obtain such consent, the individual would be likely to give it; or
(f) the organisation has taken reasonable steps to ensure that the information which it has transferred will not be held, used or disclosed by the recipient of the information inconsistently with the National Privacy Principles.

10 Sensitive information

10.1 An organisation must not collect sensitive information about an individual unless:
  (a) the individual has consented; or
  (b) the collection is required by law; or
  (c) the collection is necessary to prevent or lessen a serious and imminent threat to the life or health of any individual, where the individual whom the information concerns:
    (i) is physically or legally incapable of giving consent to the collection; or
    (ii) physically cannot communicate consent to the collection; or
  (d) if the information is collected in the course of the activities of a non-profit organisation—the following conditions are satisfied:
    (i) the information relates solely to the members of the organisation or to individuals who have regular contact with it in connection with its activities;
    (ii) at or before the time of collecting the information, the organisation undertakes to the individual whom the information concerns that the organisation will not disclose the information without the individual's consent; or
  (e) the collection is necessary for the establishment, exercise or defence of a legal or equitable claim.

10.2 Despite subclause 10.1, an organisation may collect health information about an individual if:
  (a) the information is necessary to provide a health service to the individual; and
  (b) the information is collected:
    (i) as required or authorised by or under law (other than this Act); or

(ii) in accordance with rules established by competent health or medical bodies that deal with obligations of professional confidentiality which bind the organisation.

10.3 Despite subclause 10.1, an organisation may collect health information about an individual if:

(a) the collection is necessary for any of the following purposes:
  (i) research relevant to public health or public safety;
  (ii) the compilation or analysis of statistics relevant to public health or public safety;
  (iii) the management, funding or monitoring of a health service; and
(b) that purpose cannot be served by the collection of information that does not identify the individual or from which the individual's identity cannot reasonably be ascertained; and
(c) it is impracticable for the organisation to seek the individual's consent to the collection; and
(d) the information is collected:
  (i) as required by law (other than this Act); or
  (ii) in accordance with rules established by competent health or medical bodies that deal with obligations of professional confidentiality which bind the organisation; or
  (iii) in accordance with guidelines approved by the Commissioner under section 95A for the purposes of this subparagraph.

10.4 If an organisation collects health information about an individual in accordance with subclause 10.3, the organisation must take reasonable steps to permanently de-identify the information before the organisation discloses it.

10.5 In this clause:

non-profit organisation means a non-profit organisation that has only racial, ethnic, political, religious, philosophical, professional, trade, or trade union aims.

Index